MASS OUTFLOWS FROM STARS AND GALACTIC NUCLEI

ASTROPHYSICS AND SPACE SCIENCE LIBRARY

A SERIES OF BOOKS ON THE RECENT DEVELOPMENTS
OF SPACE SCIENCE AND OF GENERAL GEOPHYSICS AND ASTROPHYSICS
PUBLISHED IN CONNECTION WITH THE JOURNAL
SPACE SCIENCE REVIEWS

Editorial Board

R.L.F. BOYD, *University College, London, England*

W. B. BURTON, *Sterrewacht, Leiden, The Netherlands*

L. GOLDBERG[†], *Kitt Peak National Observatory, Tucson, Ariz., U.S.A.*

C. DE JAGER, *University of Utrecht, The Netherlands*

J. KLECZEK, *Czechoslovak Academy of Sciences, Ondřejov, Czechoslovakia*

Z. KOPAL, *University of Manchester, England*

R. LÜST, *European Space Agency, Paris, France*

L. I. SEDOV, *Academy of Sciences of the U.S.S.R., Moscow, U.S.S.R.*

Z. ŠVESTKA, *Laboratory for Space Research, Utrecht, The Netherlands*

VOLUME 142
PROCEEDINGS

MASS OUTFLOWS FROM STARS AND GALACTIC NUCLEI

PROCEEDINGS OF THE SECOND TORINO WORKSHOP,
HELD IN TORINO, ITALY, MAY 4-8, 1987

Edited by

LUCIANA BIANCHI

*Osservatorio Astronomico di Torino,
Pino Torinese, Italy*

and

ROBERTO GILMOZZI

*ESA/VILSPA, IUE Groundstation,
Madrid, Spain*

KLUWER ACADEMIC PUBLISHERS

DORDRECHT / BOSTON / LONDON

Library of Congress Cataloging in Publication Data

```
Mass outflows from stars and galactic nuclei : proceedings of the
   Second Torino Workshop, held in Torino, Italy, May 4-8, 1987 /
   edited by Luciana Bianchi and Roberto Gilmozzi.
        p.     cm. -- (Astrophysics and space science library ; 142)
   Includes index.
   ISBN 9027726981
   1. Mass loss (Astrophysics)--Congresses.  2. Stars--Congresses.
 3. Galactic nuclei--Congresses.    I. Bianchi, Luciana, 1954-   .
 II. Gilmozzi, Roberto, 1954-    .  III. Series: Astrophysics and
 space science library ; v. 142.
 QB460.M37 1988
 523.01--dc19                                               88-3065
                                                              CIP
```

ISBN 90-277-2698-1

Published by Kluwer Academic Publishers,
P.O. Box 17, 3300 AA Dordrecht, Holland.

Kluwer Academic Publishers incorporates
the publishing programmes of
D. Reidel, Martinus Nijhoff, Dr W. Junk and MTP Press.

Sold and distributed in the U.S.A. and Canada
by Kluwer Academic Publishers,
101 Philip Drive, Norwell, MA 02061, U.S.A.

In all other countries, sold and distributed
by Kluwer Academic Publishers Group,
P.O. Box 322, 3300 AH Dordrecht, Holland

All Rights Reserved
© 1988 by Kluwer Academic Publishers
No part of the material protected by this copyright notice may be reproduced or
utilized in any form or by any means, electronic or mechanical
including photocopying, recording or by any information storage and
retrieval system, without written permission from the copyright owner

Printed in The Netherlands

TABLE OF CONTENTS

Foreword   xi

List of Participants   xiii

Part I: INVITED PAPERS   1

QUASI-STEADY OUTFLOWS FROM ASTROPHYSICAL OBJECTS: COMPLEMENTARY
VIEWS FROM DIFFERENT EXPERIENCES.
A.Ferrari, R.Rosner   3

MASS OUTFLOW FROM THE SUN - OBSERVATIONS AND DIAGNOSTICS.
G.Noci   11

PROPERTIES OF WINDS AND CHROMOSPHERES IN G TO M GIANTS AND
SUPERGIANTS.
D.Reimers   25

MASS LOSS FROM HOT STARS: MAIN SEQUENCE AND SUPERGIANT STARS.
H.J.G.L.M.Lamers   39

RADIATION DRIVEN WINDS OF CENTRAL STARS OF PLANETARY NEBULAE.
A.Pauldrach, R.P.Kudritzki, R.Gabler, A.Wagner   63

THE THEORY OF STELLAR WINDS FROM HOT AND COOL STARS.
A.G.Hearn   79

INTERACTING BINARIES AND MASS LOSS IN HOT STARS.
F.C.Bruhweiler   95

MASSES OF PLANETARY NEBULAE AND THEIR CENTRAL STARS.
S.R.Pottasch   109

THE FORMATION OF PLANETARY NEBULAE.
S.Kwok   123

MASS LOSS AND POST-ASYMPTOTIC GIANT-BRANCH EVOLUTION.
D.Schönberner   137

COLLIMATED OUTFLOWS FROM YOUNG STARS.
R.Mundt   147

OBSERVATIONAL EVIDENCE FOR OUTFLOW IN ACTIVE GALACTIC NUCLEI AND
IN X-RAY BINARIES.
R.T.Schilizzi   151

MECHANISMS FOR OUTFLOW IN AGNs.
M.Rees 163

REQUIREMENTS FOR THEORETICAL MODELS OF OUTFLOWS.
J.L.Linsky 177

Part II: CONTRIBUTED PAPERS 193

THE LINE SCATTERING RADIATIVE FORCE AS THE REASON OF THE
ACCELERATION AND HEATING OF THE HOT STARS WINDS
E.Ya.Vil'koviskii, L.V.Tambovtseva 195

DUST FORMATION AND MASS LOSS BY WOLF-RAYET STARS.
K.A.van der Hucht, P.M.Williams, P.S.Thé 199

IMPLICATIONS OF VARIABLE MASS-OUTFLOW ON MODELING.
V.Doazan, R.N.Thomas 205

HIGHLY RESOLVED EMISSION-LINE PROFILES OF B[e]-SUPERGIANTS.
F.-J.Zickgraf 211

HD 316285: AN EXTREME P CYGNI STAR.
D.J.Hillier, P.J.McGregor, A.R.Hyland 215

SOMETHING TO DO WITH THE X-RAY EMISSION OF EARLY TYPE STARS.
A.M.T.Pollock 219

THE O6.5IIIf STAR BD+60°2522 AND ITS INTERACTION WITH THE
SURROUNDING INTERSTELLAR MEDIUM: CCD-IMAGING AND SPECTROSCOPIC
OBSERVATIONS OF THE STELLAR WIND BUBBLE NGC 7635.
C.Jäger, C.Leitherer, C.Chavarria K. 223

STELLAR WINDS FROM Of STARS FROM INFRARED AND RADIO OBSERVATIONS.
P.Persi, L.F.Rodriguez, M.Tapia, M.Ferrari-Toniolo, M.Roth 227

STELLAR WIND OF MASSIVE STARS IN M31.
L.Bianchi, J.Hutchings, P.Massey 231

HYDROMAGNETIC WINDS FROM PARTIALLY OPEN MAGNETOSPHERES.
K.Tsinganos 235

HELICOIDAL ASTROPHYSICAL OUTFLOWS.
K.Tsinganos, G.Vlastou 239

LOCAL THERMONUCLEAR RUNAWAYS ON WHITE DWARFS.
M.Orio, G.Shaviv 245

IMPROVED FIRST ORDER MOMENT METHOD FOR DETERMINATION OF
MASS-LOSS-RATES.
D.Hutsemékers, J.Surdej 253

SHELL FORMATION AND MASS LOSS IN THE PLANETARY NEBULA A78.
A.Manchado, A.Mampaso, S.R.Pottasch   257

NARROW-BAND IMAGING OF PLANETARY NEBULAE WITH THE C.F.H.
TELESCOPE PHOTON COUNTING CAMERA.
C.T.Hua   259

V645 CYGNI AND THE DUCK NEBULA.
R.W.Goodrich   263

MASS LOSS FROM PLANETARY NEBULAE: IC418.
S.Cerrato, M.Baessgen, L.Bianchi, M.Grewing   267

THE KINEMATICAL PROPERTIES OF PLANETARY NEBULAE.
L.Bianchi, C.Falcetta   273

THE "S" PROCESS NUCLEOSYNTESIS IN LOW MASS STARS AND THE
NEUTRON SOURCE $^{13}C(\alpha,n)$ ^{16}O.
G.Picchio, M.Busso, R.Gallino   279

SPECTROSCOPIC STUDY OF GAS OUTFLOWS IN THE BIPOLAR NEBULA S106.
A.Riera, A.Mampaso, P.J.Phillips, J.M.Vilchez, A.Manchado   285

CCD IMAGES OF THE SERPENS BIPOLAR NEBULA.
A.I.Gomez de Castro, C.Eiroa (abstract)   289

FORMATION OF PLANETARY NEBULAE.
A.Finzi   291

BOW SHOCK STRUCTURES NEAR YOUNG STELLAR OBJECTS.
T.P.Ray, R.Mundt   293

WINDS FROM COLD PRE-MAIN SEQUENCE STARS: IONIZATION STRUCTURE
AND LINE INTENSITY.
A.Natta, C,Giovanardi, F.Palla   295

HAVE THE ENERGETICS OF MOLECULAR LINE FLOWS BEEN EXAGGERATED?
J.E.Dyson, J.Cantó, L.F.Rodriguez   299

INFRARED EMISSION FROM OUTFLOWS ASSOCIATED WITH YOUNG STARS.
F.O.Clark, R.J.Laureijs, C.Y.Zhang, G.Chlewicki, P.R.Wesselius   303

INDICATION OF OUTFLOWS FROM YOUNG STARS IN THE SERPENS
MOLECULAR CLOUD.
C.Y.Zhang, F.O.Clark, R.J.Laureijs   309

THE JET OF HH34: NEW RESULTS.
Th.Bührke   313

HIGH-RESOLUTION SPECTROSCOPY OF JETS FROM YOUNG STARS.
J.Solf   317

A THIN SHELL MODEL FOR MOLECULAR OUTFLOWS.
M.Robberto, G.Silvestro   321

EVIDENCE OF THIN DUST SHELLS IN SOME RS CVn STARS.
F.Scaltriti, M.Busso, M.Robberto, P.Persi, G.Silvestro   325

NEAR INFRARED - IRAS CANDIDATES FOR MOLECULAR OUTFLOWS IN
LYNDS CLOUDS.
A.Mampaso, J.P.Phillips, R.Gomez-Reñasco, R.Carballo   329

MASS OUTFLOWS AND JETS IN SYMBIOTIC STARS.
R.Viotti, G.B.Baratta, L.Piro, A.Altamore, A.Cassatella   333

IRAS AND NEAR-INFRARED OBSERVATIONS OF PECULIAR NEBULOSITIES.
P.Persi, M.Busso, M.Ferrari-Toniolo, L.Origlia, M.Robberto,
F.Scaltriti, G.Silvestro.   337

RADIO MAPPING OF TYPE I POST-MAIN-SEQUENCE NEBULAE.
J.P.Phillips, A.Mampaso.   343

$J = 3-2$ AND $J = 2-1$ CO MAPPING OF HIGH VELOCITY OUTFLOW SOURCES.
J.P.Phillips, G.J.White   349

STELLAR MASS LOSS BY TURBULENT ALFVÉN WAVES.
R.Opher, V.J.S.Pereira   355

MASS LOSS FROM THE PROGENITOR OF KEPLER'S SUPERNOVA:
CHARACTERISTICS AND EVOLUTIONARY CONSTRAINTS.
R.Bandiera   359

OBSERVATIONS OF SN1987A WITH I.U.E.
R.Gilmozzi   365

NEUTRINO OUTFLOW FROM SN1987A DETECTED IN THE MONT BLANC
OBSERVATORY.
M.Aglietta et al   367

GEOMETRICAL SIMULATIONS OF ASTROPHYSICAL JETS.
L.Zaninetti   373

ON THE STABILITY OF ROTATING SHEAR FLOWS.
W.Glatzel   377

ON THE STABILITY OF SUPERSONIC SHEAR LAYERS.
W.Glatzel   381

SELF-SUSTAINING GALACTIC WINDS IN SPIRAL GALAXIES WITH
STARBURST NUCLEI.
J.E.Beckman   385

OBSERVATIONS OF OUTFLOW AND ITS CONSEQUENCES IN CIRCUM-NUCLEAR
ZONES OF SPIRALS.
C.Muñoz-Tuñon, J.E.Beckman   391

EVOLUTION OF PERTURBATIONS AND SHOCK FORMATION IN STELLAR
WINDS AND JETS.
E.Trussoni, A.Ferrari, R.Rosner, K.Tsinganos   397

THE FLOWING INTERSTELLAR MEDIUM IN THE HOST GALAXIES OF QUASARS.
J.E.Dyson, J.J.Perry   401

NEW ARGUMENTS SUPPORTING THE INTRINSIC ORIGIN OF BAL QSOs:
FORMATION OF THE COMPLEX Lyα+NV LINE PROFILE.
D.Hutsemékers, J.Surdej   405

Part III: FUTURE SPACE MISSIONS   411

THE LYMAN MISSION.
M.Grewing   413

THE SOHO PROJECT.
E.Antonucci   419

QUASAT - A SPACE VLBI OBSERVATORY.
R.T.Schilizzi   425

THE INTERNATIONAL EUV/FUV HITCHHIKER (IEH).
IEH Team   433

Subject Index   437

FOREWORD

The objective of this workshop was to put together observational and theoretical works on outflows from different kinds of astrophysical objects, occurring on different scales and at various evolutionary phases, and to discuss the impact of observations from future space missions. For the stars, we thought to follow throughout the evolution the relevance (rates and dynamical modes) of the mass loss phenomenon, e.g. to explain how and when massive stars loose most of their initial mass to end up with typical WD masses. The observations of the solar wind were included for being a unique case where the origin and propagation of the outflow can be resolved. We thought that the comparison with similar phenomena occurring in galactic outflows would be fruitful, as demonstrated by recent works on galactic winds and jets.

The interest of having this workshop in Torino came because there are groups in this area, at the Astronomical Observatory and at the Institute of Physics of the University, involved in the theoretical and observational studies of outflows from astrophysical objects.

The members of the Scientific Organizing Committee were: V.Castellani, C.Cesarski, P.Conti, A.Ferrari, A.Gabriel, M.Grewing, Y.Kondo, H.Lamers, V.Manno, M.Rees and R.Schilizzi. The Local Organizing Committee was: L.Bianchi, G.Massone and E.Antonucci.

During the workshop the following topics were treated: the solar wind, the mass loss from cool stars and from hot stars (m.s. stars, supergiants and evolved stars like Planetary Nebulae nuclei), the formation of Planetary Nebulae, the mass flows in interacting binary systems, and the cases of collimated outflows: young stellar objects and galactic nuclei. Recent results on these topics were presented in 16 invited reviews and 61 contributed and poster papers. The future space projects of the European Space Agency which could provide relevant observations for the study of astrophysical outflows were illustrated in 7 invited presentations.

This workshop has been realized thanks to the financial support of: Consiglio Nazionale delle Ricerche, European Space Agency, Cassa di Risparmio di Torino, Citta' di Torino, Provincia di Torino, Regione Piemonte, Unione Industriale and A.M.M.A. During the workshop, the help of several members of the staff of the Astronomical Observatory: M. Marini, G.Chiumiento, E.deZanet, M.Lattanzi, R.Manzi, F.Racioppi and M.Sarasso has been extremely valuable.

The editors.

LIST OF PARTICIPANTS

E. ANTONUCCI, University of Torino, Italy
R. BANDIERA, Osservatorio di Arcetri, Firenze, Italy
G. B. BARATTA, Osservatorio di Roma, Italy
P. BENVENUTI, STECF / ESO Garching, W Germany
L. BIANCHI, Osservatorio di Torino, Italy
G. BODO, Osservatorio di Torino, Italy
J. A. M. BLEEKER, LSR, Utrecht, Netherlands
F. C. BRUHWEILER, NASA GSFS, Mariland, U.S.A.
T. BUHRKE, MPI Heidelberg, FRG
M. BUSSO, Osservatorio di Torino, Italy
M. CAMENZIND, Landessternwarte Konigstuhl Heidelberg, FRG
S. CERRATO, Astronomisches Institut Tuebingen, FRG
C. CESARSKY, CEN Saclay, Gif-sur-Yvette, France
C. CHIOSI, Osservatorio di Padova, Italy
F.O. CLARK, LSR, Groningen, Netherlands
D. DAL FIUME, Istituto TESRE, CNR, Bologna, Italy
T. DE JONG, University of Amsterdam, The Netherlands
V. DOAZAN, Observatoire de Paris, France
J. E. DYSON, University of Manchester, UK
R. FARAGGIANA, Universita di Trieste, Italy
A. FERRARI, Osservatorio di Torino, Italy
A. FINZI, TECHNION, Haifa, Israel
R. GALLINO, Universita di Torino, Italy
R. GILMOZZI, IUE/VILSPA, Madrid, Spain
C. GIOVANARDI, Osservatorio di Firenze, Italy
W. GLATZEL, MPI Garching, FRG
A.I. GOMEZ DE CASTRO, Observatorio Nacional, Madrid, Spain
R. GOODRICH, Lick Observatory, U.S.A.
M. GREWING Astronomisches Institut Tuebingen, FRG

A. HEARN, Sterrewacht "Sonnenborg", Utrecht, Netherlands
J. HILLIER, SAO, Cambridge, Ma, USA
J. HRON, Institut fur Astronomie, Wien, Austria
K. A. VAN DER HUCHT, SRL, Utrecht, Netherlands
D. HUTSEMEKERS, Universite de Liege, Belgium
C. JAGER, Landessternwarte Koningstuhl, Heidelberg, FRG
S. KWOK, University of Calgary, Canada
H. LAMERS, LRS, Utrecht, Netherlands
J. LINSKY, JILA, Boulder, Colorado, USA
V. MANNO, Scientific Directorate, ESA, Paris, France
A. MASANI, Universita di Torino, Italy
D. MAROCCHI, Universita di Torino, Italy
S. MASSAGLIA, Universita di Torino, Italy
G. MASSONE, Osservatorio di Torino, Italy
V. JATENCO SILVA PEREIRA, Universitade de Sao Paulo, Brasil
C. MOROSSI, Osservatirio di Trieste, Italy
R. MUNDT, MPI Heidelberg, FRG
L. NOBILI, Universita di Padova, Italy
G. NOCI, Osservatorio di Firenze, Italy
M. ORIO, TECHNION, Haifa, Israel
G. PAULDRACH, Universitats-Sternwarte, Munich, FRG
P. PERSI, Istituto Astrofisica Spaziale, Frascati, Italy
J. P. PHILLIPS, Queen Mary College, London, U.K.
A.M.T. POLLOCK, EXOSAT Observatory, ESTEC, Netherlands
T. P. RAY, Dunsik Observatory, Dublin, Ireland
M. REES, Institute of Astronomy, Cambridge, U.K.
D. REIMERS, Sternwarte Hamburg, FRG
A. RIERA, Universidad de Canarias, Tenerife, Spain
M. ROBBERTO, Osservatorio di Torino, Italy
R. RUFFINI, Universita di Roma, Italy
F. SCALTRITI, Osservatorio di Torino, Italy
A. SCARMATO, Universita di Bologna, Italy
R.T. SCHILIZZI, Radiosterrenwacht, Dwingeloo, Netherlands
J. SCHMID-BURGK, MPI Bonn, FRG

D. SCHONBERNER, Universitat Kiel, FRG
E. SZUSZKIEWICZ, SISSA, Trieste, Italy
E. TANZI, Istituto di Fisica Cosmica, Milano, Italy
R. N. THOMAS, Radiophysics INC., Boulder, Co, USA
K. TSINGANOS, University of Crete, Greece
E. TRUSSONI, Universita di Torino, Italy
R. TUROLLA, Universita di Padova, Italy
P. VANBEVEREN, Vrije Universiteit, Brussel, Belgium
M.J. WILSON, University of Leeds, U.K.
L. ZANINETTI, Universita di Torino, Italy
C. Y. ZHANG, LSR, Groningen, Netherlands
F. J. ZICKGRAF, Landessternwarte Konigstuhl, Heidelberg, FRG

PART I: INVITED PAPERS

QUASI-STEADY OUTFLOWS FROM ASTROPHYSICAL OBJECTS: COMPLEMENTARY VIEWS FROM DIFFERENT EXPERIENCES

Attilio FERRARI, Istituto di Fisica Matematica, Università di Torino,
and Osservatorio Astronomico di Torino, Italy

Robert ROSNER, Harvard-Smithsonian Center for Astrophysics, Cambridge, USA

ABSTRACT. As a general introduction to the Workshop, we discuss the motivation for joining the perspectives of solar, stellar and extragalactic astronomers, both theorists and experimentalists, on the problem of mass outflows from astrophysical objects. We illustrate some of the principal physical similarities and differences between outflows in various astrophysical systems, and show how experiences derived from different subject areas can help to build a complete picture of the wind phenomenon.

PHENOMENOLOGY

The discovery that most astrophysical objects are associated with quasi-steady mass outflows is very recent; in the case of the Sun, this discovery dates from the direct sensing of the solar wind as the Pioneer probes cruised through the wind itself. However, the existence of an outflowing solar wind had in fact been suggested earlier on the basis of theoretical considerations, including the presence of the extended hot solar corona, intensity modulation of low-energy cosmic rays, and (most classically) cometary tail behaviour.

More recently, optical and ultraviolet observations of stellar atmospheres have allowed us to detect (and diagnose) mass outflows from both hot and cool stars; high resolution images of radiogalaxies have shown the presence of collimated jets, prompting the idea that galactic nuclei drive highly relativistic flows; and radio and infrared observations of star forming regions have shown that outflows exist from the very onset of stellar evolution. The general impression is thus that virtually all 'active' and relatively inactive astrophysical objects are associated with quasi-steady plasma outflows, independent of the actual processes – release of gravitational potential energy, wave damping, photon escape and scattering – which provide the direct source of the energy driving the flows.

The list of objects known to be associated with quasi-steady mass outflows ranges from the Earth (the terrestrial polar wind), to the Sun and most other stars (solar and stellar winds), to regions of star formation (protostellar bipolar outflows), pulsars (relativistic polar winds), compact binaries and SS433 (jets), and galaxies and galactic nuclei (galactic winds and jets).

The underlying motivation for this meeting is to discuss the rationale for constructing a comprehensive model of mass outflows. The obvious first question to answer is whether different types of mass outflows have any physics in common, given the large variety of observed morphologies and distinct physical environs. What is remarkable is that despite these differences, it nevertheless seems to be the case that such outflows share several common physical elements: a coronal plasma at the 'surface' of the astrophysical object; and a process which drives this 'coronal' gas to supersonic velocities and, ultimately, to escape from gravitational confinement. In order to understand this ubiquity of flow acceleration, it will be important to define the physical conditions which govern the acceleration processes, as well as the ultimate wind parameters once an outflow has escaped the gravitational potential well from whence it came.

A rather different set of questions concerning outflows focuses on their interaction with (i) the ambient external medium (in which the wind morphological must play an important role), and (ii) the 'parent' object itself, whose evolution may be influenced by the mass loss itself.

Finally, we note in passing that the basic physics of mass outflows has a close correspondence with the physics of mass inflows, which are also fundamental in astrophysics. It may be that an exchange of ideas between these two otherwise rather different flow regimes will also be fruitful, given the relatively rapid evolution of the respective fields within the past decade.

THE BASIC PHYSICS OF OUTFLOWS

The basic physical elements defining a (quasi-)steady mass outflow from a gravitationally bound system are (i) a driving mechanism and (ii) a restraining force, which combine to form a more or less complicated 'de Laval' nozzle, in which the transition from subsonic to supersonic flow occurs. In order to focus the discussion on the essentials, consider the form of the force balance equation governing a steady outflow, in the simple limit of a 1-D model,

$$\rho \mathbf{v} \cdot \nabla \mathbf{v} = -\nabla p + \rho \mathbf{g} + \frac{1}{c}\mathbf{J} \times \mathbf{B} + \text{other forces} + \text{geometric terms.} \quad (1)$$

Here $\mathbf{v}$ is the outflow velocity, p the gas pressure, ρ the mass density, $\mathbf{J}$ the current density, $\mathbf{B}$ the magnetic field, and $\mathbf{g}$ the gravitational field. The only relevant terms on the right hand side of (1) are the gravitational force, which typically provides the restraining force, and the pressure gradient, which provides the (thermal) driving force; this is the form in which the solar wind problem was first discussed by E.N. Parker (1958), but, remarkably enough, we know today that the solar wind is probably the one case of mass outflow for which this simplest of models is entirely inadequate. Thus, we now know that additional terms on the right hand side of (1) – which reflect departures of the flow physics from simple thermally driven, radially-symmetric outflow – figure prominently in defining the ultimate wind conditions. These modifications of the simplest wind model include, for example, electrodynamic forces, which can be exerted by waves propagating in the flow itself, and may modify the effective flow acceleration and restraint. Similarly, geometric corrections which arise because of deviations from spherically symmetry can lead to both additional driving and restraining forces on the outflow (e.g., the added geometric terms can act as either momentum addition or subtraction terms; cf. Holzer 1977).

To illustrate, consider the simplest departure from the elementary spherically-symmetric thermally-driven wind, in which (1) can be reduced to a a simple 1-D equation for an isothermal flow of the form

$$\frac{M^2-1}{2M^2}\frac{dM^2}{dr} = \frac{1}{A(r)}\frac{dA(r)}{dr} - \frac{Gm_\odot}{r_0 v_{sound}^2 r^2}, \quad (2a)$$

where

$$v_{sound} = 128\sqrt{\Gamma T_6^{(0)}} \text{ km s}^{-1}, \quad T_6 = \frac{T}{10^6 K}; \qquad (2b)$$

here M is the flow Mach number, and $A(r)$ is the flow tube cross-sectional area. All quantities with subscript and superscript '0' (i.e., $M_0, T_6^{(0)}$) are measured at the base of the flow $r = r_0$, and Γ is the gas adiabatic index (see Kopp and Holzer 1976).

In this relatively simple case, it is the term involving the derivative of the flow tube cross-sectional area, $A(r)$, which leads to the additional complications. That is, observed steady outflows by definition operate such that driving forces balance restraining forces so as to allow the transition from the subsonic to the supersonic regime ($v_{sound} \sim v_{escape}$); in equation (2), this corresponds to the point where the left hand side of (2) must vanish, so that the possibility for a transonic transition depends upon the existence of a zero of the right-hand side of (2) — it is the existence of one or more of such zeros, and their position(s), that is affected in a major way by the additional geometric term. Such flows, in which this transonic transition figures centrally, are the main focus of this Workshop. There are of course obvious alternative physical circumstances: for example, when $v_{sound} \gg v_{escape}$, the driving forces can completely dominate any possible restraining forces, and the outflow has the character of a transient explosive mass ejection, rather than a steady flow; at the opposite extreme, if $v_{sound} \ll v_{escape}$, we are in the so-called 'breeze' regime, in which the outflow has very low density, never becomes supersonic (i.e., the right-hand side of (2) has no zeros in the vicinity of the physical system under study), and is virtually impossible to observe; such flows are of little consequnce from from the astronomical point-of-view.

BASIC ACCELERATION PROCESSES

Many theoretical models have been proposed to explain steady outflows, but as just alluded to, the essence of the kind of flows we are considering is that, no matter what the acceleration process, the driving and restraining forces on the flow must come into balance somewhere within the region of interest.

We now briefly discuss some of the acceleration models; along the way, we shall also note the fact that in all probability these models apply not only to the objects for which they were originally proposed, but also have wider application.

1. *Thermal Pressure Gradient.* The conceptually simplest process for accelerating outflows is the gas pressure gradient exerted by a hot plasma at the bottom of the gravitational potential well. We have already noted that the *first* wind model, proposed for the Sun by E.N. Parker (1961), was in fact of this type. Since then, this model has been applied to stars and galactic nuclei.

2. *MHD Waves.* If the 'bottom' of the wind is a highly turbulent region, then the gas pressure gradient can be reasonably replaced by the pressure force exerted by waves propagating outwards in the wind plasma, and depositing energy and momentum as they damp in the ambient gas. This type of model is quite appropriate for the Sun, stars and galactic nuclei, for which one can identify a turbulent 'surface'; furthermore, the presence of magnetic fields in such systems allows for the generation and propagation of MHD waves, which (as just discussed) damp in propagating outwards, and release their momentum in the surrounding plasma, which is thereby accelerated.

3. *Radiation Pressure.* For sufficiently high temperatures at the bottom of the 'well', the photon density and the opacity of the plasma may be so high that radiation pressure becomes the principal driving mechanism which competes with the restraining influence of the gravitational pull. The stability properties of models incorporating this mechinism are only now being investigated; results by Owocki and Rybicki (1984) suggest that such models are quite unstable.

4. *Centrifugal Force.* Plasma entrained on 'open' magnetic field lines tied to rapidly rotating objects experiences a 'sling-shot' effect, which is produced by the Lorentz force, and which acts as the driving body force.

5. *Pulsational Instabilities.* For specific evolutionary conditions, the stellar layers immediately below the photosphere may become pulsationally unstable. The photosphere then acts as a piston, driving an outward flow of the overlying layers.

WIND MODELS FOR SPECIFIC ASTROPHYSICAL SYSTEMS

The above processes have been applied to a variety of specific astrophysical systems, resulting in a rich panoply of wind models. In the present section, we summarize the present situation for the Sun and several other typical classes of objects.

1. THE SUN.

1.1 *Thermally-Driven Wind.* Recent results on coronal X-ray emission, combined with UV and in-situ solar wind measurements, have proven that the solar wind *is not* thermally driven. For instance, the so-called high-speed streams have velocities of ~ 800 km s^{-1}, which cannot possibly be obtained via thermal driving, given the observational constraints on the temperature of the underlying coronal plasma.

1.2. *MHD Wave-Driven Wind.* Alfven waves are actually observed in the solar wind; and a relatively well-developed theory exists for modeling Alfven wave propagation in the wind. These circumstances makes these waves good candidates for providing the driving momentum via wave pressure.

2. STARS.

2.1. *Thermally-Driven Winds.* There is no good supporting evidence for a thermally-driven wind for any star we are aware of; such models fail especially for giant and supergiants. To the best of our knowledge, the only reason thermally-driven winds are ever considered as viable candidates for stellar wind driving is that the failure of such models in the solar case is not widely appreciated.

2.2. *MHD Wave-Driven Winds.* Models for winds from giants and supergiants based on wave momentum deposition and flow acceleration can reproduce the main features of observations, but require fine tuning of crucial physical parameters in the theory, in particular, the ratio of the wave damping distance scale, λ_{damp}, to the coronal scale height (e.g. Cassinelli and McGregor 1984).

2.3. *Radiation Pressure-Driven Winds.* Models for hot stars based on a line-driving mechanism have been fairly successful (e.g. Lucy and Salomon 1970, Hearn 1985); but similar ideas applied to the opacity from dust grains in the atmospheres of giants and supergiants appear to be less fruitful, as dust grains fail to form close enough to the star where the outflow should be accelerated (e.g. W. Hagen et al. 1983).

2.4. *Pulsation-Driven Winds.* Models for stars with unstable photospheres, such as Mira-type variable stars, yield outflows which do provide the needed initial momentum driving; at large distances, however, an additional agent is needed to push the outflow to its asymptotic velocity, and this might be the radiation pressure on grains at large radii.

2.5. *Centrifugally-Driven Winds.* Rapidly rotating, magnetized stars can drive winds via the 'rotating-hoop' effect, as described originally by Weber and Davis (1967). Certainly rotation has an important effect on coronal heating (Rosner et al. 1985), and therefore may also play a role in the driving of outflows.

3. NEUTRON STARS, PULSARS.

3.1. *Electromagnetic Wave-Driven Winds.* In addition to the mechanisms valid for stars in general, neutron stars with high rotational velocities and large magnetic field strengths allow for yet more complex flow acceleration processes which do not operate in less 'exotic' environs. For example, charged particles tied to the pulsar magnetospheres can be accelerated via an electrodynamic process based on the emission of a large-amplitude e.m. wave at the rotation frequency (Gunn and Ostriker 1970). Such outflows can reach relativistic speeds, and can give rise to large momentum deposition in the circumstellar plasma. This model has had some success in explaining the activity of Crab-like supernova remnants (Benford et al. 1978).

3.2. *Electron-Positron Pair Winds.* The same model of a magnetic rotator for compact stars as just described can be shown to imply the operation of high-energy processes within the pulsar magnetospheres above the magnetic polar caps which lead to an electron-positron cascade via the generation of gamma rays in strong induced electric fields (Ruderman 1969). The pairs are then accelerated along magnetic field lines outside the magnetosphere proper, and in the equatorial plane of rotation.

4. ACCRETION DISKS, SS433, BINARY STARS.

4.1. *Thermally-Driven Winds.* This is the most 'popular' model for these high-energy astrophysical objects. Temperatures $\approx 10^8$ K are required to produce supersonic (and, in some cases, relativistic) outflows, and such temperatures are conceivable in the deep potential wells. Furthermore, the structure of the disk itself provides the flow collimation which is essential for explaining the jets observed in objects such as Cyg X-3 and SS433. Finally, modulations of the accretion rate may explain observed variations in the jet morphology. The weakness of this model is the fact that for the one object in which model should have worked (namely, the Sun), we now know that the model is not satisfactory.

4.2. *Radiation Pressure-Driven Winds.* This mechanism is similar to the previous case, but instead appeals to the high photon concentration which is present in the deep gravitational wells associated with these compact stars. The idea works rather successfully both for optically thin and optically thick plasmas; however, the maximum reachable asymptotic velocities are of the order of $c/3$.

4.3. *Magnetically-Driven Winds.* Disk magnetic fields can in principle easily drive an outflows centrifugally, as in the Weber and Davis model. At the same time, this process leads to a shedding of the disk angular momentum, which may be the consistent explanation why accretion starts in the first place.

5. PROTOSTELLAR OBJECTS.

5.1. *Magnetically-Driven Bipolar Jets.* Rotation and accretion in stellar formation regions are observed to be associated with the presence of collimated beams of matter oriented along two opposite preferential directions (which may be along their rotation axis. In a model presented by Pudritz and Norman (1983), the matter comes off an accretion disk, and is centrifugally driven by the Lorentz force generated by the rotating lines frozen into the disk plasma.

5.2. *Radiation Pressure-Driven Jets.* Large photon fluxes exerting their pressure on plasma (and/or perhaps grains) around the accretion disk 'punch' their way throughout the circumstellar matter. In the case of thick accretion disks, collimation can be rather good, and further provides optically-emitting jets with shocks appearing as Herbig-Haro objects.

6. GALAXIES.

6.1. *Thermally-Driven Winds.* This model is just a straightforward application of the stellar methodology to galactic conditions. Apparently, the typical conditions are such that the model is here successful; however, the observational data are certainly only weakly constraining.

6.2. *Magnetically-Driven Winds.* Spiral galaxies are well known to possess magnetic fields aligned along the arms. This situation seems very promising for driving magnetically-coupled outflows; however such a model has not as yet been carefully explored.

7. ACTIVE GALACTIC NUCLEI, RADIOGALAXIES, JETS.

7.1. *Radiation Pressure-Driven Jets.* This is the dominant model to date, and is a straightforward application of the studies of jets from compact stars with accretion disks. The general features of the flow acceleration can be explained in terms of a central 'cauldron', with temperatures $\sim 10^9$ K. Calculations have been done both for optically thin and optically thick jets, but the model cannot explain the relativistic velocities which seem to be associated with superluminal sources.

7.2. *MHD Winds.* An interesting possibility for driving relativistic flows off active galactic nuclei is related to the detailed MHD theory of matter in orbit around massive compact objects, and combines the basic elements of centrifugally-driven and MHD wave-driven outflows. This type of model requires incorporation of general relativistic effects, and is still far from providing definite results.

7.3. *Large Amplitude Electromagnetic Wave-Driven Jets.* This model is the extension of the pulsar model to the scale of active nuclei, when interpreted as highly magnetized, rotating objects. The wave has in this case a very low frequency, and therefore its propagation requires an almost empty region around the nucleus. In contrast, self-collimation effects would favour the formation of jets over large scales as in radiogalaxies.

CONCLUDING REMARKS AND SUGGESTIONS FOR DISCUSSION

This brief introductory survey of mass outflow models for various astrophysical objects has focused on the variety of methods used to interpret outflow data; and discussed the common roots which point toward the possibility of working out a general dynamics for such flows. There is as yet no direct, clear evidence of any specific physical mechanism to be preferred to the others, and this is especially due to the poor information about physical constrains.

These considerations entail two immediate conclusions:
1. Analytic and numerical studies must be pushed forward to clarify the influence of the physical parameters which arise in studies of mass outflows. In this respect, numerical simulations will possibly allow us to 'experiment', and try out suggestions in a way which the data do not allow us.

2. The acquisition of more, and more detailed, data must be insured. This requires both the continuation of present observational programs at radio, optical, IR, UV, ... frequencies, and the intervention of new sophisticated techniques which in fact require observations from space via satellites. Many of the programs which the various national space agencies have in their schedule will certainly contribute in this respect.

This issues are not likely to be resolved by this Workshop, but we hope that this occasion will at least make researchers in different fields aware of common aims, problems, and techniques.

Acknowledgements. The authors wish to thank the Consiglio Nazionale delle Ricerche (Italy), the Italian Ministero della Pubblica Istruzione (Italy) and the National Science Foundation (USA) for financial support.

References

Benford, G., Bodo, G., and Ferrari, A., 1978, *Astron. Astrophys.*, **70**, 815.
Cassinelli, J.P., and McGregor, K., 1986, in *The Phsyics of the Sun*, ed. P. Sturrock, New York: Springer Verlag.
Goldreich, P., and Julian, W.H., *Astrophys. J.*, **157**, 869.
Gunn, J.O., and Ostriker, J.P., 1970, *Astrophys. J.*, **160**, 979.
Hagen, W., Stencel, R.E., and Dickinson, D.F., 1983, *Astrophys. J.*, **274**, 286.
Hearn, A.G., 1985, in *The Origin of Non-Radiative Energy and Momentum in Hot Stars*, NASA CP-2358.
Holzer, T.E., 1977, *J. Geophys. Res.*, **82**, 23.
Kopp, R.A., and Holzer, T.E., 1976, *Solar Phys.*, **49**, 43.
Lucy, L.B., and Salomon, P.M., 1970, *Astrophys. J.*, **159**, 879.
Owocki, S., and Rybicki, G., 1984, *Astrophys. J.*, **284**, 337.
Parker, E.N., 1958, *Astrophys. J.*, **128**, 664.
Parker, E.N., 1963, *Interplanetary Dynamical Processes*, New York: Wiley Interscience.
Pudritz, R.E., and Norman, C.A., 1983, *Astrophys. J.*, **274**, 677.
Rosner, R., Golub, L., and Vaiana, G.S., 1985, *Ann. Rev. Astron. Astrophys.*, **23**, 413.
Weber, E.J., and Davis, L.D., 1967, *Astrophys. J.*, **148**, 217.

MASS OUTFLOW FROM THE SUN - OBSERVATIONS AND DIAGNOSTICS

G. Noci
Institute of Astronomy, University of Florence
Largo E. Fermi 5
50125 Florence
Italy

ABSTRACT. This review contains a brief description of the observational aspects of the main processes of mass outflow from the Sun and of their interpretation; it includes an indication of the problems still open in this area. In the last part the paper focuses on the diagnostic possibilities that arise from the observation of the corona with coronographs/spectrometers operating in the ultraviolet. It discusses the possibility of determining density, electron and ion temperatures, outflow speed.

1. INTRODUCTION

The main processes of mass loss from the Sun are the continuous particle emission called solar wind, the sporadic emission of clouds of plasma (coronal mass ejections, CMEs) and radiation. A simple calculation shows that the last one is the most important process, since, amounting to 4.3×10^{12} gr/sec, it is about three times more effective than the first, and more than one order of magnitude than the second. I will not discuss the solar luminosity, however, but will consider only the other two processes.

2. SOLAR WIND

2.1. General Picture

Since its discovery the solar wind has been extensively studied by "in situ" measuremets from spacecraft. Several review articles describe its properties and its connections with the solar sources (see, for example, Hundhausen, 1977; Crooker, 1983; Schwenn, 1983; and references therein).
   The solar wind parameters have considerable space-time variations; to show them one can use plate 1 of Feldman et al. (1979), that refers to solar wind observations in the 1971-1978 period. The main features emerging from this figure are the

complementarity between two couples of parameters (speed and proton temperature on one side, density and electron tempeature on the other side) and the 27-day periodicity. This periodicity, which manifests itself in the fast solar wind, points clearly to a solar source that lasts for several solar rotations. It is one of the reasons that make meaningful to single out the category of the fast solar wind; another reason is that there are sharp boundaries in the interplanetary space between slow wind and fast streams (Rosenbauer et al., 1977). The recurrent fast wind streams are therefore characterized, beyond the high speed (up to 800 km/sec in some cases), by high proton temperature and low density and electron temperature; they are characterized also by a considerably lower variability of the physical parameters, with respect to the slow wind (Bame et al., 1977), by a larger alpha particle content and much lower angular momentum flux (Schwenn, 1983).

Let us turn, now, to a description of the solar sources of the wind.

The observations of the low corona have shown that it is organized in coronal holes and in quiet and active regions. Extrapolations of the photospheric magnetic field indicate that its lines of force in coronal holes are open towards the interplanetary space, so that the coronal plasma can flow along them and escape the Sun. The low latitude holes are generally

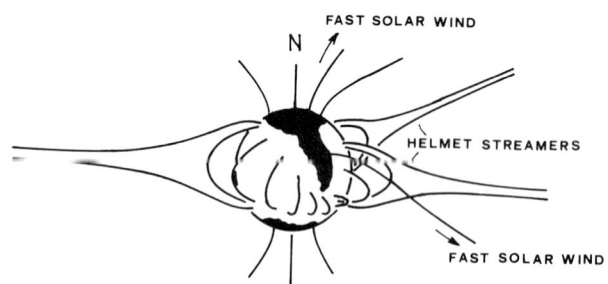

Figure 1 - A schematic view of the solar corona. The lines represent the lines of force of the magnetic field; coronal holes in black. (Adapted from Hundhausen, 1977).

connected (at least the large ones) to the polar hole of the same magnetic polarity, so that they can be thought of as extensions of this. While low latitude coronal holes last for some months and appear mainly in the declining phase of the activity cycle, the polar holes are permanent, so that the solar corona appears to be structured - troughout all the solar cycle

- in two polar caps of opposite polarity (with open magnetic field lines) and an activity zone between them. The polar caps have irregular boundaries that vary considerably with time.

Extrapolation of the photospheric field shows also that quiet and active regions are covered with very complicated closed magnetic structures, which prevent the plasma from escaping the Sun. At larger heights the structuring of the corona in these regions becomes somewhat simpler: white light pictures show the presence of "helmet streamers", which overlie active regions and have in the lower part an arch structure with the feet in regions of opposite magnetic polarity. These closed magnetic regions constitute a roughly equatorial strip separating the two polar regions of open field (fig. 1).

The width of the closed magnetic regions decreases with height up to the point where also the magnetic field inside the streamers becomes open towards the interplanetary space, so that the magnetic configuration of the corona becomes organized in two regions of open fields (coronal holes) of opposite polarity, separated by a current sheet (top of streamers) (fig. 1).

The origin of the recurrent high speed wind streams from coronal holes is now well established, but it is not quite clear yet where the source of the slow wind is. However, the picture which emerges putting together these coronal and interplanetary data is that of high speed wind streams flowing from coronal holes or from some regions inside them and slow wind from hole boundaries. The situation which results in the interplanetary space is depicted in fig. 2, where the shaded region corresponds to slow wind and the unshaded one to fast wind. Fig. 2 explains also how, in spite of an essentially dipolar structure of the magnetic field, more than two field reversals can be observed during a solar rotation from a spacecraft orbiting close to the ecliptic plane.

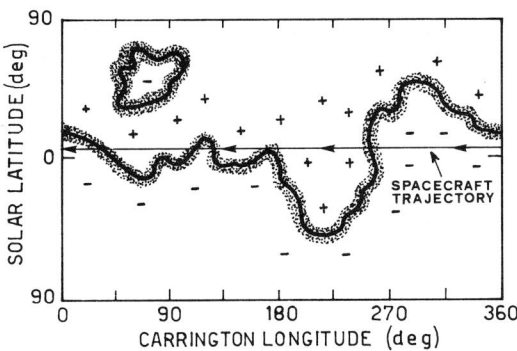

Figure 2 - Streamer belt in interplanetary space. Schematic view of its intersection with the 1 a.u. sphere: current sheet (blak curve) and low speed solar wind (shaded). (From Gosling et al., 1981).

The extreme case of slow wind is then the wind flowing from the streamers, which should be found in the interplanetary space close to the current sheet. The characteristics of the plasma around the current sheet have been studied by the Los Alamos group with the data obtained by the IMP 6, 7 and 8 satellites (1971-1978). From this study it emerges that the interplanetary signatures of streamers are opposite to those of coronal holes. In fact the statistical analysis of Borrini et al. (1981) and Gosling et al. (1981) has shown that the passage of the current sheet is accompanied, on the average, by a minimum of speed and proton temperature and a maximum of density; the He abundance has a marked minimum.

2.2. Problems

We have, therefore, a picture, probably correct in its general lines, of the Sun - solar wind connections, but many aspects of it are still poorly understood.
Let us begin with the most obvious: what causes the difference between slow and fast wind? Note that the model described above implies a geometrical difference in the low corona for the open magnetic tubes: in fact, the plasma which flows inside an open tube in a streamer runs along a bent course, in contrast with the more straight course of the plasma flowing out of the center of a coronal hole. Thus, is the difference between fast and slow wind caused by the different bending of the flow tubes in the inner corona? Or, probably related to that, is it a different increase in the cross section of the magnetic tubes where the wind flows? Theoretically it has been shown that a rapid divergence of a flow tube produces multiple critical points and the possibility of a very fast flow with a subsonic/supersonic transition very deep in the corona (Kopp and Holzer, 1977). Hence the geometry of the flow tubes in the low corona can be at the origin of the difference between fast and slow wind. Alternatively, the cause can be some difference (e.g. in energy deposition) at the coronal base between flow tubes in coronal holes and flow tubes at the sides of streamers (where the mean magnetic field is larger). On the other hand, why, in spite of the large speed differences, are the particle and energy fluxes more or less the same in fast and slow wind (Schwenn, 1983)?
This problem is an aspect of the problem of acceleration. It does not exist yet a satisfactory model of the expansion, capable of reproducing the physical parameters observed at 1 a.u. starting with realistic values at the coronal base (see, e.g., Leer et al., 1982, and references therein). It is well known that classical solar wind models are unable of attaining this result; to overcome this difficulty "ad hoc" terms of heat or momentum deposition in the wind have been added in the equations. Another approach has been to drop the Spitzer-Harm expression for the heat conduction coefficient (which assumes

transport by collisions) in the region where the collision time becomes larger than the expansion time (Holweg, 1976). In this view the heat is transported only by the electrons of the high velocity tail of the distribution (non-Maxwellian); the main result is that the reduced transport efficiency results in a decrease of the electron temperature at 1 a.u., which improves the agreement with the observations. A "correct" simple expression for the collisionless heat conduction flux does not exist, however; the expressions proposed include some factor which is simply a rough extimation. These difficulties are particularly present in the high speed streams; although, according to some authors (Olbert, 1983; Scudder and Olbert, 1983), the effect of the suprathermal electrons may be large enough to account for the high speed streams, most researcers conclude that, for the explanation of these, it is necessary to invoke energy deposition, even if one assumes inhibited thermal conduction.

Therefore it would be very interesting to get some observational data concerning these aspects, as, for example, the amount of energy deposition and its location; this could permit to discriminate between various mechanisms (MHD waves, jets of transition region material (Pneuman, 1982), others). This result would be important not only for solar and interplanetary physics, but also for fundamental plasma physics, because the conditions under which MHD waves dissipate are not clearly known.

Several other problems exist, as:
- The behaviour of the physical parameters in the interplanetary space around the current sheet, described above (a satisfactory interpretation does not exist yet).
- The large variability in the various physical parameters observed in the interplanetary space for the slow wind. Since different measurements in the interplanetary space refer necessarily to different flow tubes, what causes these differences in adjacent tubes? Are they a product of the process of transfer or are they to be traced to the solar source? Let us consider the elemental abundances: in this case particle conservation requires that the variability be due to differences in the flux of particles of a given species between adjacent tubes in the inner corona; what are the causes of it?
- The sharp boundaries existing in the interplanetary space between high speed streams and slow wind. Do these boundaries have their root in similar boundaries present in the inner corona, or their origin is to be attributed to some phenomena occurring in the interplanetary space?

2.3. Mass loss

The mass loss of the Sun due to the solar wind can be calculated assuming that the particle flux observed in the interplanetary space close to the ecliptic plane is representative also of the rest of the heliosphere at the same heliocentric distance. The observations show that the flux in the high speed streams is not much different from that in the slow wind (Schwenn, 1982); an average value of 3.5×10^8 cm^{-2} sec^{-1} at 1 a.u. can be taken as representative of both. It follows a mass loss of 1.6×10^{12} gr/sec = 2.6×10^{-14} M$_\odot$/y.

3. CORONAL MASS EJECTIONS

A third category of solar wind is more difficult to identify, namely the interplanetary extension of the coronal mass ejections (CMEs). These explosive phenomena originating in the inner corona, which consist of the ejection towards the interplanetary space of a cloud of magnetized plasma, have been identified rather recently (about fifteen years ago). Data on them have been acquired observing in the visible the continuous radiation scattered by the free electrons, both with coronographs mounted on spacecraft and with a coronograph at the Observatory of Mauna Loa. (See the reviews by Dulk, 1980; Rust and Hildner, 1980; Stewart, 1980; Dryer, 1982; Wagner, 1984).

CME observations in the corona show that the characteristics of these phenomena vary widely (speed, shape, density, associated coronal activity). For example, CMEs associated with flares have higher speeds (300-1200 km/sec above 2 r$_\odot$) than those associated with eruptive prominences (<600 km/sec) (Gosling et al., 1976), and, while the former have constant speed, the latter show substantial acceleration in the inner corona (McQueen and Fisher, 1983).

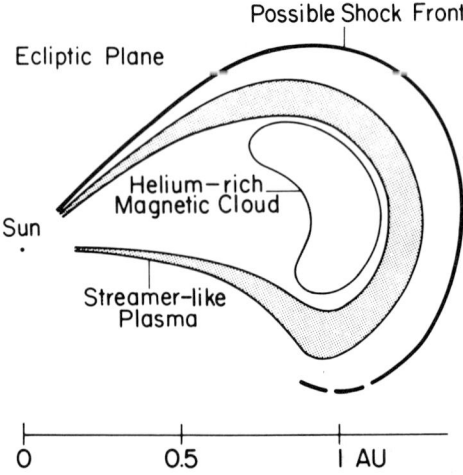

Figure 3 - Schematic picture of a coronal mass ejection (from Crooker, 1983).

In the interplanetary space and in the Earth magnetosphere several phenomena have been observed that could be due to the arrival of a CME. These can be shocks, noncompressive density enhancements, geomagnetic storms, etc.. The relationships between these phenomena are rather complicated, and controversial is their association to CMEs. It is worth mentioning, however, the rather simple scheme suggested by Crooker (1983) and shown in fig. 3. Note that the shock wave which precedes the magnetic cloud can or can not be present, according to the velocity of the cloud itself with respect to the undisturbed solar wind.

In this model the observation of the various phenomena depends on the position of the satellite or of the Earth, and on whether the shock is present or not. For example, if the satellite hits the transient marginally no helium enhancement and possibly also no shock will be observed.

Apart from the validity of this model for the CME structure in the interplanetary space, we lack the knowledge of the mechanisms (possibly more than one) originating CMEs, although theoretical effort in this direction has been made. Another fundamental point not clearly established yet is the three-dimensional shape of CMEs; there is, in fact, no agreement on whether CMEs are nearly planar or bubble shaped, with significant depth along the line of sight. (Evidence is growing in favour of the latter view, however).

The calculation of the mass loss of the Sun due to the coronal transients requires the knowledge of their frequency. According to Hundhausen et al. (1984) the obsevations gave 0.74/day at the time of Skylab (1973), 0.90/day at the time of the Solar Maximum Mission (1980), which implies a little variation with the solar cycle. Assuming that the number of CMEs which escaped obsevation (mainly because covered by the occulting disc of the coronograph) correspond to a factor of two, we can take the value of 2/day for a rough estimate. For the average mass we take 3×10^{15} gr, a value which can be in error by an order of magnitude (Wagner, 1984). It follows a mass loss of 7×10^{10} gr/sec, about twenty times smaller than that due to the solar wind (it could however be of the same order of magnitude, given the errors quoted).

4. DIAGNOSTICS

To clarify the problems listed in sections 2 and 3 one needs to measure the wind speed and other physical parameters, like density, ion and electron temperatures, etc. from the coronal base up to several solar radii of heliocentric distance.

At present we have knowledge of the physical parameters in the low corona ($r = 1 - 1.2\ r_\odot$) mainly from the UV, X-ray and radio observations. The informations we have on the more extended corona come from white light and radio observations.

The first ones concern the proper emission of an optically thin plasma, therefore they give information on its electron temperature, on its density and on the abundances. They can give information also on the velocity fields, for which there are very little data, however, concerning the corona (there are more data on the transition region chromosphere-corona, but difficult to interpret because the velocity pattern of this region is very complicated).

As for the extended corona, the white light emission comes from the photospheric continuum scattered by the free electrons and by dust particles; the former component concerns therefore the electron density. Also the radio emission comes from the free electrons and thus carries informations concerning the electron gas. Being mainly of non-thermal origin it does not concern the electron temperature of the undisturbed corona. Furthermore, being a continuum, it contains nothing about velocity, except when the source of radiation moves through the corona, if enough space resolution is available. (These velocity data, therefore, apply to coronal transients, as CMEs, not to the continuous pocess of mass loss which is the solar wind.)

To conclude, the "classical" observations of the extended corona do not yield data that concern the wind velocity and the proton, ion and electron temperatures.

Recently tecniques have been devised to accomplish this kind of measurements in the coronal range ~1.2 - ~10 $r_\odot$, by observing the radiation scattered by coronal ions (Kohl and Withbroe, 1982; Withbroe et al., 1982a).

The coronal spectrum obtained with a slitless spectrograph, launched with a rocket during the solar eclipse of March 3, 1970, showed that the Ly$_\alpha$ line is quite intense in the solar corona. The analysis of the data indicated that this line is due to scattering of the chromospheric Ly$_\alpha$ by the residual neutral H atoms in corona (Gabriel et al., 1971).

The line total (i.e. integrated over the line profile) intensity in the direction toward the observer (unit vector $\underline{n}$) is therefore given by

$$I = (h \lambda_o B_{12}/4\pi) \int_{-\infty}^{\infty} N_1 dx \int_{\Omega} p(\psi) d\omega' \int_0^{\infty} I_{ex}(\lambda,\underline{n}') \emptyset(\lambda-\lambda_o) d\lambda , \quad (1)$$

where h is Plank constant, B_{12} the Einstein coefficient for the line, N_1 the number density of hydrogen atoms in the ground level, I_{ex} the intensity of the exciting chromospheric Ly$_\alpha$, $\emptyset$ the coronal absorption profile, λ the wavelength and λ_o its laboratory value for the Ly$_\alpha$ transition. $p(\psi)d\omega'$ is the probability that a photon travelling along $\underline{n}$ was travelling in the solid angle $d\omega'$ (around $\underline{n}'$) before scattering, Ω the solid angle subtended by the chromosphere at the point of scattering and x a coordinate along $\underline{n}$. (Gabriel et al., 1971; Noci et al., 1987). Hence the intensity of the radiatively excited coronal lines depends linearly on the density, while that of the collisionally excited lines depend on its square, which implies a slower decrease with the heliocentric distance in the former case.

The continuum intensity scattered by the free electrons

(quoted above) has the expression:

$$I_\lambda = \sigma \int_{-\infty}^{\infty} N_e \, dx \int_\Omega p_e(\varphi) I_{ex}(\lambda, \underline{n}') \, d\omega' ,\qquad (2)$$

where σ is the scattering cross section, N_e the electron density, p_e the analogous of p and I_{ex} the exciting photospheric intensity.

Eqs (1) and (2) show that the intensity ratio gives the N_1/N_e ratio, i.e. the electron temperature (T_e), once the exciting intensities are measured.

It is also possible to deduce the proton temperature by measuring the profile of the scattered Ly$_\alpha$; this gives the kinetic temperature of the neutral H atoms, but the exchange time between H atoms and protons is shorter than the expansion time up to r = 0.1 a.u., so that the H kinetic temperature is in fact also the proton kinetic temperature (Holzer, 1977).

These Ly$_\alpha$ measurements (total intensity and profile) have been accomplished a few times by the Center for Astrophysics group, with the use of a coronograph/spectrometer on board a rocket, in the coronal interval 1.5 - 4 r$_\odot$ (Kohl et al., 1980; Withbroe et al., 1982b; Withbroe et al., 1985).

Another determination of N_e, independent of that based on the white light (eq. (2)), would be possible with an instrument sensible enough to detect the Ly$_\alpha$ scattered by the free electrons of the corona,

$$I_\lambda = \phi_e(\lambda-\lambda_0) \sigma \int_{-\infty}^{\infty} N_e \, dx \int_\Omega p_e(\varphi) \, d\omega \int_0^{\infty} I_{ex}(\lambda, \underline{n}') \, d\lambda ,\qquad (3)$$

where I_{ex} is now again the intensity of the chromospheric Ly$_\alpha$ and the profile ϕ_e reproduces the velocity distribution of the free electrons, so that it is expected to have a Gaussian shape with parameter proportional to $\sqrt{2kT_e/m}$, being m the electron mass and k Boltzmann constant. The electron scattered Ly$_\alpha$ will therefore be much wider than the Ly$_\alpha$ scattered by the H atoms (roughly by a factor equal to the square root of the proton/electron mass ratio). With no need of having the profile, the total intensity of the line would give the density. The ratio between this intensity and that of the Ly$_\alpha$ scattered by the H atoms will then be independent of the exciting intensity and of the instrument calibration, the intensities being determined with the same instrument; independent of these quantities will then be also the electron temperature determined by the ratio.

It would be very interesting to check this value of T_e by determining the width of the electron scattered Ly$_\alpha$, a method which, like the previous one, does not require the knowledge of the exciting intensities.

These diagnostic possibilities make it very interesting the use of a coronograph/spectrometer operating in the ultraviolet, and, in fact, coronographs planned for future space missions will have these capabilities. Furthermore, an UV coronograph would also make it possible to determine the physical parameter which is probably the most important one in the source region of the solar wind, namely the coronal expansion speed. This possibility arises from the occurrence of the phenomenon called Doppler dimming (Hyder and Lites, 1970; Beckers and Chipman, 1974). In fact, the considerations based on eq. (1) are no longer valid if the plasma that scatters the chromospheric or transition region radiation moves away from the Sun, as in the solar wind. In this case the source of exciting radiation is seen, in the frame of the absorbing plasma, moving away, and hence the exciting line shifted towards the red, which causes a misplacement with respect to the coronal absorbing profile. In other words, $I_{\ell_x}(\lambda)$ should be substituted, in eq. (1), with $I_{\ell_x}(\lambda-\delta\lambda)$, $\delta\lambda$ being the red shift due to the component of the solar wind velocity in the direction $\underline{n}'$. The corresponding decrease of the intensity of the scattered Ly_α gives the possibility of measuring the solar wind speed.

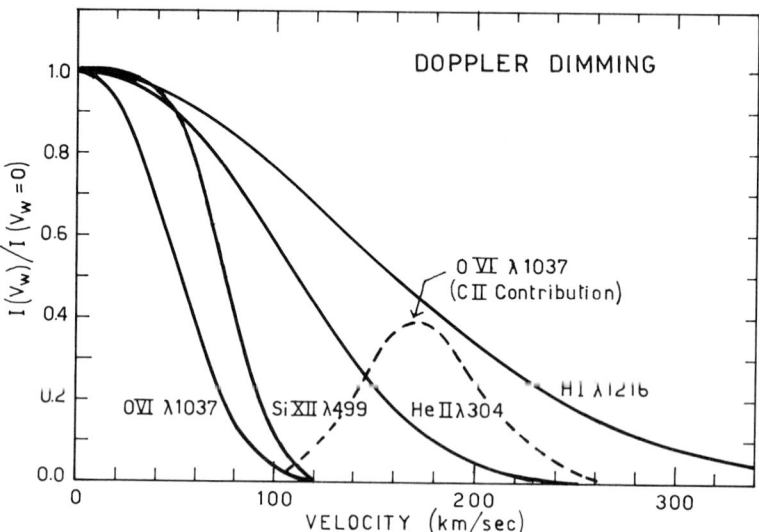

Figure 4 - Doppler dimming of the radiative component of coronal lines. (Adapted from Kohl and Withbroe, 1982, and from Noci et al., 1987).

Ly_α is the only line of the UV coronal spectrum wich is radiatively excited, also at low heights (Noci, 1971). Other UV resonance lines produced in the corona have radiative components which, although less important than the collisional components at low heights, become dominant at greater heights for the reason given below eq. (1). These lines also can be used to determine the wind speed. Since the width of the absorption profile of these lines (due to the random motions of the absorbing ions) decreases proportionally to the square root of the mass of the ion, it is possible, using lines of different ions, to vary the interval of the velocity where the effect is important. Doppler dimming of some coronal lines, calculated with an isothermal corona where the kinetic temperatures are 1.5×10^6 K, is shown in fig. 4.

Thus in regions of low outflow speed (streamers or coronal holes at low heights) one can use the Ly_α radiation (combined with the white light continuum if the electron scattered component is not available) to get the electron density and temperature, and one of the other lines to deduce the wind speed. In regions of faster wind the radiatively excited components are suppressed by Doppler dimming (except for Ly_α and for some other line which will be discussed below); in this case to determine the wind speed one should have some information on T_e (if possible, from the width of the electron scattered Ly_α, or from the collisional components) and on N_e (as before), and then apply eq. (1) to the coronal Ly_α.

Among the coronal UV lines, very promising, beyond Ly_α, are those of the OVI resonance doublet at 1031.9 and 1037.6 Å, because their radiatively excited components become dominant at low heights. The fact that we have a doublet makes it easy to separate the radiatively excited from the collisionally excited component, since the importance of the two components is different for the two lines of the doublet (Kohl and Withbroe, 1982). The collisional component will then give the OVI abundance, the radative component the outflow speed. Still better, the outflow speed can also be obtained by the doublet ratio, with no need of abundance information, if the collisional components are not negligible (Noci et al., 1987).

Furthermore for the line at 1037.6 Å it is possible to have radiative excitation, with sufficient Doppler shift, from the photons in the nearby chromospheric line at 1037.0 Å produced by CII. Accordingly the velocity sensitivity of the line is not limited to the low velocities, but goes above 200 km/sec, as shown in fig. 4 (dashed curve). For more details on the use of the resonance doublets, chiefly that of OVI, see Noci et al. (1987).

It must be pointed out that the outflow speed deduced from ion lines is not necessarily the same as that of hydrogen (and protons), particularly in the low corona (Geiss et al., 1970; Joselyn and Holzer, 1978; Borrini and Noci, 1979); it is rather a lower limit to the solar wind speed. Although this means that the information on this speed is smaller, it will make extremely interesting to measure ion and hydrogen outflow speeds in the same region using the velocity interval where the Ly_α and ion line sensitivity overlap. This could clarify the mechanism by which the heavy ions are dragged in the solar wind. (See Geiss, 1982, for a discussion of this problem).

Before finishing, it is worth touching the subject of another field of diagnostic possibilities which would open if one could also measure the ion line widths. The kinetic temperature deduced would permit to separate thermal from non-thermal (mass independent) components (comparing with hydrogen), yielding information on turbolence, wave motions, ion heating etc..

It is proper to conclude with the auspice that the diagnostic possibilities briefly described here will be realized in future space missions, in order to make it possible to determine the physical parameters of the extended corona and so throw light on the problems pointed out in the first part of this review.

References

- Bame, S.J., Asbridge, J.R., Feldman, W.C., and Gosling, J.T.: 1977, J. Geophys. Res. 82, 1487.
- Beckers, J.M., and Chipman, E.: 1974, Solar Phys. 34, 151.
- Borrini, G., and Noci, G.: 1979, Solar Phys. 64, 367.
- Borrini, G., Gosling, J.T., Bame, S.J., Feldman, W.C., and Wilcox, J.M.: 1981, J. Geophys. Res. 86, 4565.
- Crooker, N.U.: 1983, in "Solar Wind V" (NASA Conf. Publ. 2280), M. Neugebauer ed., p. 303.
- Dryer, M.: 1982, Space Sci. Rev. 33, 233.
- Dulk, G.A.: 1980, in "Radio Physics of the Sun" (IAU Symp. 86), M.R. Kundu and T.E. Gergely eds., p. 419. (Reidel, Dordrecht).
- Feldman, W.C., Asbridge, J.R., Bame, S.C., and Gosling, J.T.: 1979, J. Geophys. Res. 84, 7371.
- Gabriel, A.H., Garton, W.R.S., Goldberg, L., Jones, T.J.L., Jordan, C., Morgan, F.J., Nicholls, R.W., Parkinson, W.J., Paxton, H.J.B., Reeves, E.M., Shenton, C.B., Speer, R.J., and Wilson, R.: 1971, Astrophys. J. 169, 595.
- Geiss, J.: 1982, Space Sci. Rev. 33, 201.
- Geiss, J., Hirt, P., and Leutwyler, H.: 1970, Solar Phys. 12, 458.
- Gosling, J.T., Borrini, G., Asbridge, J.R., Bame, S.C., Feldman, W.C., and Hansen, R.T.: 1981, J. Geophys. Res. 86, 5438.
- Gosling, J.T., Hildner, E., McQueen, R.M., Munro, R.H., Poland, A.I., and Ross, C.L.: 1976, Sol. Phys. 48, 389.
- Hollweg, J.V.: 1976, J. Geophys. Res. 81, 1649.
- Holzer, T.E.: 1977, Rev. Geophys. Space Phys. 15, 467.
- Hundhausen, A.J.: 1977, in "Coronal Holes and High Speed Wind Streams", J.B. Zirker ed., p. 225. (Colo. Assoc. Univ. Press, Boulder).
- Hundhausen, A.J., Sawyer, C.B., House, L., Illing, R.M.E., and Wagner, W.J.: 1984, J. Geophys. Res. 89, 2639.
- Hyder, C.L., and Lites, B.W.: 1970, Solar Phys. 14, 147.
- Joselyn, J., and Holzer, T.E.: 1978, J. Geophys. Res. 83, 1019.
- Kohl, J.L., Weiser, H., Withbroe, G.L., Noyes, R.W., Parkinson, W.H., Reeves, E.M., Munro, R.H., and MacQueen, R.M.: 1980, Astrophys. J. 241, L117.
- Kohl, J.L., and Withbroe, G.L.: 1982, Astrophys. J. 256, 263.
- Kopp, R.A., and Holzer, T.E.: 1976, Solar Phys. 49, 43.

- Leer, E., Holzer, T.E., and Fla, T.: 1982, Space Sci. Rev. 33, 161.
- McQueen, R.M., and Fisher, R.R.: 1983, Sol. Phys. 89, 89.
- Noci, G.: 1971, in "Physics of the Solar Corona", C.J. Macris ed., p. 13. (Reidel, Dordrecht).
- Noci, G., Kohl, J.L., and Withbroe, G.L.: 1987, Astrophys. J. 315, 706.
- Olbert, S.: 1983, in "Solar Wind V" (NASA Conference Publ. 2280), M. Neugebauer ed., p. 149.
- Pneuman, G.W.: 1982, Astrophys. J. 265, 468.
- Rosenbauer, H., Schwenn, R., Marsch, E., Meyer, B., Miggenrieder, H., Montgomery, M.D., Muhlhauser, K.H., Philipp, W., Voges, W., and Zink, S.M.: 1977, J. Geophys. 42, 561.
- Rust, D.M., and Hildner, E.: 1980, in "Solar Flares", P.A. Sturrock ed., p 273. (Colo. Assoc. Univ. Press, Boulder).
- Schwenn, R.: 1983, in "Solar Wind V" (NASA Conference Publ. 2280), M. Neugebauer ed., p. 489.
- Scudder, J.D., and Olbert, S.: 1983, in "Solar Wind V" (NASA Conference Publ. 2280) M. Neugebauer ed., p. 163.
- Stewart, R.T.: 1980, in "Solar and Interplanetary Dynamics" (IAU Symp. 91), M. Dryer and E. Tandberg-Hanssen eds., p 333. (Reidel, Dordrecht).
- Wagner, W.J.: 1984, Ann. Rev. Astron. Astrophys. 22, 267.
- Withbroe, G.L., Kohl, J.L., Weiser, H., and Munro R.H.: 1982a, Space Sci. Rev. 33, 17.
- Withbroe, G.L., Kohl, J.L., Weiser, H., and Munro R.H.: 1985, Astrophys. J. 297, 324.
- Withbroe, G.L., Kohl, J.L., Weiser, H., Noci, G., and Munro, R.H.: 1982b, Astrophys. J. 254, 361.

PROPERTIES OF WINDS AND CHROMOSPHERES IN G TO M GIANTS AND SUPERGIANTS

Dieter Reimers
Hamburger Sternwarte, Universität Hamburg
Gojenbergsweg 112, 2o5o Hamburg 8o
Federal Republic of Germany

ABSTRACT. Our present knowledge of winds and chromospheres of G, K and
M giants is reviewed which in recent years has grown fast, mainly due
to the advent of X-ray and UV satellites. In particular, it is demonstrated that in Zeta Aur type eclipsing binaries, where a B star moves within the wind of a K or G supergiant, the B star can be used as a light
source which probes the wind and extended chromosphere of the cool supergiant with spatial resolution. It is shown that with more accurate wind
velocities and mass-loss rates having become available, systematic trends
of wind properties with stellar parameters become apparent.

I. INTRODUCTION

In the first two decades of research on mass-loss of cool stars, the
main emphasis was on determining from optical spectra as accurately as
possible the rate of mass-loss with the aim to help to understand late
stages of stellar evolution. Mass-loss rates remained fairly uncertain,
although systematic trends were first recognized around 1974 (Reimers,
1975 a, b). A more recent overview over optical data can be found in
Dupree (1986).
 With the advent of X-ray and UV-satellites a new era of wind
studies of cool giants began. For more detailed reviews of UV and X-ray
data see Linsky (1985), Dupree and Reimers (1987) and Reimers
(1987 a, b).
 In this overview, the main emphasis will be on the exploitation of
the binary technique with IUE and is an attempt to find trends in cool
stellar wind data on the basis of more accurate results compared to
what was available 1o years ago.

2. MASS-LOSS INDICATORS AND INCIDENCE OF MASS-LOSS IN THE HR DIAGRAM

Circumstellar CaII, MgII and Ly α : All stars cooler and more luminous
than on a line in the HR diagram defined roughly by (K5, M_V = o),
(K4, -1), (K2, -1.8), (G5, -4), (Go, -4.5) have far-shifted CaII H+K

absorption components (Reimers, 1977a), cf. Fig.1. In case of the spectroscopic binaries μ UMa (MoIII) and ξ Cyg (K5Ib) it could be shown that these K4 absorption lines remain stationary - while the photospheric lines move back and forth due to orbital motion - and thus indicate circumstellar matter.

Similar far-shifted absorption lines have been found in the MgII resonance lines (Hartmann et al., 1981 ; Reimers, 1982, Dupree and Reimers, 1987). Due to larger optical depths of the MgII lines (higher Mg abundance) compared to CaII lines the MgII visibility limit is at slightly lower luminosity and higher temperatures. Line asymmetries most probably caused by winds have also be seen in Ly α, in α Aur, and α TrA (cf. Dupree and Reimers, 1987). In the CaII and MgII resonance doublets, the ratio of the violet to the red emission peak (V / R) provides more indirect evidence for outflow of matter from red giants. Stencel (1978) and Stencel and Mullan (1980) have shown that the lines bounding those regions in the HR diagram where V/R > 1 (no outflow) changes to V/R < 1 (outflow) are near to and parallel to the line in the HR diagram beyond which stars have CS lines. It is not clear whether V/R < 1 values are just caused by blue-shifted absorption by escaping material (far away from the star) or are caused by expanding chromospheres, or both. Without detailed modelling, the origin of the asymmetry of the selfreversed CaII and MgII profile remains unclear. Even the conclusion that V/R < 1 means mass-loss appears still speculative.

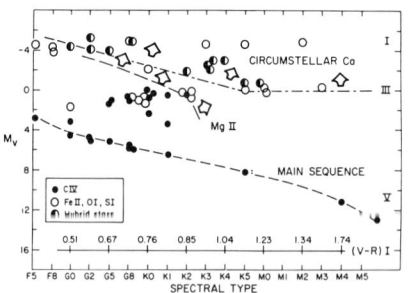

Figure 1. The presence of various spectral lines in stars of different spectral types and luminosity classes. The broken line denotes the boundary above which variable CS CaII lines appear (Reimers, 1977 a). The position of the MgII asymmetry change is also indicated (Stencel and Mullan, 1980), from Dupree and Reimers (1987).

Other means to detect mass-loss in red giants are far shifted HeI 10830 Å components in hybrid stars, shifted CaII H+K emission, radio emission of partially ionized winds in nearby giants, thermal emission from dust and molecules in highly evolved stars - which will not be discussed here - , and last not least the spectra of hot binary companions of red giants.

3. WIND VELOCITIES

A survey of stellar wind velocities in the whole red giant region of the HR diagram was first conducted by means of CaII K_4 shifts (Reimers,

1977 a).

Detailed measurements of wind velocities by means of high-resolution IUE spectra taken at various binary phases of ζ Aur, 31 Cyg, 32 Cyg, 22 Vul, and δ Sge confirm the values found more simply from K_4 shifts.

The massive stellar winds of cool stars are characterized by low asymptotic flow speeds v_w, usually less than the surface escape velocity v_{esc}. The energy required to drive the winds is $\dot{E}_w = \frac{1}{2} \cdot \dot{M} \cdot (v_w^2 + v_{esc}^2)$, i.e. the sum of the kinetic energy $\dot{E}_{kin}$ of the wind and the energy $\dot{E}_{pot}$ required to lift the wind out of the gravitational field of the stars. Usually the second term dominates in large, low gravity stars. The distribution of wind energy between kinetic energy and potential energy can be seen in more detail in Fig.2 where we plot wind velocity v_w versus escape velocity v_{esc} (data from Table 1). Virtually all cool stars are placed between the lines $v_w = v_{esc}$ and $v_w = 10^{-1} v_{esc}$ which corresponds to $\dot{E}_{kin} = \dot{E}_{pot}$ and $\dot{E}_{kin} = 10^{-2} \cdot \dot{E}_{pot}$.

In the range $v_{esc} \sim$ 100 to 160 km/s stars occupy the whole range between $v_w/v_{esc} = 1$ and 10^{-1}. G supergiants (e.g. α Aqr and 22 Vul) and KII hybrid atmosphere stars are close to $v_w = v_{esc}$, M supergiants like α Ori are near $v_w = 10^{-1} v_{esc}$, while MII giants and K supergiants like 32 Cyg occupy the intermediate regime.

Although the data basis is still too small it appears that above $v_{esc} = 160$ km/s and below 80 km/s the stars are close to $v_w = v_{esc}$ and $v_w = 10^{-1} v_{esc}$, respectively. We have yet no interpretation for the transition from the "high velocity mode" (Solar type winds and hybrid star winds) to the "low velocity mode" (M giants and supergiants) where nearly all the driving energy of the wind is used to lift the escaping matter out of the potential well. The way of transition from one type to the other (Fig.2) may be an indication for a transition from one dominating wind acceleration mechanism to another mechanism while in the intermediate regime both (or several) mechanisms operate simultaneously.

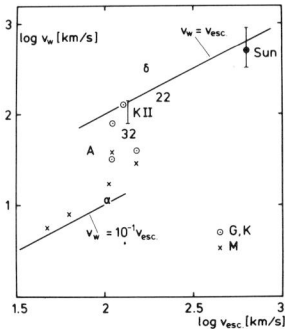

Figure 2. Stars with well determined wind velocities v_w (= asymptotic flow speeds) and gravitational escape velocities v_{esc} from Table 1. $\alpha = \alpha$ Ori, $A = \alpha$ Boo, $\delta = \delta$ And, $22 =$ Vul, $32 = 32$ Cyg.
The upper and lower lines correspond to $\dot{E}_{kin} = \dot{E}_{pot}$ and $\dot{E}_{kin} = 10^{-2} \cdot \dot{E}_{pot}$ for the winds.

TABLE 1. Compilation of data for stars with reliable wind studies (for Figs. 2 and 7)

Star	Sp.	Stellar data T_e	$\frac{R}{R_\odot}$	$\lg \frac{L}{L_\odot}$	$\frac{M}{M_\odot}$	Wind data $\dot{M}$ [$M_\odot$ yr^{-1}]	v_w [km/s]	v_{esc}	$\frac{\dot{E}_{wind}}{4\pi R^2}$ [10^5 erg cm^{-2} s^{-1}]	Source for wind data
α Ori	M 2 I ab	3500	740	4.87	~20	$2 \cdot 10^{-6}$	11	102	2	1
α Sco	M 1.5 ab	3540	625	4.68	~18	$2.2 \cdot 10^{-6}$	17	105	1.5	2
α¹ Her	M 5 II	3300	178	3.29	1.7	$1 \cdot 10^{-6}$	8	61	0.7	3
32 Cyg	K 5 I ab	3800	188	3.82	8	$1.6 \cdot 10^{-6}$	60	128	0.84	4
31 Cyg	K 4 I b	3800	202	3.91	6.2	$0.7 \cdot 10^{-7}$	80	109	0.96	5
ζ Aur	K 4 I b	3950	140	3.41	8.3	$1.1 \cdot 10^{-7}$	40	153	0.38	6
δ Sge	M 2 II	3600	140	3.43	8	$2.8 \cdot 10^{-8}$	28	149	1.3	7
22 Vul	G 3 I b-II	5200	~40	2.99	4.3	$4 \cdot 10^{-8}$	160	203	~13	7
α Boo	K 1 III p	4250	28	2.36	0.5	$6 \cdot 10^{-9}$	40	83	0.11	7
						$2 \cdot 10^{-10}$:*				8
						$6 \cdot 10^{-10}$				9
Sun	G 2 V	5780	1	1	1	(<7 $\cdot 10^{-9}$)			(<3.8)	10
						$2 \cdot 10^{-14}$	470	620		
HR 8752	G 0 I a	5000	1000	5.5	30	$2 \cdot 10^{-5}$	30	108	0.63	11
α Aqr	G 2 I b		120		~5	10^{-5}	127	130	6.7	12
Hybrids	K 3, 4 II		85		~4		80-180	~135		13
δ And	K 3 III		22		1.8		300	~170		14
o Cet			210		1.2		5.6	47		15
HR 3153	M 1 II		162		5		38	110		16,17

(*) for 30 % hydrogen ionization (upper limit is $7 \cdot 10^{-9}$ with H$^+$/H = 0.01)

(1) Mauron (1985) (2) Bowers and Knapp (1986) (3) Hagen et al. (1987) (4) Hjellming and Newell (1983)
(5) Kudritzki and Reimers (1978) (6) Reimers (1977b) (7) Che et al. (1983) (8) Reimers and Schröder (1983)
(9) Reimers and Che-Bohnenstengel (1986) (10) Linsky (1986) (11) Zirker (1984) (12) Lambert and Luck
(1978) (13) Reimers (1987 b)
(14) ibid. (15) Judge et al. (1987) (15) Reimers and Cassatella (1985) (17) Reimers (1977 c)

4. MASS-LOSS RATES

Optical CS lines of single stars: While the winds of late type giants are easily detected as violet shifted cores - at high resolution P Cyg profiles superimposed upon photospheric line cores - of strong resonance lines of neutral or singly ionized metals like CaII, MgII, TiI and TiII, BaII, SnII, NaI, FeI, ... it has turned out to be impossible to determine mass-loss rates quantitatively from these lines. The reason is that while it is possible to measure ion column densities N_{ion} and wind velocities v_w from a theoretical analysis of the P Cyg type lines, it is not possible to infer from spectroscopic observations where in the line of sight the CS lines are formed. Since the mass-loss rate $\dot{M} \propto N_{ion} \cdot v_w \cdot R_i$, where R_i is the inner shell radius, and $R_i \neq R_{star}$ is not known, $\dot{M}$ cannot be determined from CS lines of single M giants and supergiants.

The only technique for measuring mass-loss of single (nonbinary) stars appears to be spatially resolved imaging of CS shells in scattered resonance line photons like KI 7699 Å or NaD (Mauron et al., 1984; 1986). In case of α Ori, the KI line was seen as 5o" from the star.

However, since KI and NaI are minor ionization species, reliable knowledge of the ionization of metals and of the formation of CO (carbon can be a major electron donor) is necessary. Nonequilibrium effects (flow time ∿ recombination time scale in the outer envelope) could be shown to be negligible for KI while large effects have been found for α Ori for CaI, CaII and MgI at distances $\geq 10^3$ stellar radii ($\gtrsim$ 2o" apparent distance) from detailed ionization calculations including outflow effects (Robel, 1987).

Microwave continuum emission: The partly ionized winds of K and M giants reveal thermal ff-emission. Four nearby giants have been positively detected at 6 cm and 2 cm with the VLA with spectral indices close to the o.6 ($S_\nu \sim \nu^{0.6}$) as predicted for an optically thick wind (Drake and Linsky, 1986). The rate of loss of ionized matter for α Her is ∿ 1 % of the total rate as determined by the binary technique (Reimers, 1977 b). If similar ionization degrees are valid for K giants like α Boo, a mass-loss rate of ∿ $7 \cdot 10^{-9}$ $M_\odot$/yr is derived. This is certainly an upper limit, since ionization degrees could be higher than in α Her (M5II). From chromospheric modelling using MgII emission, Linsky (1986) gives $\dot{M} = 2 \cdot 10^{-10}$ $M_\odot$/yr and an ionization fraction of ∿ 5o % for α Boo.

Binary technique: The rate of mass-loss can be determined from CS absorption lines of a predominant stage of ionization seen in the spectrum of either a visual companion or of a hot companion which can be separated from the red giant in the UV due to its complementary spectral energy distribution. In both cases one avoids the difficulty of locating the shell, since the geometry of the visual system or the binary system with known orbital elements permits to locate the origin of the CS lines.

The visual binary technique has been applied to α Her (Deutsch, 1956 ; Reimers, 1977 b), α Sco (Kudritzki and Reimers, 1978) and to o Cet (Reimers and Cassatella, 1985 b).

With the launch of the IUE, the UV technique could be applied to a number of ζ Aur systems (eclipsing binaries) and VV Cep systems (like

α Sco, Boss 1985). At IUE wavelengths, the optical separation of red
giants with hot companions can be replaced by a separation through complementary energy distribution of the components. At IUE wavelengths, in
particular in the short-wavelength range, one observes a pure B star
spectrum upon which numerous CS P Cygni type lines formed in the extended wind and chromospheric absorption lines (near eclipse) of the
red giant are superimposed.

The B star serves as an astrophysical light source (a "natural
satellite") which moves around in the wind of the red giant. However,
compared to widely separated visual binaries, a number of additional
difficulties arise

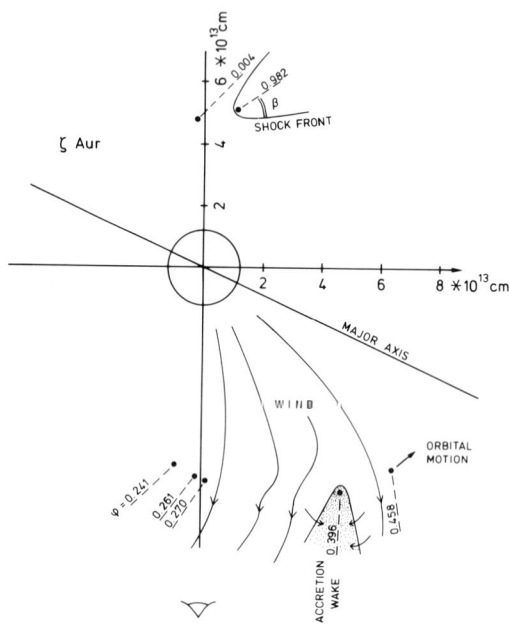

Figure 3. A roughly to-scale presentation of the accretion shock front
and wake as observed for ζ Aur (from Che and Reimers, 1986).

- a non-spherical, 3-dimensional line transfer problem has to be solved since the light source (B star) is excentric from the wind symmetry center. Computer codes that solve this problem have been developed in the 2-level approximation by Hempe (1982, 1984) and for the multilevel case by Baade (1986)
- the wind is disturbed in the immediate surrounding of the B star as it moves supersonically through the wind and forms an accretion shock front (Chapman, 1981). However, a detailed study of the accretion shocks has shown that their geometrical size is very small compared to the CS shell and can be neglected in line transfer calculations (Che-Bohnenstengel and Reimers, 1986), c.f. Fig.3.
- the hot B star ionized the wind, i.e. an HII region is formed within the red giant wind. In 31 Cyg and α Sco, in particular, the size of the HII region is large, and it has to be taken into account quantitatively since ions like SiII and FeII which are used for the mass-loss rate determination may be doubly ionized within the HII regions.

On the other hand, ζ Aur binaries are the only stars besides the Sun where the winds and extended chromospheres can be studied with

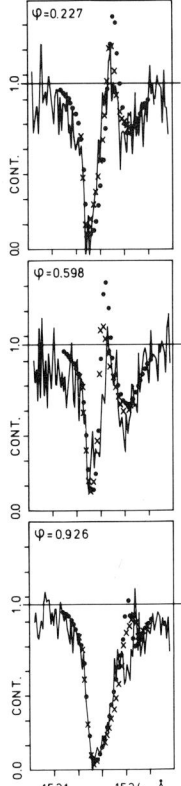

Figure 4. Observed and theoretical SiII UV Mult.1 line at different phases for 22 Vul (G3Ib-II), from Reimers and Che-Bohnenstengel (1986).

spatial (height) resolution.

In ζ Aur/VV Cep systems, the wind is visible at all phases in P Cyg type profiles (during total eclipse of B star pure emission lines) of ions like FeII, SiII, SII MgII, CII, AlII, and OI. These lines are formed by scattering of B star photons in the wind of the red giant. A few wind lines like FeII Mult.9 ($\sim$1275 Å) are seen in pure absorption due to the branching ratios of the upper levels which favour reemission as FeII UV Mult.191 photons (Hempe and Reimers, 1982; Baade, 1986).

Theoretical modelling of wind line profiles and of their phase dependency has yielded accurate mass-loss rates and wind velocities for a number of systems (Table 1). It has turned out that a good mass-loss determination requires both phases with the B star in front of the red supergiant (which yields wind turbulence v_t) and phases with the B star behind the red supergiant (which yield the wind velocity v_w). Typically, $v_w \approx 2 v_t$. Further details can be found in Che et al. (1983). It turned out that it was possible to match the circumstellar line profiles at all phases with one set of parameters v_w, v_t and - within a factor of 2 - one mass-loss rate $\dot{M}$ (Fig.4). This means that at least in the orbital plane the envelope asymmetries (in density) are within a factor of 2 on a length-scale of several K giant radii.

In several binary systems consisting of a red giant and a hotter companion, the B star is early enough (earlier than $\sim$ B 4) to ionize the wind in part. The HII region within the wind produces thermal Bremsstrahlung which has been detected in a number of systems. Mass-loss rates have been determined for α Sco (Hjellming and Newell, 1983) and HR 8752 (Lambert and Luck, 1978). A compilation of the most accurate mass-loss rates of red giants is given in Table 1 and Fig.7.

5. EXTENDED CHROMOSPHERES AND WIND EXPANSION

There is growing evidence that closely related to strong mass-loss in red giants is the existence of geometrically extended chromospheres.

During a lunar occultation of 119 Tau (M2Iab) it was observed that H_α light comes from a region having at least twice the diameter of that which produces the continuum. In α Ori it was observed by means of speckle interferometry in H_α that the detectable H_α diameter is about 5 times the stellar diameter (Goldberg et al., 1981 ; Hege et al., 1986). It was also found that the extended H_α chromosphere has an elongated shape. Recall that during a solar eclipse bright red features (H_α) are seen here and there around the obscured disk. This was the origin of the name "chromosphere" of the Sun.

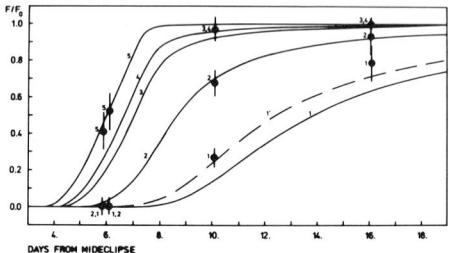

Figure 5. Light curves at 1350 Å (1), 1513 (2), 1983 Å (3), 1960 Å (4) and 2992 Å (5) during eclipse of 32 Cyg. Solid line represents best fit with model chromosphere (from Schröder, 1986).

In Zeta Aurigae type eclipsing binary systems, the extended chromospheres can be studied with height resolution when the B star moves behind the extended atmosphere of the K supergiant. IUE observations with high spectral resolution during chromospheric eclipse phases offer several advantages over optical studies as performed in the 1950s (cf. Wilson, 1960 b): (i) in the UV at wavelengths $\lambda \lesssim 2800$ Å, the B star provides a smooth continuum on which the chromospheric absorption lines are superimposed (ii) in strong UV lines the chromosphere can be seen up to projected heights of about one stellar radius (iii) the steep increase of continuous opacity - Rayleigh scattering at neutral hydrogen - provides a means to derive a density model of the inner chromosphere from the wavelength and time dependence of totality in UV below 2000 Å.

Schröder (1985 a, 1985 b, 1986) has used both absorption lines and continuum data (Fig.5) for constructing empirical model chromospheres for 32 Cyg, ζ Aur and 31 Cyg. Chromospheric densities could be represented by power laws of the form $\rho \sim r^{-2} \cdot h^{-a}$ with a $\approx$ 2.5 where r is the distance from the center of the star and h is the height above the photosphere. The empirical density distribution shows that after a steep decrease in the inner chromosphere already in the upper chromosphere (height > 1/2 to 1 giant star radii R_*) expansion starts ($\rho \sim r^{-2}$), a typical density at a height of 2 R_* is 10^7 cm^{-3} (Fig.6).

Since one observes total particle densities in the expanding chromosphere up to h $\approx$ 1.5 R_K , and in addition the wind density and

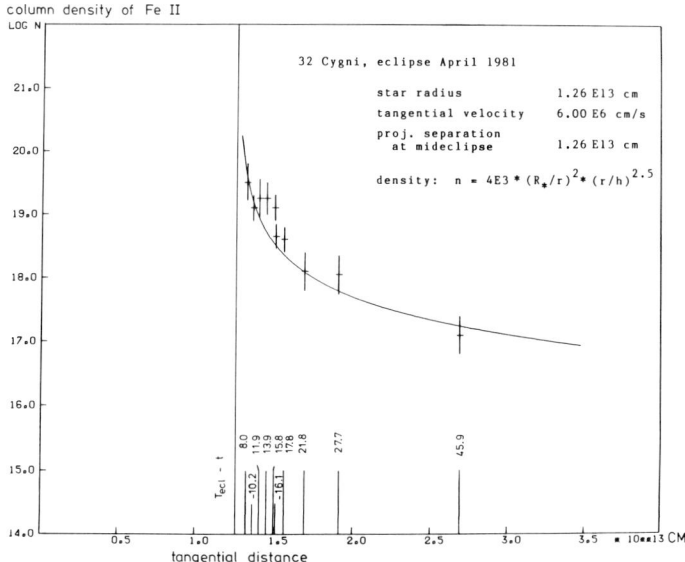

Figure 6. Chromospheric density distribution obtained from curve of growth analysis of ultraviolet FeII lines (Schröder, 1985 a, b).

velocity outside of 5 R_K, one can try to look for consistency by assuming a steady wind, i.e. to apply the equation of continuity. Using $\dot{M} = 4\pi r^2 \cdot \rho(r) \cdot v(r)$ and $\rho(r) = \rho_0 \cdot (R_*/r)^2 \cdot (r/(r-R_*))^a$ for the chromosphere, we find $v(r) = \dot{M}(4\pi\rho_0 R_*^2)^{-1}(1 - R_*/r)^a$ and a wind terminal velocity $v_\infty = \dot{M}(4\pi\rho_0 \cdot R_*^2)^{-1}$ which can be checked with observed values for ρ_0, $\dot{M}$ and v_∞ for consistency. For 32 Cyg and 31 Cyg Schröder (1985a) found consistency, which means that the empirical density distribution (when extrapolated by the equation of continuity to the outer wind) yields the correct mass-loss rate. In case of ζ Aur, the chromospheric density distribution was far too steep - at least at that particular limb position during eclipse - to give the mean mass-loss rate, which might be a stellar analogue to a solar coronal hole.

According to the semiempirical velocity law v(r), the velocity increases steeply close to the star (steeper than in the Sun, cf. Reimers, 1987b), due to the fact that the driving energy of the wind is mainly potential energy (cf. Fig.2). In giants like 32 Cyg, the relatively low asymptotic flow speed ($v_\infty/v_{esc} = 0.4$) constrains the amount of energy and momentum added to the flow in the supersonic regime and requires a way of "fine tuning", a self-regulating mechanism (cf. Holzer and McGregor, 1985).

The empirical model chromosphere of ζ Aur also provides a test of a method developed to derive the electron density, electron temperature and the geometrical extent of giant star chromospheres from the CII (UV 0.01) λ 2325 Å multiplet and the CII 1335 Å resonance doublet (Stencel et al., 1981; Carpenter et al., 1985; Brown and Carpenter, 1985).

Using the CII method, it was claimed that coronal stars, i.e. below the wind/corona 'dividing line', have a small geometrical extent ($R/R_* < 1.001$) of the CII emitting region, while noncoronal stars have typically $R/R_* \approx 2$ (1.4 to 5).

Application of the CII 2325 Å method to a double shift high-resolution spectrum taken of ζ Aur during total eclipse of the B star - at other phases the wavelength range around 2300 Å is dominated by the B star - yielded the following results (Schröder et al., 1987): i) the CII 2325 Å flux is matched with the empirical model chromosphere obtained from eclipse data (Schröder, 1985), ii) most CII 2325 Å is emitted by the innermost chromosphere, iii) line intensity ratios within the CII 2325 Å multiplet are affected by optical depth effects, at least in supergiants, and iv) the method to determine the geometrical extent of giant star chromospheres from CII emission gives results which are quantitatively incorrect.

6. PHYSICAL PROPERTIES

In case of 32 Cyg and 22 Vul, the wind electron temperature T_e has been estimated from the observed population of excited FeII levels.

In the wind of 32 Cyg, at distances of more than 5 R_K (K supergiant radii), Che-Bohnenstengel (1984) derived a wind electron temperature $T_e \approx 4800$ K for 1 % hydrogen ionization ($n_e/n_H = 0.01$) and $T_e \approx 10^4$ K for smaller electron densities.

In the chromosphere of 32 Cyg, Schröder (1986) estimated $T_e \approx 8500$ K at 0.2 R_K and an increase to about 11 000 K at 0.5 R_K height. Hydrogen ionization appears to increase over the same range from about $n_e/n_H = 10^{-3}$ to 10^{-2}. This means that strong nonradiative heating occurs in heights above the photosphere where the wind starts and most of the wind driving energy - mainly potential energy, cf. Table 1 - is deposited. It can be shown that for the semiempirical velocity law discussed above, the energy deposition into the wind per unit mass reaches its maximum at ~ 1 R_K above the photosphere.

The results for 32 Cyg are consistent with radio observations of α Ori which imply an extended chromosphere with a temperature in the range 7000-9000 K (Wischnewski and Wendker, 1981). In the semiempirical model of the outer atmosphere of α Ori by Hartmann and Avrett (1984), which matches line profiles of CaII, MgII, H_α and SiII, microwave emission, and a mass-loss rate of 10^{-6} $M_\odot$/yr, chromospheric temperatures of 5000-8000 K extend outwards to about 10 R_*.

In 22 Vul, a further G2Ib-II Zeta Aur type eclipsing binary, which is in several respects (high wind velocity, rotation, intermediate mass) similar to hybrid stars like α Aqr, a wind electron temperature of $30 \pm 10 \cdot 10^4$ K was estimated with the assumption of pure electron collision excitation (Reimers and Che-Bohnenstengel, 1986). However, since radiative excitation via high levels cannot be excluded at present, it is highly desirable to prove or disprove the existence of a high-temperature wind in a G supergiant. The location of 22 Vul in Fig.2 is also not inconsistent with a relatively high wind temperature.

There is little direct evidence for wind temperatures at large distances from the stars. Only in the case of α Her, from the observed absence of lines from excited fine structure levels of TiII, one can exclude that T >> 100 at a distance of 300 M giant radii (Reimers, 1977b), consistent with adiabatic cooling of the wind at large distances ($T \sim r^{-4/3}$).

7. IMPLICATION FOR STELLAR EVOLUTION

Table 1 summarizes the most accurately determined mass-loss rates and wind velocities for G, K and M giants and supergiants determined mainly by means of the binary technique. In addition, mass-loss rates from the KI imaging technique and the 21 cm line emission are included. The two stars (α Ori, α Sco) for which good rates are available from completely independent techniques indicate the accuracy which can be achieved (factor ~ 2). Mass-loss rates for more stars and in particular more accurate values cannot be expected in the near future for normal giants. The situation is probably more favourable for highly evolved stars with molecular line emission. However, for most of these stars (like OH-IR stars) stellar parameters like mass are poorly known.

I shall therefore briefly discuss the implications for stellar evolution. Nearly all stellar evolution calculations which included mass-loss in the red giant stage have applied the semiempirical scaling law
$\dot{M}$ ($M_\odot$/yr) = $\eta \cdot 4 \cdot 10^{-13}$ L/g $\cdot$ R   (1) proposed on dimensional arguments and calibrated with empirical mass-loss rates for a number of Pop I

giants and supergiants (Reimers, 1975). The dimensionless factor η (1/3 ≤ η ≤ 3) has been introduced in order to take into account the then considerable uncertainty of mass-loss rates.

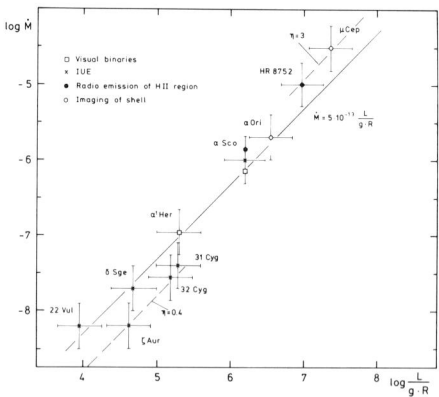

Figure 7. Mass-loss rates $\dot{M}$ $(M_\odot/yr)$ for stars from Table 1 versus L/g·R (in solar units).

For this reason we test the simple scaling law with the improved empirical mass-loss rates.

In Fig. 7, the mass-loss rates from Table 1 are plotted versus L/g·R . Two conclusions are evident
1) All observed mass-loss rates can be represented within realistic error bars (factors of 2 in both $\dot{M}$ and L/g·R to either side) by a relation $\dot{M} = 5 \cdot 10^{-13}$ L/g·R .   (η = 1.25)
2) The new empirical rates seem to indicate, however, a tendency for a steeper dependence of mass-loss on luminosity, resp. L/g·R.

A comparison of observed late stages of stellar evolution with evolutionary calculations applying red giant mass-loss in the simple parametrized form imposes constraints on η
i) Horizontal branch star masses and the observed upper luminosity limit of globular cluster AGB stars require η ≈ 0.4   (Renzini, 1977) for Pop. II stars
ii) The observed maximum initial mass for white dwarf progenitors M_{WD} as determined from white dwarf members of the intermediate age cluster NGC 2516 is M_{WD} ≈ 8 $M_\odot$ or even higher (Reimers and Koester, 1982).
iii) Mass-loss rates of advanced AGB stars (OH-IR stars) seem to be higher by about an order of magnitude than predicted by the scaling law with η = 1 (Baud and Habing, 1983). Similarly, PN demand higher mass-loss rates if interpreted as excited winds of red giant progenitors.

Inspection of Fig.7 shows that for luminosities typical at the tip of the Pop.II AGB (log L $\simeq$ 3.2, log L/g·R $\simeq$ 5.3), η = 0.4 does not contradict observations within realistic errors.

Similarly, for high luminosity η = 3 gives a good fit to observations. According to Iben and Renzini (1983), η = 3 would yield M_{WD} $\simeq$ 1o. However, observations at high luminosities in Fig. 7 are from massive stars and should not be extrapolated deliberately to advanced evolutionary stages of intermediate mass stars. Baud and Habing have proposed a parametrization of observations of mass-loss from OH-IR stars in the form

$$\dot{M}(t) = \mu \cdot \frac{L \cdot R}{M_e(t)} \quad \text{where} \quad \mu = \frac{M_e}{M} \cdot 4 \cdot 10^{-13}$$

and $M_e(t)$ is the envelope mass instead of the total mass. This modified relation gives about the same rates for most of the AGB lifetime as relation (1), while at the end of the AGB phase the reduced envelope mass leads to a steep increase.

In conclusion, accurately determined mass-loss rates are in full agreement with stellar evolution constraints. However, a semiempirical fit of mass-loss rates with $\dot{M} \sim L/g \cdot R$ (or $\dot{M} \sim (L/g \cdot R)^{1.3}$ as might be derived from Fig.7 should not be extrapolated to stars of distinctly different properties like OH-IR stars with thick dust shells, F stars, carbon stars etc.

REFERENCES

Baade,R. 1986, Astron.Astrophys. 154, 145
Baud,B., Habing,H.J. 1983, Astron.Astrophys. 127, 73
Bowers,P.F., Knapp,G.R. 1986 in "Workshop on the Late Stages of Stellar Evolution", Calgary
Brown,A., Carpenter,K.G. 1984, Astrophys.J. 287, L43
Cacciari,C., Freeman,K.C. 1983, Astrophys.J. 268, 185
Carpenter,K.G., Brown,A., Stencel,R.E. 1985, Astrophys.J. 289, 676
Che-Bohnenstengel,A., 1984, Astron.Astrophys. 138, 333
Che-Bohnenstengel,A., Reimers,D. 1986, Astron.Astrophys. 156, 172
Che,A., Hempe,K., Reimers,D. 1983, Astron.Astrophys. 126, 225
Cohen,J. 1976, Astrophys.J. 1o3, L127
Deutsch,A.J. 1956, Astrophys.J. 123, 21o
Drake,S.A., Linsky,J.L. 1986, Astron.J. 91, 6o2
Dupree,A.K. 1986, Ann.Rev.Astron.Ap. 24, 377
Dupree,A.K., Reimers,D. 1987, in: The Scientific Accomplishments of the IUE, Reidel, Dordrecht/Holland
Gahm,G., Hultquist,L. 1972, Astron.Astrophys. 16, 329
Goldberg,L. 1979, Q.J.R.A.S. 2o, 361
Goldberg,L., Hege,E.K., Hubbard,E.N., Strittmatter,P.A., Cocke,W.J. 1981, SAO Spc.Rep. 392, 131
Hagen,H.-J., Hempe,K., Reimers,D. 1987, Astron.Astrophys. in press
Hartmann,L., Dupree,A.K., Raymond,J.C. 1981, Astrophys.J. 246, 193
Hartmann,L., Avrett,E.H. 1984, Astrophys.J. 284, 238
Hege,E.K., Hebden,J.C., Christou,J.C. 1986 in Cool Stars, Stellar Systems, and the Sun (M. Heilik, D.M. Gibson eds.), Springer Verlag Berlin etc. p.414

Hempe,K. 1982, Astron.Astrophys. 115, 133
Hempe,K., Reimers,D. 1982, Astron.Astrophys. 1o7, 36
Hjellming,R.M., Newell,R.T. 1983, Astrophys.J. 275, 7o4
Holzer,T.E., MacGregor,K.B. 1985, in Mass Loss from Red Giants
    (M. Morris, B. Zuckermann eds.) D. Reidel, p.229
Iben,I., Renzini,A. 1983, Ann.Rev.Astr.Ap. 21, 271
Judge,P.G., Jordan,C., Rowan-Robinson,M. 1987, M.N.R.A.S. 224, 93
Kraft,R.P., Preston,G.W., Woolf,S.C. 1964, Astrophys.J. 14o, 235
Kudritzki,R.P., Reimers,D. 1978, Astron.Astrophys. 7o, 227
Lambert,D.L., Luck,R.E. 1978, M.N.R.A.S. 184, 4o5
Linsky,J.L. 1986, Irish Astron.J. 17, 343
Mauron,N. 1985, Doct.Thesis, Univ.Toulouse
Mauron,N., Fort,B., Querci,F., Dreux,M., Fauconnier,T., Lamy,P. 1984,
    Astron.Astrophys. 13o, 341
Mauron,N., Cailloux,M., Tilloles,P., Lefêvre.O. 1986, Astron.Astrophys.
    165, L9
O'Brien,G.P., Lambert,D.L. 1986, Astrophys.J.Suppl. 62, 899
Reimers,D. 1975a, Mem.Soc.Roy.Sci Liège 6e Ser. 8, 369
          1975b, in Problems in Stellar Atmospheres and Envelopes (eds.
    B. Baschek, W.H. Kegel, G. Traving), Springer-Verlag, Berlin etc.
    p.229
          1977a, Astron.Astrophys. 57, 395
          1977b, Astron.Astrophys. 61, 217 (Erratum 67, 161)
          1977c, Astron.Astrophys. 54, 485
          1982, Astron.Astrophys. 1o7, 292
          1987a, in Circumstellar Matter (IAU Symp. 122). D. Reidel,
                in press
Reimers,D. 1987b, in Solar and Stellar Physics (eds. E.H. Schröter,
    M. Schüssler) Lecture Notes in Physics, Springer,Berlin etc. in press
Reimers,D., Cassatella,A. 1985, Astrophys.J. 297, 275
Reimers,D., Che-Bohnenstengel,A. 1986, Astron.Astrophys. 166, 252
Reimers,D., Koester,D. 1982, Astron.Astrophys. 116, 341
Reimers,D., Schröder,K.-P. 1983, Astron.Astrophys. 124, 241
Robel,E. 1987, Diplomarbeit Univ. Hamburg
Schröder,K.-P. 1983, Astron.Astrophys. 124, L16
              1985a, Astron.Astrophys. 147, 1o3
              1985b, Doct.Thesis, Univ. Hamburg
Schröder,K.-P. 1986, Astron.Astrophys. 17o, 7o
Schröder,K.-P., Reimers,D., Carpenter,K.G., Brown,A. 1987,Astron.Astro-
Stencel,R.E., 1978, Astrophys.J. 223, L37   phys. subm.
Stencel,R.E., Mullan,D.J. 198o, Astrophys.J. 238, 221
Stencel,R.E., Linsky,J.L., Brown,A., Jordan,C., Carpenter,K.G., Wing,
    R.F., Czyzak,S. 1981, M.N.R.A.S. 196, 47
Wilson,O.C. 196oa, Astrophys.J. 132, 136
            196ob, in Stellar Atmospheres (ed. J.L. Greenstein),
    Univ.Chicago Press, Chicago
Wischnewski,E., Wendker,H.J. 1981, Astron.Astrophys. 96, 1o2
Zirker,J.B. 1984, in Effects of Variable Mass Loss on the Local Stellar
    Environment (R. Stalio, R.N. Thomas eds.), Trieste, p.25

MASS LOSS FROM HOT STARS: MAIN SEQUENCE AND SUPERGIANT STARS

Henny J. G. L. M. Lamers
Joint Institute for Laboratory Astrophysics
National Bureau of Standards and University of Colorado
Boulder, CO 80309-0440 and
SRON Laboratory for Space Research*
Beneluxlaan 21, 3527 HS Utrecht, The Netherlands

ABSTRACT. The determination of mass loss from early type stars by various methods is critically discussed. The reliability of the mass loss rates published in the literature is less than generally adopted.
   The role of rotation on the mass loss is discussed in relation to the rapidly rotating B and Be-stars. The coupling between rotation and pulsation plays a major role in determining the structure and the variability of the winds.
   The winds of stars near the Eddington limit are discussed in relation to the Luminous Blue Variables. Their winds are very different from those of normal supergiants, although both are due to radiation pressure. Small initial changes, possibly due to nonradial pulsations, can trigger large effects in the wind.

1. INTRODUCTION

The discovery of strong P Cygni profiles in the spectra of three early type supergiants in Orion by a rocket UV experiment of Morton (1967) provided the first evidence that hot stars are losing mass. Since then, many observations with UV spectrographs such as S59, Copernicus, BUSS, the UCL balloon experiment and IUE have shown that mass loss from early type stars is not restricted to the supergiants, but that it also occurs during the main sequence phase for all stars with $M_{bol} \lesssim -6^m$, which corresponds to $M \gtrsim 15\ M_\odot$.
   The mass loss rates can be derived from UV lines, Balmer line emission, radio fluxes and the IR excess. The methods are described in §§2, 3 and 4, and the accuracy of the results is discussed critically.
   The mass loss from early type stars was thought to be stationary and could be explained successfully by the radiation driven wind models. However, repeated observations of a small sample of stars have shown that the winds are variable on timescales as short as hours

*Permanent address

and that they are inhomogeneous on linear scales substantially smaller than the stellar radius.

Rotation may play an important role in the mass loss from stars, if the rotation rate exceeds some critical value. This is most noticeable in the Be-stars which are rapidly rotating stars. Their mass loss seems to be concentrated in the equatorial region with a high density and low outflow velocities. The mass loss from the polar regions is characterized by low density and high velocities, similar to the winds of normal stars. The most enigmatic characteristics of the Be stars are the very large changes in mass loss on timescales of decades. This shows that rotation is not the only agent which enhances the mass loss in these stars, but that another mechanism can trigger large variations.

Stars near the Eddington limit show mass loss characteristics very different from normal stars. This is demonstrated most clearly by the Luminous Blue Variables (LBV's) which are early type hypergiants in the helium core burning phase. Their mass loss is variable on all timescales from centuries to months. The most famous examples for large outbursts are P Cygni and η Car. In this case internal instabilities may trigger large mass ejections from the atmosphere which has a very small effective gravity.

In this paper I will concentrate on the determination of the mass loss rates with its uncertainties and on the variability of normal stars. The role of rotation is discussed in connection with the Be stars and the effects of low gravity are discussed in connection with the LBV's.

For earlier reviews the reader is referred to
- Conti (1978), Lamers and Cassinelli (1987) and Hearn (1987): for normal stars.
- Abbott et al. (1986), Henrichs (1987) and Lamers and de Loore (1987): for variability.
- Doazan (1982), Snow and Stalio (1987) and Lamers (1987b): for Be-stars.
- Appenzeller (1986), Davidson (1987) and Lamers (1986a, 1987a): for Luminous Blue Variables.

2. MASS LOSS STUDIES FROM UV LINES

2.1 The Method of Line Fitting

Mass loss rates can be determined from the analysis of the P Cygni profiles of UV resonance lines and lines from metastable levels. This analysis implies the comparison between the observed profiles and those calculated for spherical symmetric winds with an assumed velocity law. The profiles are usually calculated with the Sobolev approximation (e.g. Castor and Lamers, 1979) because of its simplicity and accuracy for winds with large velocity gradients. Since the Sobolev approximation is not valid at low velocities, this linefitting does not give information about the conditions at the base of the wind. The Comoving Frame Method (Mihalas et al., 1975) can predict the lineprofiles more accurately since it does not depend on the Sobolev approximation and the transition from the photosphere to the wind is treated properly. However due to the complexity of this method, compared to the Sobolev method, it has been used in the analysis of stellar winds of only a small number of stars. The recently published SEI method (Lamers et al., 1987), based on the calculation of the source function in the Sobolev approximation but with the exact integration of the radiative transfer equation may be a good compromise for the study of lines formed in stellar winds since it has a high accuracy and can be used efficiently on small size computers.

Castor and Lamers (1979) have shown that the profiles of the absorption part of the UV resonance lines are very sensitive to the radial optical depth $\tau(v)$ but insensitive to the velocity law, $v(r)$. The radial optical depth is defined by

$$\tau(v) = (\pi e^2/mc) f \lambda_o n_i'(r)(dv/dr)^{-1} \qquad (1)$$

where f is the oscillator strength, λ_o is the rest wavelength and $n_i'(r)$ is the number density of the absorbing atoms. So, by fitting the observed profiles, the value of $\tau(v)$ can be derived. If the velocity law is known or can be estimated the value of $n_i'(r)$ can be determined. This number density is related to the mass loss rate by

$$\dot{M} = 4\pi r^2 v(r) \; n_i'(r)/\{(n_H/\rho)Aq_i(r)E_i(r)\} \qquad (2)$$

where (n_H/ρ) is the number of H-atoms per unit mass; $A = n_E/n_H$ is the abundance of the element E relative to H; $q_i = n_i/n_E$ is the ionization fraction of the observed ion and $E_i = n_i'/n_i$ is the excitation fraction of the observed transition. For resonance lines one can assume $n_i' \simeq n_i$. A combination of Eq. (1) and (2) gives an expression for $\dot{M}$ as a function of $\tau(v)$.

If the abundances and $q_i(r)$ were known with a high degree of accuracy, the function $\tau(v)$ would only depend on $\dot{M}$ and $v(r)$ which could both be derived from linefitting (see Hamann, 1980). In general this is not the case: one has to assume a velocity law and make an educated guess of the ionization fraction to derive $\dot{M}$. One usually assumes a velocity law of the type

$$v(r) = v_o + (v_\infty - v_o)(1-R*/r)^\beta \qquad (3)$$

first proposed by Lamers and Rogerson (1978); where v_o is the initial velocity of the wind and v_∞ is the terminal velocity which can be derived from the observed short-λ edge of the P Cygni profiles. The initial velocity is typically on the order of the sound velocity, i.e. ~10 km/s, whereas $v_\infty \sim$ 2000 to 3500 km/s for O-stars. The studies of UV line profiles indicate $\beta \simeq 0.5$ to 1 (Gathier et al., 1981; Garmany et al., 1981) in agreement with the predictions for radiation driven winds (Abbott, 1982; Pauldrach et al., 1986).

Adopting the velocity law of Eq. (3) one finds

$$\tau(v) \simeq 3.38 \times 10^{18} f \lambda_o A q_i(r) \dot{M} R_*^{-1} v_\infty^{-2} \beta (R*/r)^2 (v/v_\infty)^{2\beta-1} \quad (4)$$

with λ_o in Å, $\dot{M}$ in $M_\odot/yr$, $R*$ in $R_\odot$ and V_∞ in km/s. We have neglected terms of the order v_o/v_∞. For a typical O star with $R* = 10\ R_\odot$, $v_\infty = 2000$ km/s and $\beta = 1$ we find that a P Cygni profile of a typical UV resonance line with $f \simeq 0.3$ and $\lambda_o \simeq 1300$ Å is saturated ($\tau \geqslant 4$ and $e^{-\tau} \leqslant 0.02$) at $v = 0.5\ v_\infty$ if

$$\dot{M} A q_i(2R_*) \geqslant 9.7 \times 10^{-13}\ M_\odot/yr \ . \quad (5)$$

This equation shows the <u>fundamental problem of mass loss studies from hot stars</u>: in stars with $\dot{M} \gtrsim 10^{-6}\ M_\odot/yr$ the most easily observed UV resonance lines of C IV, N V and Si IV are either saturated or they represent minor ionization fractions so that large and uncertain correction factors for the ionization have to be applied. In either case the mass loss rates are uncertain. The study of lines from the excited levels of O IV, O V and N IV does not solve this problem since this requires a large and uncertain correction for the excitation. The problem can only be solved by studying resonance lines in the wavelength range of $900 \lesssim \lambda \lesssim 1200$ Å where the lines of many more ions can be observed (see Snow and Morton, 1976; Lamers and Morton, 1976).

Mass loss studies of hot stars by means of line fitting of UV lines have been made by various groups: Lamers and Morton (1976) and Lamers and Rogerson (1978) (first mass loss determination of OB stars by linefitting); Olson and Castor (1981) and Gathier et al. (1981) (based on spectra between 930 and 1400 Å from Copernicus satellite); Garmany et al. (1981), Garmany and Conti (1984) and Prinja and Howarth (1986) (based on IUE observations); Hamann (1980, 1981) (first detailed studies with Comoving Frame Code); Lamers et al. (1978) and Praderie et al. (1980) (first studies of late B and A supergiants). An example of the linefits is shown in Fig. 1.

2.2 The Ionization Problem

All studies of mass loss suffer from the uncertainty in the degree of ionization. Since the first observations of the UV spectra by the Copernicus satellite showed the presence of O VI lines in stars as cool as B0, N V in stars as cool as B3 and C IV in stars as cool as B8 (Snow and Morton, 1976), it is clear that the ionization in the winds of early type stars cannot be explained by a simple equilibrium with the photospheric radiation. Several explanations have been proposed, involving, e.g., a warm wind of $T \sim 2.10^5$ K, and Auger ionization by X-rays created in a thin corona at the base of the wind or in shocks. The discovery of X-rays from hot stars by the Einstein satellite (Harnden et al., 1979) and the good correlation between the observed presence of O VI and the predicted presence of O IV, as expected for two-photon ionization (Cassinelli and Olson, 1979), has made Auger ionization the most likely candidate. The discovery of instabilities in the winds of early type stars from detailed studies

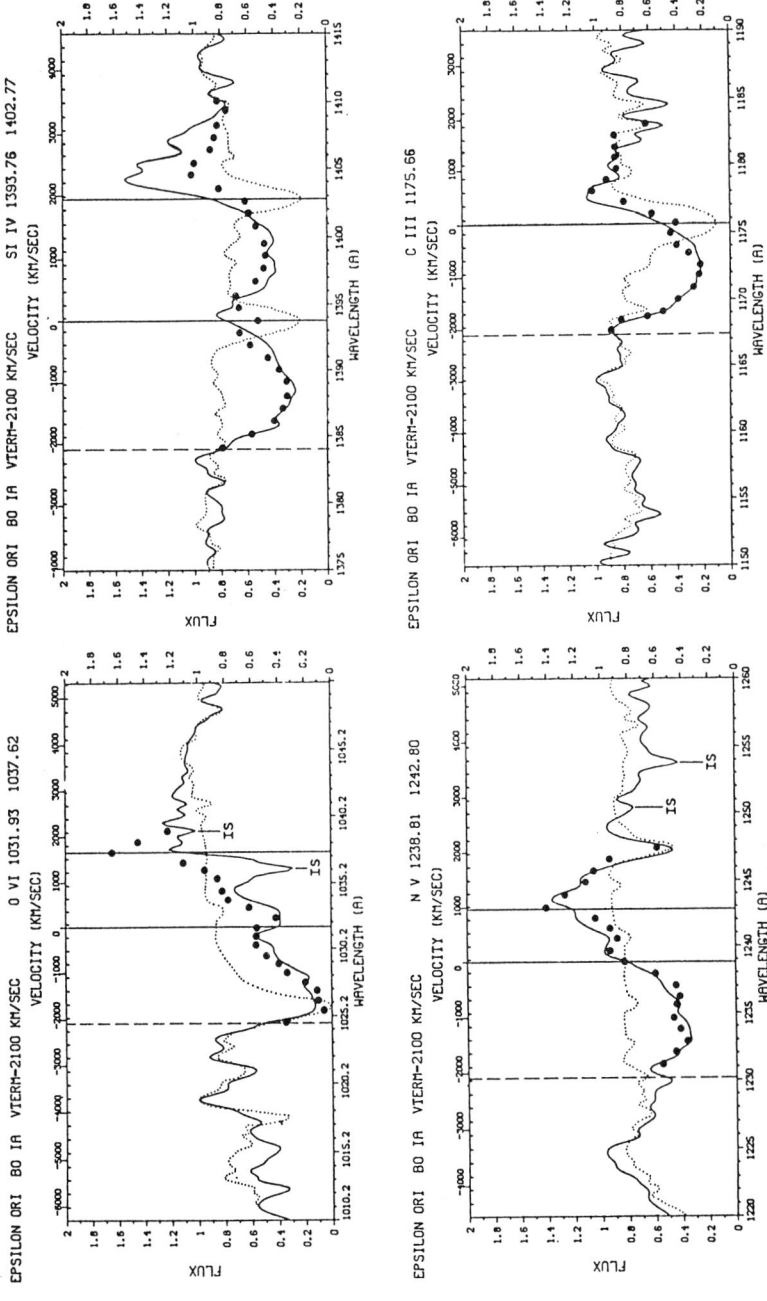

Fig. 1. Linefitting of the resonance lines of O VI, Si IV, N V and C III of ε Ori (B0 Ia). The full line is the observed spectrum, the dotted line is the adopted photospheric spectrum, the big dots are the fitted theoretical profile. The rest wavelengths are indicated by full vertical lines and the terminal velocity of 2050 km/s by dashed vertical lines (from Gathier et al., 1981). These linefits are considerably more accurate than those from other groups (e.g. Garmany et al., 1981).

of UV lineprofiles (Lucy, 1984; Lamers et al., 1988) and from X-ray flickering in X-ray binaries (White et al., 1983) suggests that shocks do occur in the winds and that the shock-generated X-rays are responsible for the superionization. The recently derived very stringent upper limit for the emission measure of a corona around ζ Pup by Baade and Lucy (1987) and the analysis of the X-ray spectrum of τ Sco (Cassinelli, 1987) have eliminated the possibility that early type stars have coronae as the basis of their wind. This leaves the Auger ionization from shocks in the wind as the best candidate for the explanation of the superionization.

Recently Pauldrach (1987) has shown, by means of very detailed calculations of the ionization balance in winds, that the superionization can also be explaned by photoionization from excited levels in a wind which is in radiative equilibrium with the photosphere. This means that there are now two explanations for the superionization. Both mechanisms must be operative. The relative importance of these two mechanisms in stars of different spectral types is presently unknown. The solution will require observations of more ions than can be obtained with IUE or ST.

Unfortunately, the analysis of the line profiles by the various groups mentioned in 2.1 did not include either one of these two explanations for the superionization. Lamers and Morton (1976), Lamers and Rogerson (1978), and Hamann (1980, 1981) assumed that the superionization was due to a warm wind, which is now known to be wrong. Olson and Castor (1981), Garmany et al. (1981), and Prinja and Howarth (1986), assumed Auger ionization from a corona at the base of the wind, which is now also known to be wrong. This may explain the curious fact that they derived an electron temperature in the wind of about 34000 K, independent of the spectral type of the stars, from O4 to B0. Lamers et al. (1978) and Praderie et al. (1980) assumed radiative equilibrium without ionization from excited levels. Gathier et al. (1981) assumed empirical ionization fractions, derived from the comparison between the UV linefitting results and independent mass loss estimates from radio data of a small number of stars. Since the radio data can only be used for stars with very high mass loss rates (§3), the extrapolation of the empirical ionization to winds of lower density is suspect. This is the reason for the large differences between the mass loss rates of the lower luminosity OB stars derived by Garmany et al. (1981) and Gathier et al. (1981).

These comments show that the mass loss rates of early type stars which are derived from UV lines only are not very reliable. If the rates are derived from the IUE spectra only, which limits the determination to essentially C IV (usually saturated), N V (a superionized atom) and Si IV (very low ionization fraction in most O stars), then the errors in mass loss rates can easily amount to a factor 10. This may partly explain the large scatter in the $\dot{M}$ versus L diagrams of the less luminous O-stars (e.g. Conti and Garmany, 1984).

This rather pessimistic situation for mass loss determinations of early type stars from UV lines can be improved in the future by observations of more ions at shorter wavelength, or by more reliable calculations of ionization fractions in which photoexcitation from excited levels and Auger ionization by X-rays from shocks are taken into account.

3. MASS LOSS STUDIES FROM RADIO AND IR RADIATION

3.1. The Method

The free-free emission from the stellar winds can be measured at IR and radio wavelengths if the stars have a high mass loss rate. The absorption coefficient for free-free radiation is

$$\kappa_\lambda = 1.98 \times 10^{-23} Z^2(g+b)\lambda^2 n_i n_e T^{-3/2} \tag{6}$$

(Allen, 1973) where Z is the rms charge per atom, g and b are the free-free and free-bound gaunt factors, λ is wavelength in cm, n_e and $n_i = n_e/\gamma$ are the electron and ion densities, and T is the electron temperatures. Notice that $\kappa_\lambda \sim \lambda^2$ so the effects of free-free radiation will be most noticeable at long wavelength. The optical depth for a line of sight directed towards the center of the star through a spherically symmetric wind is

$$\tau_\lambda = 5.3 \times 10^{31} Z^2 \gamma \dot{M}^2 T^{-3/2} \mu^{-2} v_\infty^{-2} R_*^{-3} \lambda^2 (g+b) \int_x^\infty w^{-2} x^{-4} dx \tag{7}$$

where $\dot{M}$ is in $M_\odot$/yr, μ is the mean atomic weight in units of m_H, R_* is in $R_\odot$, v_∞ is in km/s, $w = v/v_\infty$ and $x = r/R_*$. For a typical early type star with $R_* = 10$, $v_\infty = 2000$, $T = 3 \times 10^4$, $g + b \simeq 2$, $Z \simeq \gamma \simeq \mu \simeq 1$ and $w(x) \simeq (1-1/x)^\beta$, from Eq. (3) with $v_0 = 0$ and $\beta = 1$, we find

$$\tau_\lambda \simeq 5.1 \times 10^{15} \dot{M}^2 \lambda^2 (w^{-1} - w + 2 \ln w) . \tag{8}$$

Table 1 gives an indication of the depth in the wind, where $\tau_\lambda = 0.3$ is reached for various wavelengths and mass loss rates.

Table 1. $r(\tau_\lambda=0.3)/R_*$

$\dot{M}$	λ=10cm	1 cm	1 mm	100 μ	10 μ	1 μ
10^{-5}	384.2	83.3	18.34	4.37	1.44	1.015
10^{-6}	83.3	18.3	4.37	1.44	1.01	<1.010
10^{-7}	18.3	4.4	1.44	1.01	<1.01	<1.010

The radio or IR energy distribution of a star will be affected by free-free emission from the wind if $\tau_\lambda \gtrsim 0.3$ corresponding to an excess of $\sim 0.1^m$. The data in Table 1 show that an O-star with $\dot{M} \sim 10^{-5}$ $M_\odot$/yr will show detectable IR excess at $\lambda > 1$ μ, a star with 10^{-6} $M_\odot$/yr will have an excess at $\lambda > 10$ μ, and a star with 10^{-7} $M_\odot$/yr will have an excess at $\lambda > 100$ μ. The table also shows that the free-free radiation at cm-wavelengths from luminous O-stars is generated at distances $r \gtrsim 10^2$ R_*.

The amount of radiation absorbed and emitted by free-free processes in stellar winds can be calculated by solving the transfer equations in a spherically symmetric wind. For an isothermal wind the flux from a star with a wind can be expressed conveniently as

$$F_\nu = F_\nu^* \left[Z_1(E_\nu) + \{B_\nu(T_w)/I_\nu^*\} Z_2(E_\nu) \right] \qquad (9)$$

where

$$Z_1 = \int_0^1 \exp\{-\tau_{max}(q)\} 2qdq \qquad (10a)$$

$$Z_2 = \int_0^\infty \left[1-\exp\{-\tau_{max}(q)\} \right] 2qdq \qquad (10b)$$

$$\tau_{max}(q) = E_\nu \int_1^\infty w^{-2} x^{-3} (x^2-q^2)^{-1/2} dx \quad \text{if } q<1 \qquad (11a)$$

$$\tau_{max}(q) = 2E_\nu \int_q^\infty w^{-2} x^{-3} (x^2-q^2)^{-1/2} dx \quad \text{if } q \geqslant 1 \qquad (11b)$$

$$E_\nu = 5.3 \times 10^{31} z^2 \gamma \dot{M}^2 T_w^{-3/2} \mu^{-2} v_\infty^{-2} R_*^{-3} \lambda^2 (g+b) \qquad (12)$$

with the same units as above (Lamers and Waters, 1984). The first term in Eq. (9) is the fraction of the stellar flux absorbed by the wind ($Z_1 < 1$), and the second term is the radiation emitted by the wind. This equation shows that in general the wind will produce a net emission (since $Z_2 + Z_1 > 1$ except when the wind has a low temperature below the brightness temperature of the photosphere. In that case $B_\nu(T_w)/I_\nu^* < 1$ and F_ν may be smaller than the photospheric flux F_ν^*. Waters and Lamers (1984) have published tables of Z_1 and Z_2 as a function of E_ν for various velocity laws.

The mass loss rates of early type stars can be derived from their IR excess or radio excess by means of Eq. (9). If the velocity law of the wind is known and its temperature can be estimated, the comparison between the measured flux F_ν and the expected flux from the photosphere F_ν^* indicates the value of E_ν in which $(\dot{M}/v_\infty)^2$ is the only unknown. If the terminal velocity is known from UV line profile observations, the value of $\dot{M}$ can be derived.

If the excess can be measured at various wavelengths, the velocity law can be derived by comparing the observed spectral dependence of F_ν/F_ν^* with the predictions for different velocity laws. This is most easily done with the curve-of-growth method for IR excess, introduced by Lamers and Waters (1984). In this method the measured excess in magn. is plotted versus $\log \lambda^2 (g+b)$. This plot is then compared with theoretical curves of growth expressed in excess

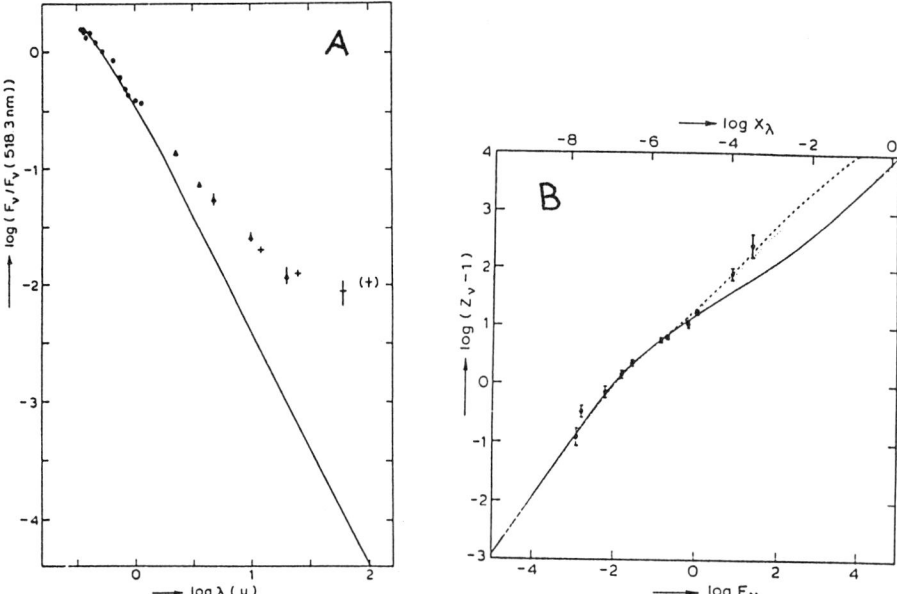

Fig. 2. (A) The IR energy distribution of the hypergiant P Cygni
(B1 Ia$^+$). The full line is the expected photospheric energy predicted
by model atmospheres. (B) The curve of growth for the IR excess of
P Cygni. The vertical scale is log $(Z_\nu-1)$ = log F_ν (excess)/F_ν
(photosph.). The full line is the calculated curve of growth for a
very slow linear velocity law. The dashed line indicates the
contribution of a shell at r ~ 15 to 200 R$_*$ (from Waters and
Wesselius, 1986).

versus log E_ν, calculated for different velocity laws. By matching
the observed and predicted curves, the shape of the c.o.g. gives the
velocity law and the horizontal shift to be applied gives the mass
loss rate. As an example Fig. 2 shows the energy distribution and the
IR curve of growth of the star P Cygni (from Waters and Wesselius,
1986).

3.2 Mass Loss Studies from Radio Flux Measurements

The radio flux of about 15 early type stars has been measured at 2 and
6 cm by Abbott (1985) whereas upper limits exist for about 50 more
stars. Since the radio flux comes from large distances from the
star, r ≳ 10 R$_*$ (see Table 1), where the velocity can be taken
as v_∞, Eqs. (9)-(12) have a very simple solution:

$$S_\nu = 23.2 \; (Z\dot{M}/\mu v_\infty)^{4/3} (\nu\gamma g)^{2/3} D^{-2} \quad \text{Jy} \quad (13)$$

(Wright and Barlow, 1975; Panagia and Felli, 1975) where S_ν is the flux observed at the earth and D is the distance in kpc. Apart from the slight dependence of the Gaunt factor on T, the temperature does not enter into this expression, because the temperature dependence of $B_\nu(T)$ and $\tau(T)$ cancel one another. This is a very important characteristic of the mass loss studies from radio fluxes: the uncertainty in the ionization or temperature of the wind is of minor importance, provided that H is fully ionized. The major uncertainty in these mass loss determinations is in general due to the distance, which enters as $D^{-1.5}$.

Unfortunately, the radio flux of typical O-stars is small and below the present detection limit of 0.1 mJy at 6 cm. For a typical O-star with v_∞ = 2000 km/s at a distance of 2 kpc the flux, in mJy, as a function of wavelength is as shown in Table 2.

Table 2. S_ν(mJy)

$\dot{M}$	λ = 21 cm	6 cm	1 cm	1 mm
10^{-5}	0.24	0.48	1.56	6.19
10^{-6}	0.01	0.02	0.07	0.29

We see that only stars with $\dot{M} \gtrsim 10^{-5}$ can be detected at radio wavelengths.

The possibilities of observing the winds of hot stars with mm-telescopes are better than at 6 cm since $S_\nu \propto \nu^{2/3}$. Moreover one might expect to see the effects of the velocity law at these wavelengths, since the mm radiation originates deeper than the radio radiation (see Table 1).

Several early type stars, for which the radio flux has been measured, turn out to be non-thermal emitters. They are recognized by a spectral index $\alpha \lesssim 0$ instead of $\alpha \simeq 0.6$, by intrinsic polarization and considerable variability (Abbott, 1985). The presence of non-thermal radio sources implies that a reliable determination of a mass loss rate from an early type star requires flux measurements at two frequencies.

3.3 Mass Loss Studies From Infrared Excess

The first determinations of the mass loss from IR observations of a sample of early type stars were made by Barlow and Cohen (1977). Other mass loss studies based on the IR excess are published by Leitherer et al. (1982), Groot and Thé (1983), Castor and Simon (1983), Abbott et al. (1984), and Berthout et al. (1985). In total the mass loss rates of about 80 normal early type stars have been determined from ground-based IR observations.

Since the IR free-free radiation originates very close to the photosphere, the IR excess is very sensitive to the density distribution close to the photosphere. If the wind is homogeneous and stationary, the velocity law can be used to describe the density distribution. However, neither the density nor the velocity distribution at layers just above the photosphere and in the lower part of the wind are known with sufficient accuracy.

Barlow and Cohen (1977) adopted the velocity law of P Cygni, which is now known to be very different from normal stars, for their analysis of the IR excess from O, B and A stars. Therefore their mass loss rates are now considered to be unreliable. Other authors adopted velocity laws of the type of Eq. (3), with v_∞ derived from UV lines and used the spectral dependence of the IR excess to derive β and the mass loss rates. Berthout et al. (1985) found indications that the velocity law is "softer" (i.e. larger values of β) for the B supergiants than for the O-stars. However, the effect is marginal and may also be due to small errors in the predicted adopted fluxes.

An additional problem with the IR excesses has been discovered recently on the basis of the IRAS data. The photospheric IR fluxes predicted by the model atmospheres of Kurucz (1979) are smaller than the observed fluxes for normal stars without mass loss by about 10% at 12 μ. The reason for this discrepancy is not known, but it seems to be real and not due to calibration errors (Waters et al., 1987). Since most studies up to now have used the Kurucz models for the determination of the excess flux, and since the excesses are generally small and on the order of 0.1 to 0.5 magn, this error in the predicted photospheric flux will result in a drastic error in the mass loss rate. This effect, and possible inhomogeneities in the wind, may explain why Castor and Simon (1983) and Abbott et al. (1984) found that the mass loss rates derived from the IR excess are poorly correlated with those derived from radio and UV data of the same stars.

Considering these problems, we conclude that at present the ground-based IR observations do not provide reliable mass loss rates for normal stars. In fact, it may be more useful to adopt the mass loss rates derived from the radio fluxes and from the UV line profiles (with its uncertainties!) and use the IR excess to derive information on the density structure of the lower parts of the wind.

4. MASS LOSS STUDIES FROM Hα EMISSION

The high density layers in the lower part of the stellar winds produce Hα emission which can be measured in high S/N optical spectra. The region where this emission is formed is about the same as that where the IR excess is generated. The study of these layers by means of Hα has a clear advantage over the IR studies, because the Hα profile provides information about the density and the velocity. The difficulty of the mass loss studies based on Hα profiles is due to the fact that accurate photospheric profile calculations are needed in order to estimate the contribution of the wind to the profile. Moreover, since the emission depends strongly on the population of the energy levels of atomic hydrogen, the statistical equilibrium equations have to be solved for the conditions in the wind.

Klein and Castor (1978) have calculated the expected Hα emission from the winds of O-type stars for a velocity law of the type of Eq. (3) with β = 0.5 and derived mass loss rates for O-type stars. Olson and Ebbets (1981) used these calculations to extend the sample of stars. However, we now know that the velocity law with β = 0.5 is too steep to match the UV or IR observations, so the resulting mass loss rates are not very reliable.

Recently Leitherer (1988) has reinvestigated the possibility that Hα can be used as a tracer of mass loss from OB stars. To circumvent the problem of the unknown velocity law in the lower part of the wind he adopted the UV mass loss rates to derive the velocity law as a function of spectral type, and found that β (see Eq. (3)) increases from β = 0.9 at types O3 to β = 2.4 at type B9. Using this spectral type dependence of the velocity law, he derived mass loss rates for the individual stars, which then of course agree with the UV rates. Since Hα observations can be obtained more easily than UV spectra, the method is, in principle, very powerful. In practice, however, the results depend on the accuracy of the UV mass loss rates used for the calibration, which is not very high.

It is interesting that Leitherer found β increasing with spectral type, since a similar result was found from the IR excess studies by Berthout et al. (1985). Since the methods are independent, this seems to indicate that the density structure in the lower part of the winds of B supergiants decreases slower outwards than in the O-stars.

Unfortunately studies of the Hα _profiles_, in which the advantage of the density and velocity information can be used, have been restricted to only very few stars (see e.g. Kunasz and Morrison, 1982) because this requires extensive calculations with the Comoving Frame Code. It would be extremely interesting to perform such an analysis for a number of O, B and A stars of different types and luminosity in order to derive the density laws and velocity laws in the lower layers of the wind as a function of temperature and gravity. This might also indicate the presence of clumping on inhomogeneities.

5. MASS LOSS RATES AND VELOCITIES FOR NORMAL STARS

The terminal velocities of the winds of early type stars scale approximately with the escape velocity at the stellar surface. The scaling factor is temperature dependent. Figure 3 shows the ratio v_∞/v_{esc} as a function of T_{eff}, determined by Abbott (1982). The dashed line indicates a best fit through the data:

$$v_\infty/v_{esc} = 24.0 - 4.6 \log T_{eff} \quad \text{for } T_{eff} \geq 25000 \text{ K} \quad (14)$$

$$v_\infty/v_{esc} = -25.3 + 6.6 \log T_{eff} \quad \text{for } 10000 < T_{eff} \leq 25000 \text{ K}$$

with

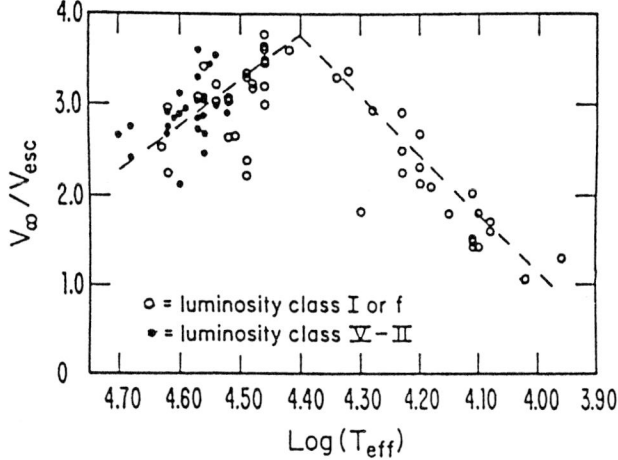

Fig. 3. The relation between the terminal velocity of the wind, normalized to the escape velocity, v_∞/v_{esc}, versus T_{eff} (from Abbott, 1982). The dashed line is a best linear fit through the data.

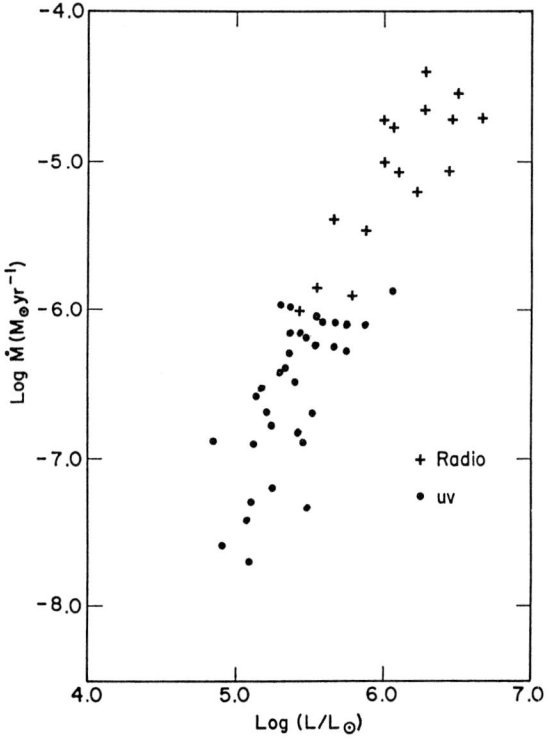

Fig. 4. The correlation between the mass loss rates and luminosity for OB stars. The "radio" rates are from Abbott (1985) and the "UV" rates are from Garmany et al. (1981) and Garmany and Conti (1984).

$$v_{esc} = (2\ GM/R)^{0.5}(1-\sigma_e L/4\pi GMc) \qquad (15)$$

where the second factor is the correction term for the effective escape velocity due to the radiation pressure by electron scattering.

The mass loss rates scale roughly as the luminosity. Figure 4 shows the mass loss rates determined from the radio data (Abbott, 1985) and those from UV data by Garmany et al. (1981) and Garmany and Conti (1984). The data show a very large scatter, especially the rates derived from the UV lines. This is not surprising, considering the fact that these rates are derived from C IV (saturated) and N V (superionized) only! I suspect that the scatter in the UV rates is largely due to the uncertainties in the ionization rate of N V, and to the fact that many of the N V lines studied are close to saturation.

Various attempts have been made to express the mass loss rates as a function of the basic stellar parameters. A least-squares fit through the data in Fig. 3 gives the following dependence

$$\log \dot{M} = -6.87\ (\pm 0.13) + 1.62\ (\pm 0.19) \log (L/10^5) \qquad (16)$$

(Garmany and Conti, 1984). Other, more elaborate fits, involving also the radius or mass have not been very successful, mainly because of the uncertainties in the mass loss rates. For instance, attempts to express the mass loss as $\dot{M} \sim L^\alpha R^\beta M^\gamma$ have resulted in the following parameters:

	α	β	γ
Lamers (1981)	1.4± 0.4	0.6± 0.1	-1.0± 1.0
Chiosi and Olson (1984)	1.7± 0.1	0.7± 0.1	-0.7± 0.1
Garmany and Conti (1984)	1.0± 0.8	0.8± 0.6	0.6± 1.2

All three groups agree that the mass loss depends on the radius to a power of about 0.7. This is important, because it implies that the mass loss rate will increase when the star moves away from the zero age main sequence. The dependence of $\dot{M}$ on the mass is poorly determined. This is mainly due to the fact that the sample of stars shown in Fig. 3 covers only a factor 2 or 3 in mass. (The accurate determination of γ by Chiosi and Olson is due to the fact that they assumed $\gamma = -\beta$.)

Alternative fits have been tried by various authors (see Chiosi and Maeder, 1986). However the fits are no more accurate than the mass loss rates upon which they are based, so their reliability is questionable.

For a comparison between the predicted and observed values of the mass loss rates and the terminal velocites, the reader is referred to Hearn (1987).

6. VARIABILITY OF THE WINDS OF NORMAL STARS

The P Cygni profiles of the UV resonance lines of early type stars often show small variable absorption components, superposed on the more stable profile. They were first discovered by Morton (1976) and by Snow (1977), who realized that their variability implies that these so-called narrow absorption components are formed in the wind, rather than in the photosphere. The narrow components and their variability have been studied extensively by, e.g., Lamers et al. (1982), Henrichs et al. (1983), and Prinja and Howarth (1986). For recent reviews see Cassinelli and Lamers (1987) and Henrichs (1987).

The narrow absorption components (NAC) are usually found in the velocity range of about -0.5 v_∞ to -0.9 v_∞. However, they can occur occasionally at velocities as small as -0.3 v_∞. The width of the NAC can range from typically 30 to 300 km/s, with the widest ones found at the lowest velocities.

When the NAC are observed at irregular epochs they seem to occur mainly near $v \simeq -0.7$ v_∞ and seem to be stable over timescales of many years. However, when the spectra are observed in rapid succession with intervals on the order of hours, it turns out that the NAC are not stationary, but show a very definite evolution (e.g. Henrichs et al., 1980; Prinja et al., 1988). The NAC first appears at $v \simeq -0.3$ v_∞ as a wide absorption. Within about 10 hrs it shifts to $v \simeq -0.7$ v_∞, meanwhile getting narrower and deeper. After about one day the feature is very narrow and only a few tens of km/s wide and then starts to get weaker. It might disappear after a few days, but by that time usually a new NAC has developed which also appeared first at $v \simeq -0.3$ and then shifts to about -0.7 v_∞. This behavior is shown in Fig. 5 for the star 68 Cyg (O7.5 III (f)). The column density of the NAC are on the order of 10^{14} to 10^{15} cm^{-2} for the observed ions, which corresponds to hydrogen column densities between 10^{19} and 10^{22} cm^{-2}. For comparison, the total column density of a wind of 10^{-6} M$_\odot$/yr, $v_\infty \simeq 2000$ km/s and $R_* \simeq 10$ R$_\odot$ is $N_H \simeq 7 \times 10^{22}$ cm^{-2}.

These characteristics indicate the growth and development of <u>density enhancements</u> in the wind. The NAC are not due to increased ionization fractions for the observed ions, since they are observed simultaneously in both high (e.g. O VI, N V) and low ionization stages (C IV, Si IV). The density enhancements first appear at $v \simeq 0.3$ v_∞, i.e. in the accelerating part of the wind. The large velocity gradient in these layers may explain the large width of the NAC. Once they appear, the density enhancements are then accelerated up to about 0.7 v_∞. The acceleration of these enhancements as derived from their velocity versus time diagrams, is generally much slower by a factor 5 to 20 than that of the wind.

Henrichs et al. (1980, 1983) have interpreted this behavior in terms of shells which are ejected by the star and are accelerated by radiation pressure. The slow acceleration and their lower terminal velocity is due to the fact that the radiative acceleration (i.e. force per unit mass) is smaller than in the wind because of the higher density. Recent observations of fast variations in the UV lines of α Cam (O9.5 Ia) by Lamers et al. (1988) and simultaneously observed variations in the polarization (Hayes, 1984) indicate that the enhancements are not due to spherical shells, but to blobs with a

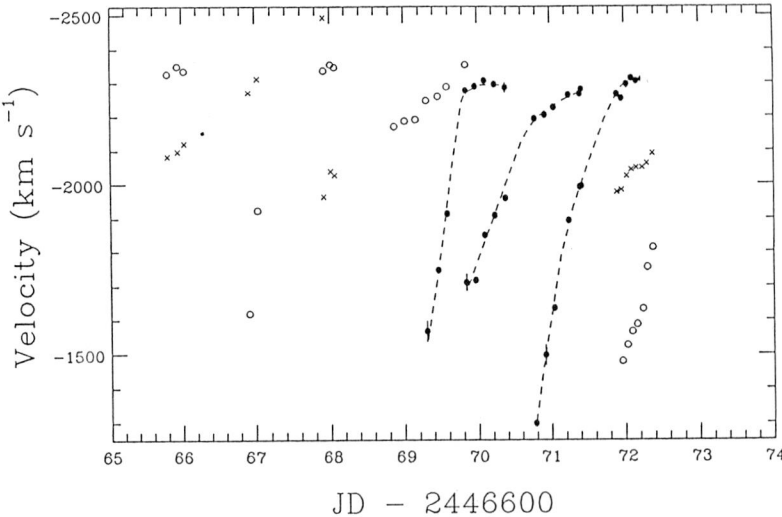

Fig. 5. The velocities of the narrow absorption components in the UV resonance lines of the star 68 Cyg O7.5 III ((f)). The velocity evolution of three components are indicated by dashed lines. Notice the acceleration and the rapid succession of the components (from Prinja et al., 1988).

typical size on the order of a stellar radius. Based on these observations Lamers et al. suggest that the winds of early type stars are inhomogeneous with about $\sim 10^{-1}$ of its mass in clumps of higher density and lower velocity than the quiescent wind. This picture is confirmed by the X-ray flickering observed in the X-ray binary 4U 1700-37 which consists of an O6 f star and a compact star. White et al. (1983) have shown that the X-ray spikes can be explained by accretion of density concentrations in the wind of the O-star with a characteristic scale of ~ 1 R_* and a column density of $N_H \lesssim 6 \times 10^{21}$ cm^{-2}.

The observed clumps may be due to the growth of radiation driven instabilities although the theoretical predictions of such instabilities can not yet be compared with the observations (see e.g. Castor, 1987).

An alternative explanation for the NAC has been proposed by Mullan (1986), who suggested that the winds of hot stars consist of fast and slow streams, in analogy with the solar wind, and that the NAC are due to the interaction between those streams. This interpretation has also been adopted by Prinja et al. (1988) for the variations observed in 68 Cyg. The model proposed by Mullan requires strong magnetic fields in the winds of hot stars, the presence of which has not been proven yet.

At present it seems clear that the winds of early type stars are not homogeneous, but that there are density enhancements of dimension $\sim R_*$ which move considerably more slowly than the wind. The interaction between the fast wind and these slower blobs may be responsible for the generation of X-rays which in turn could produce the superionization.

7. MASS LOSS AND ROTATION

The effects of rotation on mass loss are most easily seen in the main sequence B type stars. These stars can have high rotational velocities, up to v sin i $\simeq$ 500 km/s.

The study of the UV resonance lines of these stars does show an interesting effect: generally B main sequence stars with $M_{bol} > -6^m$ do not show the effects of mass loss in the UV lines, except if the stars are rapidly rotating (Lamers and Snow, 1978; Snow and Marlborough, 1980; Grady et al., 1987). The critical rotational velocity seems to increase towards later spectral types i.e. with decreasing luminosity. This behavior should be considered in relation to the predictions for radiation driven winds. Abbott (1982) showed that stars with $M_{bol} < -6^m$ can have self-initiating winds, but that in the less luminous stars only a wind which is started by some other mechanism can be sustained by radiation pressure. The observations thus suggest that in stars with large values of v sin i such an "other mechanism" is operative.

Additional evidence for a correlation between rotation and circumstellar matter has been found from the study of the IRAS infrared excess of B stars. Waters (1986) has shown that early B stars with v sin i $\gtrsim$ 100 km/s have a larger infrared excess than those with lower velocity, and that late B stars with v sin i $\gtrsim$ 200 km/s have a larger excess than those of lower velocity.

We know that rotation alone cannot be responsible for the mass loss from rapidly rotating B stars. The main reason to suspect the presence of an additional mechanism is the variability of the mass loss in rapidly rotating stars. The best examples are the Be stars, characterized by high rotational velocities and Balmer emission lines.

It has become clear during the last decennium that the mass loss from Be stars is varying tremendously and that these stars can change their spectral types from Be to normal B (without emission) in timescales of years or decades (Doazan, 1982). This indicates that the mass loss rates can change by orders of magnitude whereas the rotation obviously remains the same.

The mechanism that is responsible for the mass loss of rapidly rotating stars and for its variability produces circumstellar matter which is highly concentrated along the equator region. The high densities on the order of 10^{12} cm^{-3} derived from the Hα profiles and from the IR excess are much higher than those derived from the UV lines, which are $\sim 10^8$ cm^{-3}. This indicates that the UV lines are formed in a different region of the wind. Poeckert and Marlborough (1978) and Lamers and Waters (1987) have argued that these observations can be explained by a wind which consists of two regions: a low velocity, high density equatorial region where the Hα emission and the IR excess is generated, and a high velocity, low density region outside the equator, where the UV lines are formed. Such a model is supported by the observed polarization of Be stars and by the characteristics of the X-ray variability in Be X-ray binaries (Waters et al., 1987).

There is increasing evidence that the mass loss from Be stars and from other rapidly rotating B stars is due to nonradial pulsations (NRP). See Baade (1987) for an excellent review. The rapidly rotating stars can have NRP modes of both high and low azimuthal order, m ≃ 2, whereas slow rotators only have modes of higher order. The amplitudes of the low-m modes are variable. The large variations of the winds of the rapidly rotating stars may be due to the variations in the NRP modes. For low order sectorial modes of pulsation the amplitudes are strongly peaked at the equator. This may explain the high density equatorial discs around Be stars.

The picture which emerges for rapidly rotating B stars is that of nonradial pulsating stars in which the rotation results in a preference for the excitation of the low-m modes. These low-m modes produce an enhanced mass loss, especially in the equatorial region. The low mass flux from the polar regions can be accelerated to high velocities, ~2000 km/s, by radiation pressure, but the high mass flux from the equatorial region can be accelerated to low velocities, ~200 km/s, only. The combined effects of a difference in mass flux and a difference in velocity between the polar and equatorial regions produce a very large density difference between the polar wind and the equatorial disc.

8. MASS LOSS NEAR THE EDDINGTON LIMIT

The characteristics of the stellar winds from stars near the Eddington limit are very different from those of normal stars. The best example of this can be found in the winds of the Luminous Blue Variables (LBV's). These are early type hypergiants of spectral types O to A with luminosities of $M_{bol} < -8^m$ which are highly variable in visual magnitudes and in mass loss. The LBV's are close to the luminosity upper limit in the HR diagram. Lamers and Fitzpatrick (1988) have shown that the effective gravity in the atmospheres of LBV's are very small because they are close to their Eddington limit for radiation pressure by numerous metal lines in the Balmer continuum.

The main characteristics of the winds of the LBV's are: high mass loss rates of $1-5 \times 10^{-5}$ $M_\odot$/yr; low terminal velocity ~100 to 300 km/s; very slow acceleration, reaching half the terminal velocity at about 5 - 10 R_*; large variability in mass loss (Lamers, 1987a). The variability in the mass loss occurs on all timescales from months to centuries. See Fig. 6.

The low terminal velocity can be understood in terms of the low effective gravity and the small escape velocity. We showed in Fig. 3 that v_∞ scales with v_{esc}. The high mass loss rate may also be a result of the low effective gravity and high luminosity. In fact, Pauldrach et al. (1986) have predicted a mass loss rate of about

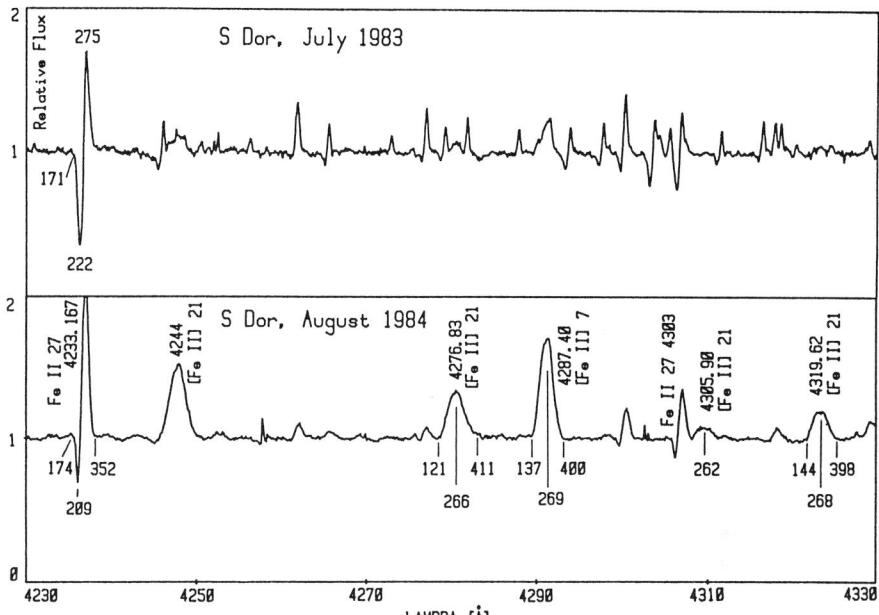

Fig. 6. Changes in the optical spectrum of the Luminous Blue Variable S Dor. The same part of the spectrum taken at two epochs one year apart is shown. The first spectrum was taken at maximum visual brightness and shows many narrow P Cygni profiles. The second spectrum was taken during the subsequent minimum, and is dominated by strong [Fe II] lines formed in the shell which was ejected during the maximum.

3×10^{-5} $M_\odot$/yr and a terminal velocity of 400 km/s for the LBV P Cygni (B1 Ia$^+$) using the standard radiation driven wind theory. These values are close to the observed ones of 2×10^{-5} $M_\odot$/yr and $v_\infty \simeq 300$ km/s. However this theory cannot explain the very slow acceleration of the winds of LBV's as it predicts that 0.5 v_∞ is reached at $r \simeq 2$ R_* instead of the observed value of about 5 - 10 R_*.

I suspect that the very slow acceleration of the winds of LBV's is due to the fact that the radiation pressure is produced by other lines than in the case of normal supergiants. This is suggested by the observed UV spectrum of LBV's: the winds of normal early-B supergiants are ionized to C IV, Si IV and even N V, whereas the winds of early-B LBV's are ionized to Mg II, Fe II, Fe III etc. This lower degree of ionization is probably due to the higher density in the winds (higher $\dot{M}$ and lower v_∞). If the LBV winds have a lower degree of ionization than the normal stars the winds will be driven by different lines, which may produce a slower velocity law. In fact Lamers (1986b) has shown that radiation pressure due to a large number ($\sim 10^2 - 10^3$) of optically thick lines of singly ionized metals can explain the very "soft" velocity law in the wind of P Cygni. A very slow velocity law will produce a high density in the wind, since $\rho \sim \dot{M} v^{-2} r^{-2}$, and a high density will result in a low degree of ionization.

These considerations suggest that the characteristics of stellar winds may change very drastically if a star gets close to its Eddington limit: high mass loss ($\dot{M} \gtrsim 10^{-5}$ M$_\odot$/yr) produces low ionization in the wind which implies slower acceleration. The high density of the wind produces P Cygni profiles in the Balmer lines, which is a general characteristic of most LBV's.

The mechanism of the large variations in the mass loss rates of LBV's is not known. Maeder (1983) has suggested that the very large outbursts of P Cygni in AD 1600 and of η Car in AD 1840 in which about ~1 M$_\odot$ was ejected might be due to an evolutionary effect when the star reached its stability limit against turbulent pressure. However, this does not explain the sudden outbursts which seem to require some energy storage and release mechanism (see Davidson, 1987 for a review).

The smaller variations in the mass loss of about a factor 2 or 3 on a timescale of years to decades might be explained by dynamical effects in the atmosphere. Appenzeller (1986, 1987) proposed that a small increase of $\dot{M}$ of a star close to its Eddington limit might lead to a runaway situation, because this will increase the radius of the pseudo-photosphere ($\tau_c \sim 0.3$), thus decreasing T_{eff}, which decreases the Eddington limit and makes the atmosphere even more unstable. The resulting runaway on a dynamical timescale will stop only when the temperature has dropped to $T_{eff} < 10000$ K. At such low temperatures the Eddington limit (L_{Edd}) would <u>increase</u> with decreasing T_{eff}, so the star is in dynamic equilibrium again. But, due to the rapid expansion of the outer layers, it will be out of thermal equilibrium and the layers just below the photosphere will now adjust on a thermal timescale. The main effect probably will be a heating of the layers at the base of the photosphere, which may bring the star back into its initial equilibrium. A small change in $\dot{M}$ might trigger a new outburst.

To illustrate how drastic the effects of an increase in mass loss by about a factor 3 can be, Table 3 presents the characteristics of the star S Dor during minimum and maximum visual brightness (Leitherer et al., 1985)

<u>Table 3</u>. S Dor during Minimum and Maximum

Phase	Year	M_v	M_v	M_{bol}	BC	T_{eff}	$R_*/R_\odot$	$\dot{M}$
Min	1965	11.3	−7.3	−9.3	2.0	22000	44	?
Max	1983	9.3	−9.3	−9.3	0	8000	330	5×10^{-5}

Notice that the radius of the "photosphere" has increased from 44 to 330 R$_\odot$, i.e., by a factor of 8! Recent calculations by Davidson

(1988) have shown that such an increase in radius requires an increase in the mass loss rate of only a factor 2 or 3.

We conclude that the winds of stars near their Eddington limit have very different characteristics from those of normal stars (low velocity, slow acceleration, low ionization) and that these winds are highly variable, because small effects might trigger large changes. It is not clear at present what the cause of the "small effects" are. Nonradial pulsations may play an important role, similar to the case of the Be-stars. In fact the photometric "microvariations" observed in LBV's and in other luminous early type supergiants by van Genderen (1986) suggest the presence of nonradial pulsations of the underlying stars.

9. SUMMARY

The accuracy of the determinations of the mass loss rates for early type stars is lower than generally assumed. The mass loss rates derived from the UV lines are uncertain because the ionization fractions of the observed ions are not well known. This is particularly true if the mass loss rates are derived from IUE data only, which for the O-stars are limited to the N V lines, (the C IV lines are usually saturated). The mass loss rates derived from the $H\alpha$ emission and the IR excess are uncertain because they depend very heavily on the density structure or velocity law in the transition between the photosphere and the wind. The structure of these layers is not accurately predicted by the present theories of radiation driven winds. The only reliable mass loss determinations at present are those derived from radio data. Unfortunately this method can only be used for ~15 of the most luminous stars.

The situation can be improved in the future in several ways: reliable calculations of the ionization equilibrium in stellar winds, analysis of the profiles of $H\alpha$ emission lines, observation of the free-free emission at mm-wavelengths, and observations of UV resonance lines in the wavelength range from 912 to 1250 Å.

Rotation plays a critical role in the mass loss from rapidly rotating main sequence B-type stars and Be stars. The mass loss is concentrated in the equatorial regions. This high concentration is probably due to the effects of nonradial pulsations in sectorial low-m modes. The large changes in mass loss from these stars could be due to changes in the nonradial pulsation modes of the star. It is expected that the next few years will bring new insight in this field, when the relation between pulsation and mass loss has been observed in more stars.

Stars near the Eddington limit have winds with high mass loss rates, low velocities and a low degree of ionization. Small initial changes in mass loss, possibly due to nonradial pulsations of the stars, can trigger large effects. In the near future we can expect to see great improvements in our understanding of these phenomena, since substantial observational and theoretical programs are presently being carried out by various groups.

Acknowledgement

The author is grateful to P. Conti and C. Garmany for hospitality at JILA (Boulder) where this review was written. I also want to thank the JILA Scientific Reports Office for preparing the manuscript. This research was supported by NSF grant AST-8520728 to the University of Colorado.

References

Abbott, D. C. 1982, Astrophys. J. 259, 282.
Abbott, D. C. 1985, in Radio Stars, eds. R. M. Hjellming and D. M. Gibson (Dordrecht: Reidel) p. 61.
Abbott, D. C., Garmany, C. D., Hansen, C. J., Henrichs, H. F., Pesnell, W. D. 1986, Publ. Astron. Soc. Pacific 98, 29.
Abbott, D. C., Telesco, C. M., Wolff, S. C. 1984, Astrophys. J. 279, 225.
Allen, C. W. 1973, Astrophysical Quantities (London: Athlone).
Appenzeller, I. 1986, in Luminous Stars and Associations in Galaxies, eds. C. W. H. de Loore et al. (Dordrecht: Reidel).
Appenzeller, I. 1987, in Instabilities in Luminous Early Type Stars, eds. H. J. G. L. M. Lamers and C. W. H. de Loore (Dordrecht: Reidel) p. 137.
Baade, D. 1987, in Physics of Be-stars, eds. A. Slettebak and T. P. Snow (Cambridge: Cambridge University Press) p. 361.
Baade, D., Lucy, L. B. 1987, Astron. Astrophys. 178, 213.
Barlow, M. J., Cohen. M. 1977, Astrophys. J. 213, 737.
Berthout, C., Leitherer C., Stahl, O., Wolf, B. 1985, Astron. Astrophys. 144, 87.
Cassinelli, J. P. 1987, in Instabilities in Luminous Early Type Stars, eds. H. J. G. L. M. Lamers and C. W. H. de Loore (Dordrecht: Reidel) p. 273.
Cassinelli, J. P., Lamers, H. J. G. L. M. 1987, in Scientific Accomplishments of IUE, eds. Y. Kondo et al. (Dordrecht: Reidel) p. 139.
Cassinelli, J. P., Olson, G. L. 1979, Astrophys. J. 229, 304.
Castor, J. I. 1987, in Instabilities in Luminous Early Type Stars, eds. H. J. G. L. M. Lamers and C. W. H. de Loore (Dordrecht: Reidel) p. 159.
Castor, J. I., Lamers H. J. G. L. M. 1979, Astrophys. J. Suppl. 39, 481.
Castor, J. I., Simon, T. 1983, Astrophys. J. 265, 304.
Chiosi, C., Maeder, A. 1986, Ann. Rev. Astron. Astrophys. 24, 329.
Chiosi, C., Olson, G. L. 1984, private communication.
Conti, P. S. 1978, Ann Rev. Astron. Astrophys. 16, 371.
Davidson, K. 1987, in Instabilities in Luminous Early Type Stars, eds. H. J. G. L. M. Lamers and C. W. H. de Loore (Dordrecht: Reidel) p. 127.
Davidson, K. 1988, Astrophys. J. (in press).

Doazan, V. 1982, in B stars with and without Emission Lines, eds. A. B. Underhill and V. Doazan (NASA SP-456).
Garmany, C. D., Conti, P. S. 1984, Astrophys. J. 284, 705.
Garmany, C. D., Olson, G. L., Conti, P. S., Van Steenberg, M. E. 1981, Astrophys. J. 250, 660.
Gathier, R., Lamers, H. J. G. L. M., Snow, T. P. 1981, Astrophys. J. 247, 173.
Grady, C. A., Bjorkmann, K. S., Snow, T. P. 1987, Astrophys. J. 320, 376.
Groot, M., Thé, P. S. 1983, Astron. Astrophys. 120, 89.
Hamann, W. R. 1980, Astron. Astrophys. 84, 342.
Hamann, W. R. 1981, Astron. Astrophys. 100, 169.
Harnden, F. R., Branduardi, G., Elvis, M., Gorenstein, P., Grindlay, J. E. 1979, Astrophys. J. (Lett.) 234, L51.
Hayes, D. P. 1984, Astron. J. 89, 1219.
Hearn, A. G. 1987, in Circumstellar Matter, eds. I. Appenzeller and C. Jordan (Dordrecht: Reidel) p. 395.
Henrichs, H. F. 1987, in O, Of and Wolf Rayet Stars, eds. P. S. Conti and A. B. Underhill (NASA/CNRS monograph in press).
Henrichs, H. F., Hammerschlag-Hensberge, G., Howarth, I. D., Barr, P. 1983, Astrophys. J. 268, 807.
Henrichs, H. F., Hammerschlag-Hensberge, G., Lamers, H. J. G. L. M. 1980, in Proc. Second European IUE Conf., eds. B. Battrick and J. Mort (ESA SP-157) p. 147.
Klein, R. I., Castor, J. I. 1978, Astrophys. J. 220, 902.
Kunasz, P., Morrison, N. D. 1982, Astrophys. J. 263, 226.
Kurucz, R. L. 1979, Astrophys. J. Suppl. 40, 1.
Lamers, H. J. G. L. M. 1986a, in Luminous Stars in Associations and Galaxies, eds. C. W. H. de Loore et al. (Dordrecht: Reidel) p. 157.
Lamers, H. J. G. L. M. 1986b, Astron. Astrophys. 159, 90.
Lamers, H. J. G. L. M. 1987a, in Instabilities in Luminous Early Type Stars, eds. H. J. G. L. M. Lamers and C. W. H. de Loore (Dordrecht: Reidel) p. 99.
Lamers, H. J. G. L. M. 1987b, in Physics of Be stars, eds. A. Slettebak and T. P. Snow (Cambridge: Cambridge University Press) p. 219.
Lamers, H. J. G. L. M., de Loore, C. W. H. (eds.) 1987, Instabilities in Luminous Early Type Stars (Dordrecht: Reidel).
Lamers, H. J. G. L. M., Fitzpatrick, E. 1988, Astrophys. J. (in press).
Lamers, H. J. G. L. M., Gathier, R., Snow, T. P. 1982, Astrophys. J. 258, 186.
Lamers, H. J. G. L. M., Morton, D. C. 1976, Astrophys. J. Suppl. 32, 715.
Lamers, H. J. G. L. M., Perinotto, M., Cerruti-Sola, M. 1987, Astrophys. J. 314, 726.
Lamers, H. J. G. L. M., Rogerson, J. B. 1978, Astrophys. J. 66, 417.
Lamers, H. J. G. L. M., Snow, T. P. 1978, Astrophys. J. 219, 504.
Lamers, H. J. G. L. M., Snow, T. P., de Jager, C., Langerwerf, A. 1988, Astrophys. J. (in press).

Lamers, H. J. G. L. M., Stalio, R., Kondo, Y. 1978, Astrophys. J. 223, 207.
Lamers, H. J. G. L. M., Waters, L. B. F. M. 1984, Astron. Astrophys. 136, 37.
Lamers, H. J. G. L. M., Waters, L. B. F. M. 1987, Astron. Astrophys. 182, 80.
Leitherer, C. 1988, Astrophys. J. (in press).
Leitherer, C., Appenzeller, I., Klare, G., Lamers, H. J. G. L. M., Stahl, O., Waters, L. B. F. M., Wolf, B. 1985, Astron. Astrophys. 153, 168.
Leitherer, C., Hefele, H., Stahl, O., Wolf, B. 1982, Astron. Astrophys. 108, 102.
Lucy, L. B. 1984, Astron. Astrophys. 140, 210.
Maeder, A. 1983, Astron. Astrophys. 120, 113.
Mihalas, D., Kunasz, P. B., Hummer, D. G. 1975, Astrophys. J. 202, 465.
Morton, D. C. 1967, Astrophys. J. 150, 535.
Morton, D. C. 1976, Astrophys. J. 203, 386.
Mullan, D. J. 1986, Astron. Astrophys. 165, 157.
Olson, G. L., Castor, J. I. 1981, Astrophys. J. 244, 179.
Olson, G. L., Ebbets, D. 1981, Astrophys. J. 248, 1021.
Panagia, N., Felli, M. 1975, Astron. Astrophys. 39, 1.
Pauldrach, A. 1987, private communication.
Pauldrach, A., Puls, J., Kudritzki, R. P. 1986, Astron. Astrophys. 164, 86.
Poeckert, R., Marlborough, J. M. 1978, Astrophys. J. Suppl. 38, 229.
Praderie, F., Talavera, A., Lamers, H. J. G. L. M. 1980, Astron. Astrophys. 86, 271.
Prinja, R. K., Henrichs, H. F., Howarth, I. D. 1988, Monthly Notices R. Astron. Soc. (in press).
Prinja, R. K., Howarth, I. D. 1986, Astron. Astrophys. Suppl. 71, 357.
Snow, T. P. 1977, Astrophys. J. 217, 760.
Snow, T. P., Marlborough, M. 1980, Astrophys. J. (Lett.) 203, L87.
Snow, T. P., Morton, D. C. 1976, Astrophys. J. Suppl. 32, 429.
Snow, T. P., Stalio, R. 1987, in Scientific Accomplishments of IUE, eds. Y. Kondo et al. (Dordrecht: Reidel) p. 183.
Van Genderen, A. M. 1986, Astron. Astrophys. 157, 163.
Waters, L. B. F. M. 1986, Astron. Astrophys. 159, L1.
Waters, L. B. F. M., Coté, J., Aumann, H. H. 1987, Astron. Astrophys. 172, 225.
Waters, L. B. F. M., Lamers, H. J. G. L. M. 1984, Astron. Astrophys. Suppl. 57, 327.
Waters, L. B. F. M., Taylor, A. R., van den Heuvel, E. P. J., Habets, G. M. H. J., Persi, P. 1988, Astron. Astrophys. (in press).
Waters, L. B. F. M., Wesselius, P. R. 1986, Astron. Astrophys. 155, 104.
White, N. E., Kallman, T. R., Swank, J. H. 1983, Astrophys. J. 269, 264.
Wrights, A. E., Barlow, M. J. 1975, Monthly Notices R. Astron. Soc. 170, 41.

RADIATION DRIVEN WINDS OF CENTRAL STARS OF PLANETARY NEBULAE

A. Pauldrach, R.P. Kudritzki, R. Gabler, A. Wagner
Institut für Astronomie und Astrophysik
der Universität München
Scheinerstr. 1
8000 München 80
Germany

I. INTRODUCTION

It is well known that stellar winds are an ubiquitous feature of young, massive, very luminous stars at the upper end of the main sequence with masses in the range 20-200 $M_\odot$ and that the evolution of these objects is strongly affected by severe mass-loss connected with the stellar winds. The most promising attempt to explain these winds quantitatively is the theory of radiation driven winds, which has been established in its original form by Lucy and Solomon (1970) and Castor, Abbott and Klein (1975, "CAK"). Most of the basic observational properties of the hot massive star stellar winds can be explained by this self-consistent theory, if the substantial improvements, which have been worked out very recently (Pauldrach, Puls, Kudritzki (1986), "PPK"; Kudritzki, Pauldrach, Puls (1987); Pauldrach (1987), Puls (1987)) are taken into account.

However, for the hot stars of low mass in the advanced stage of post-AGB-evolution - the Central Stars of Planetary Nebulae (CPN) - the situation is less clear. Many of them exhibit stellar winds, whereas others do not. The observed mass-loss rates are highly uncertain, however as the wind velocities are of the same order as for the massive, young OB-stars, it is not clear, whether the mass-loss does affect the post-AGB-evolution. (According to Schönberner (1983) this would only be the case for rates with $\dot{M} > 10^{-7} M_\odot/yr$). The physical reason for the stellar winds in CPN is also not yet established convincingly.

The primary intention of this paper, therefore, is to demonstrate that the improved theory of radiation driven winds is also appropriate for CPN. For this purpose we will briefly summarize in section II the basic observational features of CPN winds. We will show that the observations reveal a systematic behaviour, which is indicative of radiation driving being the origin. In section III we will present new radiation driven wind models for CPN, which are compared with the observations. Section IV discusses the use of the terminal velocity v_∞ as a distance indicator. The indicators of stellar winds in the optical spectra of CPN will be discussed in section V.

II. OBSERVED WINDS OF CPN

After the advent of the IUE satellite it became undoubtedly clear that stellar winds are present in many CPN (Heap, 1978, Perinotto, 1982). P-Cygni profiles in the UV have been detected for a variety of objects and many attempts have been undertaken to determine terminal velocities v_∞ as well as mass-loss rates $\dot{M}$. A typical example is the Central star of NGC 3242 (see Fig. 1). Here the detailed fit of the observed NV P-Cygni profile yielded v_∞ = 2200±100 km/s and log $\dot{M}$ = -9.0±1.0 ($\dot{M}$ in $M_\odot$/yr). The large error in $\dot{M}$ reflect mainly the uncertainty of the ionization calculations, which in our eyes is typical for these objects.

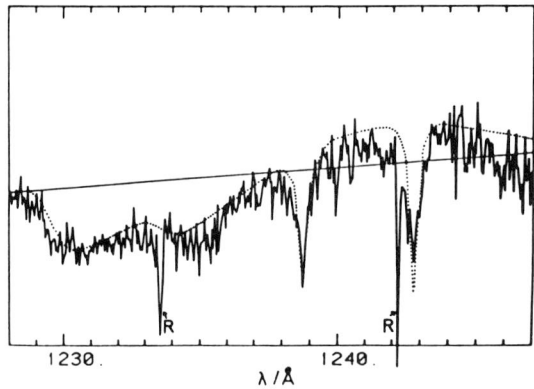

Fig. 1: IUE high resolution profile of the NV P-Cygni profile of NGC 3242. The dotted curve shows the theoretical profile fit obtained by detailed comoving frame calculations (from Hamann, Kudritzki, Méndez, Pottasch, 1984).

The general situation is best described in the paper by Cerruti-Sola and Perinotto (1985, "CSP"). Fig. 2 gives the HR-diagram of CPN observed with IUE. Although both, luminosity and effective temperature are highly uncertain for many of these objects the tendency is obvious: Winds appear for objects, which are more luminous (or have larger radii). CSP claim that for log $R/R_\odot$ > -0.5 winds are the rule and not the exception. They compare this borderline with evolutionary tracks by Schönberner (1983) and conclude that this corresponds to gravities log g ≈ 5.2. Objects with lower gravities show a wind, contrary to those with higher gravity. This conclusion is in agreement with Hamann et al. (1981), who found the same for hot subdwarfs not surrounded by a nebula.

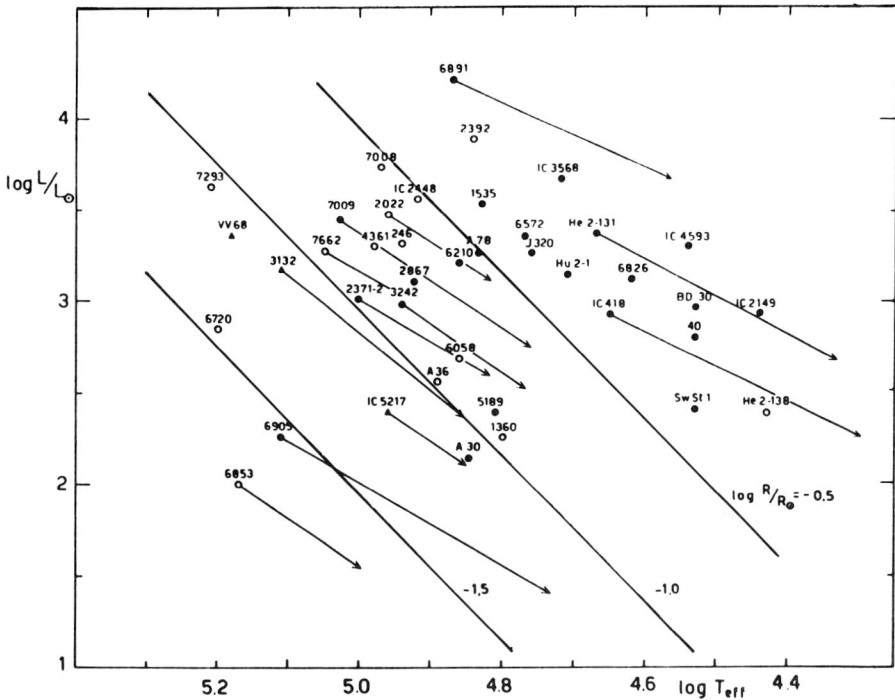

Fig. 2: The HR-diagrm of CPN showing a wind (filled circles) and no wind (open circles). Arrows reflect uncertainties of luminosity and effective temperature. Lines of constant radius are also shown (from Cerruti-Sola and Perinotto, 1985).

Mass-loss rates for CPN are displayed in Fig. 3. It is evident that the CPN do not generally fall on the extension of the relation for massive OB-stars. However, one has to bear in mind the large uncertainties of the CPN $\dot{M}$-values and the luminosities.

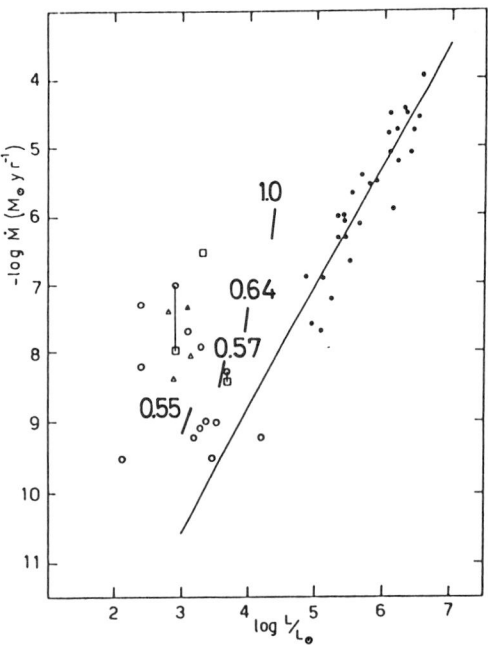

Fig. 3: Logarithm of mass-loss rate vs. log of luminosity for Pop I OB-stars (full points) and CPN (open circles, triangles and squares) (from Cerruti-Sola and Perinotto, 1985). The curves labelled by stellar mass in units of $M_\odot$ show radiation driven wind calculations described in section III.

The use of the HR-diagram for CPN has the disadvantage that it relies on distances, which are heavily debated and highly uncertain. In consequence, an alternative approach, which has been developed by Méndez, Kudritzki and colleagues (Méndez et al., 1981, 1983, 1985, 1987) has become more and more important: Here the (log g, log T_{eff})-diagram of CPN is used, where gravity and effective temperature are carefully determined by detailed NLTE model photosphere spectral analysis. The comparison with stellar evolution is then done by transforming the tracks into the (log g, log T_{eff})-plane, which gives much better constraints on the physical status of CPN - in particular independent on any assumption on distances. Fig. 4 shows the status quo of this work, now in our case with respect to the occurence of stellar winds: We see immediately, that objects with log g > 5 do not exhibit a wind. Objects with log g < 5.0 do show a wind either in the UV or in the optical (by HeII 4686 being in emission <=> Of-spectral type) or both.

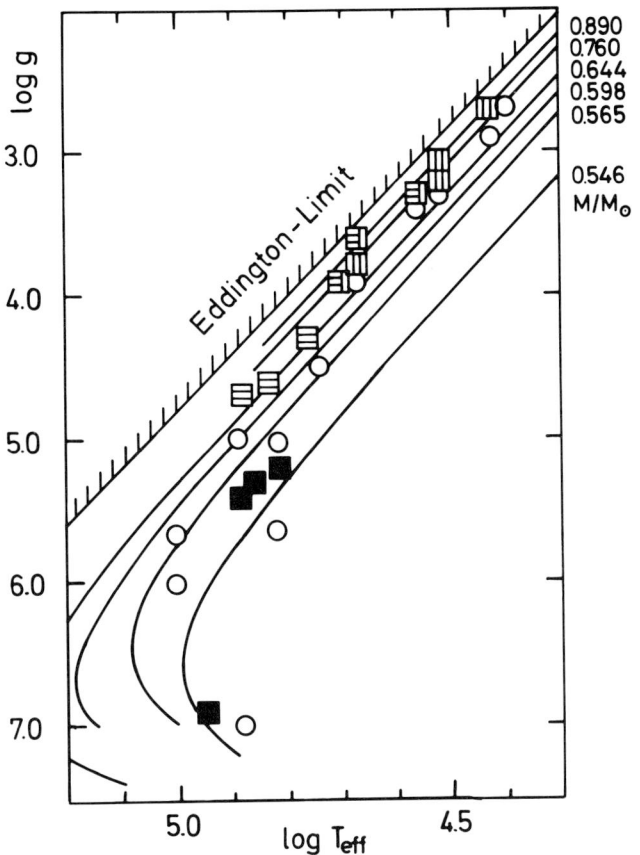

Fig. 4: The (log g, log T_{eff})-diagram of CPN according to Méndez et al. (1987, 1985). Evolutionary tracks (labelled by $M/M_\odot$) are also shown. IUE wind detections are indicated by ⊟. Wind detections by optical spectra (HeII 4686 in emission <=> Of-spectral type) are given by ⊞. Definitely no winds (IUE plus optical) is coded by ■. Objects with no winds in the optical but no IUE high resolution information yet available are described by o. The Eddington-limit is also indicated.

It is very interesting that the Of-phenomenon occurs for objects very close to the Eddington-limit. Fig. 5 shows, how HeII 4686 turns over from absorption to emission, when approaching the Eddington-limit.

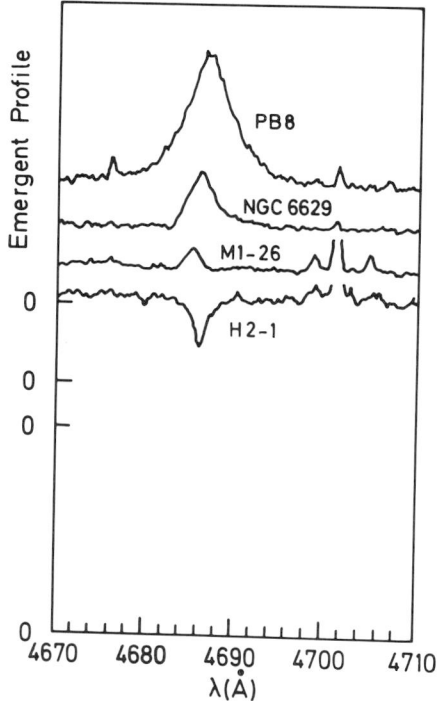

Fig. 5a: The turnover of HeII 4686 from absorption to emission, when approaching the Eddington-limit (ESO 3.6 m CASPEC-spectra from Méndez et al., 1987)

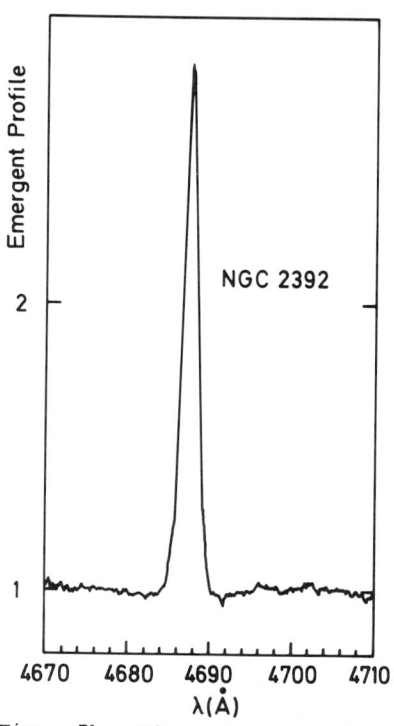

Fig. 5b: The extreme stellar HeII 4686 emission of NGC 2392, the most massive object of Fig. 4, which is closest to the Eddington-limit (from Méndez et al. 1987).

Fig. 4 and 5 allow the following conclusion: Stellar winds become less detectable (= less dense), as soon as the CPN evolve towards larger gravities. Objects closer to the Eddington-limit have stronger (= denser) winds. Fig. 6, which shows evolutionary tracks in the HR-diagram, demonstrates that this means the following: According to the core-mass luminosity relation more massive (= more luminous) objects have stronger winds. However, when evolving at constant luminosity towards higher temperature, the density or the strength of the winds is decreasing.

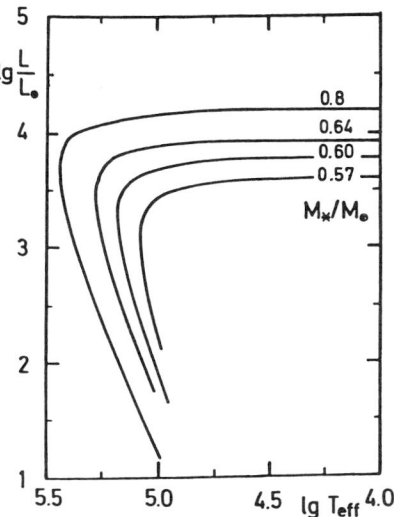

Fig. 6: The HR-diagram of CPN evolutionary tracks (Schönberner, 1983; Wood and Faulkner, 1986).

This systematic behaviour indicates already that the CPN winds are radiation driven: In this case we would expect $\dot{M}$ to increase significantly, when approaching the Eddington-limit (see CAK, Abbott, 1982, PPK, Pauldrach et al. 1985). On the other hand $\dot{M}$ should remain roughly constant with constant luminosity along such a track in Fig. 6, while the terminal velocity v_∞ <u>increases</u> with increasing escape velocity v_{esc} from the stellar surfaces. This means that for radiation driven winds of stars evolving at constant luminosity and mass towards higher temperatures, v_∞ should increase along the track, since the radius is shrinking (Abbott, 1978; PPK). Consequently, since $\dot{M} = 4\pi r^2 \rho v$, we expect the wind density to decrease along the track of constant luminosity towards higher temperature. This would explain the general behaviour encountered in Fig. 4 and 5.

A crucial observational test of this interpretation is the behaviour of the terminal velocity v_∞ of CPN as function of T_{eff}. This is shown in Fig. 7. We clearly see that v_∞ increases with T_{eff}, which strongly supports the idea of increasing wind velocities with decreasing stellar radius predicted by radiation driven wind theory.

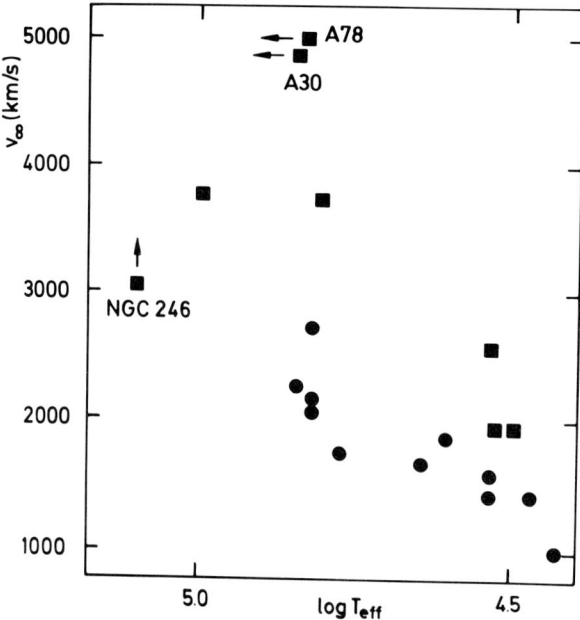

Fig. 7: Terminal wind velocities of CPN v_∞ vs. T_{eff}. Hydrogen deficient objects (■) and obejcts containing mainly hydrogen (●).

III. RADIATION DRIVEN WIND MODELS OF CPN

In this chapter we present our computations for radiation driven wind models calculated along the evolutionary tracks as displayed in Fig. 6. These calculations include the following significant improvements relative to the original CAK-theory:

- improved dynamics by dropping the radial streaming approximation (see PPK): these improvements allowed to match the observed values of v_∞ and $\dot{M}$ for a variety of massive OB stars including P-Cygni.

- detailed multi-level NLTE calculations for ionization and excitation (Pauldrach, 1987): these calculations were done for the massive O4f-star ζ Pup. They include the selfconsistent treatment for:

  - 26 elements
  - 133 ions
  - 10000 line transitions   } simultaneously
  - 100000 lines in NLTE for the line force   with
  - electron collisions   hydrodynamics
  - continuum radiative transfer

These calculations solved the "superionization" problem and produce NV, OVI as observed for cool winds, which assume a temperature $T_{wind} \approx T_{eff}$ suggested by IR observations (Lamers et al., 1984).

The selfconsistent treatment of multiple scattering as developed by Puls (1987) was not yet included. This will be done in a refined step of the calculations.

Fig. 8 shows the $(v_\infty, \log T_{eff})$-plots of these calculations along the evolutionary tracks including the observed values for CPN. The result is extremely convincing.

It allows to read off stellar masses directly and demonstrates that the masses of CPN are in a rather narrow range between 0.5 and 0.8 solar masses (Schönberner, 1981, Méndez et al. 1985, 1987).

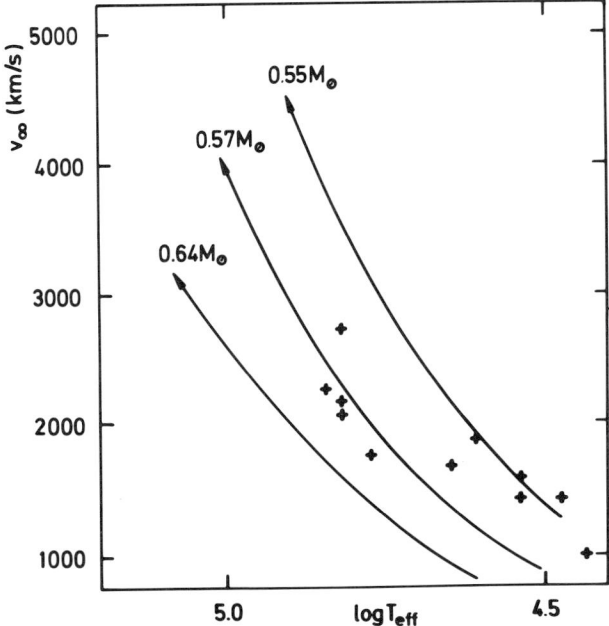

Fig. 8a: The $(v_\infty, \log T_{eff})$-diagram of radiation driven wind theory for objects with almost normal hydrogen content. Crosses give the observed values.

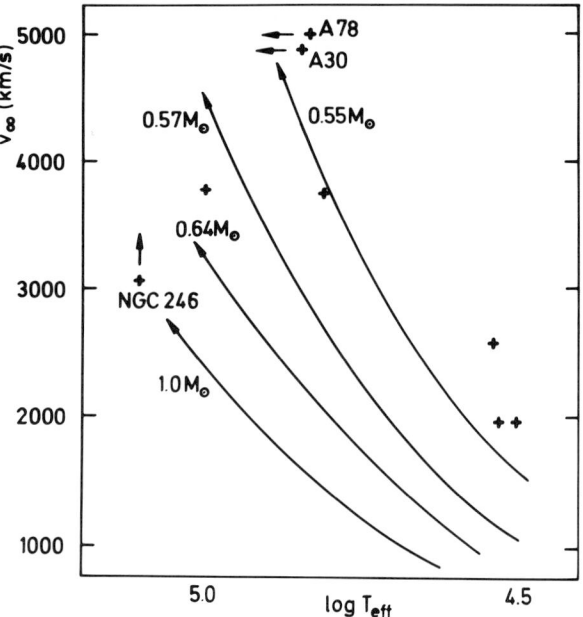

Fig. 8b: Same as Fig. 8a, but for hydrogen deficient objects. Note: only the labelled objects show absorption type photospheric spectra. The others are of spectral type WC (see Méndez et al. 1986). All T_{eff} (except for NGC 246 (see Husfeld, 1986)) are extremely uncertain, the values for A78, A30 being lower limits. The v_∞ value for NGC 246 is also a lower limit (Heap, 1986).

Fig. 9. shows the mass-loss rates computed along the evolutionary tracks. The results are extremely striking: The more massive objects, which according to the core-mass luminosity relation are more luminous, have significantly stronger winds. This is just the observed effect, as discussed in Section II. Very extreme is the track for $M = 1\ M_\odot$. Here $\dot{M}$ is almost as high as for massive O-stars, although the stellar radius is a factor of 10 smaller. This means that the winds are extremely dense in this case and must exhibit extreme emission features. NGC 2392 is obviously such an object. In Fig. 4 it lies closest to the Eddington-limit with a mass of about $0.9\ M_\odot$. Its HeII 4686 emission is extreme (Fig. 5b) thus indicating an enormous wind density. The v_∞ given in literature (Heap, 1986) is rather low: 580 km/s. We have carried out a detailed wind calculation for this object adopting $T_{eff} = 45000K$, $\log g = 3.6$, $\log L/L_\odot = 4.45$ (see Méndez et al., 1987) and obtained $v_\infty = 420$ km/s. Since we did not yet include the effects of multiple scattering (Puls, 1987), this result is quite satisfying. The theoretical mass-loss rate is very interesting: $\dot{M} = 1.4 * 10^{-6} M_\odot/yr$. Obviously, for such an object the fast evolution is dominated by the mass-loss rather than the nuclear burning time!

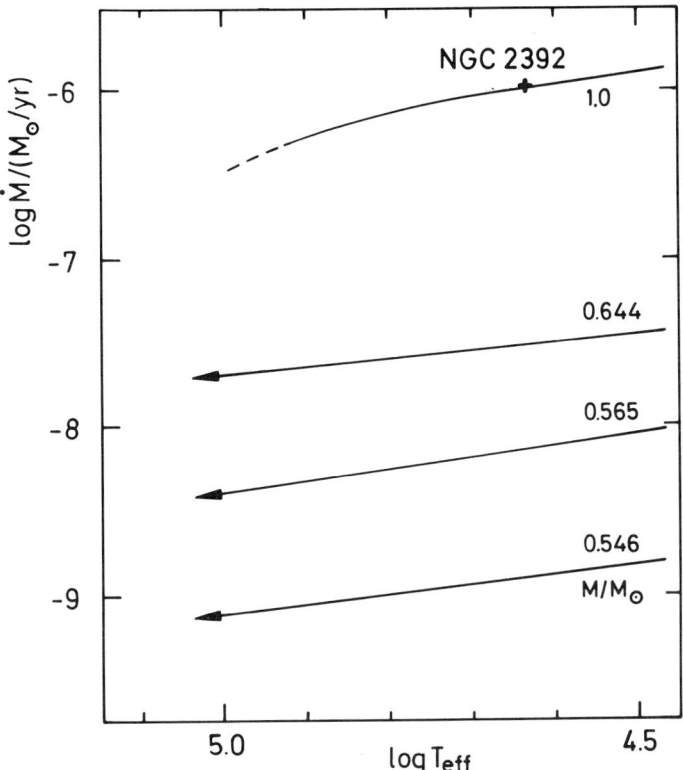

Fig. 9: The theoretical log Ṁ-log T_{eff} diagram along the evolutionary tracks.

The transformation of Fig. 9 into the log Ṁ-log $L/L_\odot$-diagram is given in Fig. 3. It shows that predictions by the improved radiation driven wind theory are not too bad, in particular when the enormous possible errors for the observed values of Ṁ and log $L/L_\odot$ are accounted for.

IV. WIND DISTANCES FOR CPN

For massive OB-stars the relation $v_\infty \approx 3 v_{esc}$ is observed, as long as $T_{eff} > 30000K$ (Abbott, 1982). Kaler et al. (1985) adopted this relation for CPN also and combined it with the core-mass luminosity relation $R_\star^2 T_{eff}^4 = f(M/M_\odot)$. Since $v_{esc}^2 = 2 G M(1-\Gamma)/R_\star$ (with $\Gamma = L_E/L$), one can determine in this way the CPN mass and radius and therefore the distance, as soon as T_{eff} is known. Kaler et al. used this method to derive "wind distances" for CPN. The crucial question is, however, whether $v_\infty = 3 v_{esc}$ holds in general for CPN. Fig. 10 shows that this is not the case. Our calculations predict that the ratio v_∞/v_{esc} depends on stellar mass, evolutionary status, photospheric helium abundance, etc.. Consequently, the method of wind distances - although intrinsically very powerful - will need some refinements to become finally quantitatively reliable.

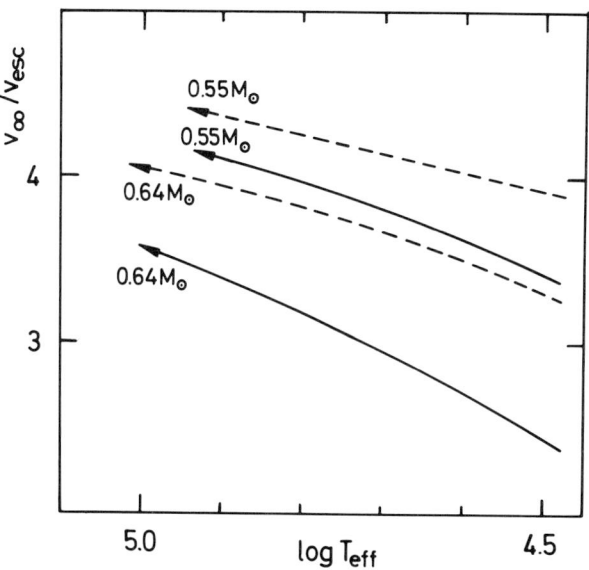

Fig. 10: The ratio v_∞/v_{esc} vs. T_{eff} for (--) hydrogen deficient objects and (-) objects containing mainly hydrogen.

V. STELLAR WINDS IN THE OPTICAL SPECTRA OF CPN

As already discussed in section III the high resolution optical spectroscopy of CPN by Méndez et al. (1987) shows clearly that stellar wind emission line features are systematically stronger pronounced for objects, which are located closer to the Eddington limit. A convincing example is given by the behaviour of HeII 4686 as displayed in Fig. 5. It is important to test, whether radiation driven wind models can reproduce this behaviour.

For this purpose a new type of "unified model atmospheres" has been developed at the Munich observatory by R. Gabler (1986) and A. Wagner (1986) in cooperation with J. Puls, A. Pauldrach and R.P. Kudritzki. These NLTE-model atmospheres are spherically extended, in radiative equilibrium and include the density and velocity distribution of radiation driven winds as described in section III. The spectra of hydrogen and helium lines are then calculated for these models by detailed non-LTE multi-level calculations in the whole atmosphere, thus treating the contribution of subsonic deeper and supersonic outer layers to the emergent line profile in the correct selfconsistent unified way including Stark-effekt broadening and velocity fields.

We have calculated a sequence of such models for T_{eff} = 50000K and $M/M_\odot$ = 0.55, 0.57, 0.64, 1.0. The corresponding luminosities for the CPN (or the gravities or the radii) were obtained from the evolutionary tracks mentioned already in the previous sections. In units of the Eddington luminosity L_E we obtain respectively: L/L_E = 0.06, 0.16, 0.34, 0.74. The 1.0 $M_\odot$ object is therefore already rather close to the Eddington limit. Its theoretical HeII 4686 profile should look similar to the one of NGC 2392 as displayed in Fig. 5b. Fig. 11 shows that this is really the case. In addition, the turnover from photospheric absorption to wind emission with increasing L/L_E as observed in Fig. 5 is well reproduced by the theoretical computations. This demonstrates that the concept of radiation driven winds is very useful also for CPN. Moreover, it renders the possibility to use the 4686 emission as a powerful luminosity (= mass, = distance)-indicator.

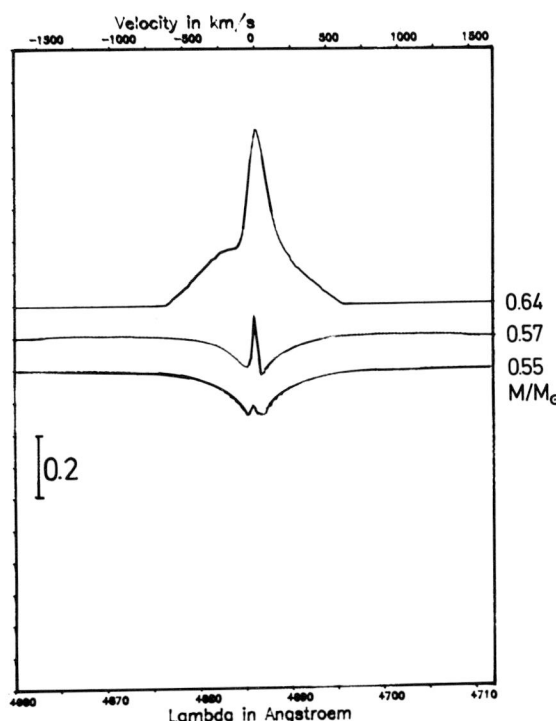

11a: objects from 0.55 to 0.64 $M/M_\odot$

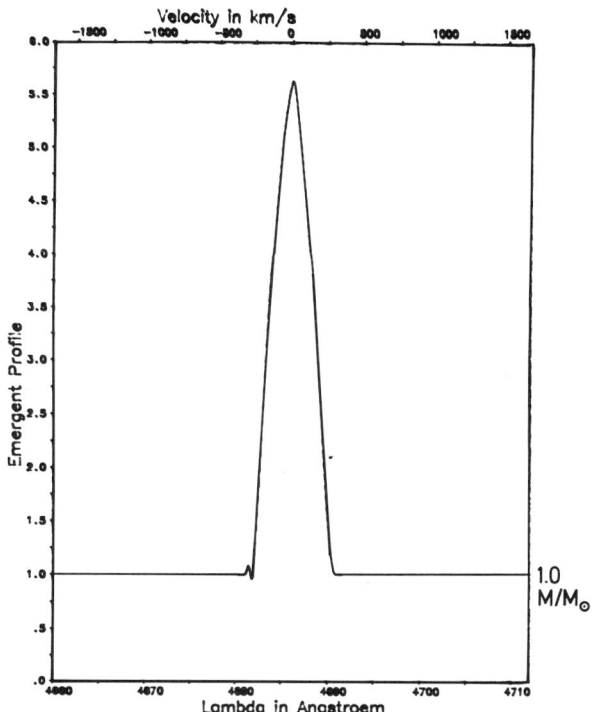

11b: an object close to the Eddington-limit $M/M_\odot = 1.0$

Fig. 11: The turnover of theoretical HeII 4686 profiles from absorption to emission when approaching the Eddington-limit, calculated for $T_{eff} = 50000K$.

VI. CONCLUSIONS

We conclude that radiation driven wind models are appropriate also for the observed winds of Central Stars. They reproduce in principle the observed systematic behaviour of mass-loss rates and terminal velocities as function of stellar mass (luminosity) and effective temperature. Consequently, we will continue to improve the details of our radiation driven wind models for these objects by including multi-line effects (see Puls, 1987) and NLTE metal line blanketing in the radiative equilibrium of the corresponding unified atmosphere models. We will test the theory by detailed spectroscopic analyses of the wind lines (both optical and UV) of individual selected objects, where the stellar parameters are already reliably determined by the analysis of photospheric lines (see Méndez et al. 1987). We will also start to analyse the spectra of those numerous CPN, which are of spectral type WC (see Méndez et al. 1986). Encouraged by the results discussed in section V, we will follow the working hypothesis that the WC Central Stars are helium rich objects close to the Eddington-limit.

ACKNOWLEDGEMENTS

We wish to thank our colleague R.H. Méndez for his support and the allowance to use part of his recent results. The detailed discussions with J. Puls, D. Husfeld, K. Butler and A. Herrero are also gratefully acknowledged. This work was supported by the DFG under grant Ku 474/11-1 and Ku 474/13-1.

REFERENCES

Abbott, D.C., 1978, Ap. J. **225**, 893
Abbott, D.C., 1982, Ap. J. **259**, 282
Castor, J., Abbott, D.C., Klein, R., 1975, Ap. J. **195**, 157
Cerruti-Sola, M., Perinotto, M., 1985, Ap. J. **291**, 237
Gabler, R., 1986, Diplomarbeit, Munich University
Hamann, W.-R., Gruschinske, J., Kudritzki, R.P., Simon, K.P., 1981, Astron. Astrophys. **104**, 249
Hamann, W.-R., Kudritzki, R.P., Mendez, R.H., Pottasch, S.R., 1984, Astron. Astrophys. **139**, 459
Heap, S.R., 1978, in IAU Symp. 83, p. 99
Heap, S.R., 1986, in ESA SP-263, p. 291
Husfeld, D., 1986, Ph.D. Thesis, Munich University
Kaler, J.B., Mo, J.E., Pottasch, S.R., 1985, Astrophys. J. **288**, 305
Kudritzki, R.P., Pauldrach, A., Puls, J., 1987, Astron. Astrophys., **173**, 293
Lamers, H.J.G.L.M., Waters, L.B., Wesselius, P.R., 1984, Astron. Astrophys. **134**, L17
Lucy, L.B., Solomon, P., 1970, Ap. J. **159**, 879
Mendez, R.H., Kudritzki, R.P., Gruschinske, J., Simon, K.P., 1981, Astron. Astrophys. **101**, 323
Mendez, R.H., Kudritzki, R.P., Simon, K.P., 1983, in IAU Symp. 103, p. 343
Mendez, R.H., Kudritzki, R.P., Simon, K.P., 1985, Astron. Astrophys. **142**, 289
Mendez, R.H., Miguel, C.H., Heber, U., Kudritzki, R.P., 1986, in IAU Colloquium 87 "Hydrogen deficient stars and related objects", ed. K. Hunger et al., Reidel, Astrophys. Sp. Sci. Library **128**, p. 323
Mendez, R.H., Kudritzki, R.P., Herrero, A., Husfeld, D., Groth, H.G., 1987, Astron. Astrophys., submitted
Pauldrach, A., Puls, J., Hummer, D.G., Kudritzki, R.P., 1985, Astron. Astrophys. **148**, L1
Pauldrach, A., Puls, J., Kudritzki, R.P., 1986, Astron. Astrophys. **164**, 86
Pauldrach, A., 1987, Astron. Astrophys., in press
Perinotto, M., 1982, in IAU Symp. 103, p. 323
Puls, J., 1987, Astron. Astrophys., in press
Schönberner, D., 1981, Astron. Astrophys. **103**, 537
Schönberner, D., 1983, Ap. J. **272**, 708
Wagner, A., 1986, Diplomarbeit, Munich University
Wood, P.R., Faulkner, D.J., 1986, Ap. J. **307**, 659

THE THEORY OF STELLAR WINDS FROM HOT AND COOL STARS

A.G. Hearn
Sonnenborgh Observatory
Zonnenburg 2
3512 NL Utrecht
The Netherlands

1. INTRODUCTION

Nearly all types of stars show continuous mass loss. For some stars the mass loss is very large, $10^{-5} M_\odot$/year for Wolf Rayet stars and M supergiants. For some stars the mass loss is very modest, $10^{-14} M_\odot$/year for the Sun.

It is becoming increasingly clear that at various stages of stellar evolution a substantial mass loss from the star is necessary to bring evolutionary calculations into agreement with observations. Examples are the mass loss necessary to form Wolf Rayet stars from O type stars and to form white dwarfs from asymptotic giant branch stars instead of supernovae.

This review deals only with stationary theories of mass loss. Mass loss from stars often shows variation with time. In some cases the variations of mass loss are just perturbations and a stationary theory probably gives a good description of the flow averaged over time. In other cases the variations of mass loss are so great that one must put question marks against the validity of a purely stationary theory. Little work has been done on time dependent theories of global mass loss. Local mass loss resulting from explosive stellar flares etc. fall outside the scope of this review.

Mass loss from a star needs energy to overcome gravity and to accelerate the stellar wind to supersonic velocities. The theories of stationary mass loss may be classified according to the driving force. These categories are shown in Table 1 with an illustration of the type of star to which the theory has been applied

The list of types of stars is illustrative and is in no way a complete list, nor is there always complete agreement that the theory is correctly applied.

In what follows each of these theories is summarized and the successes and difficulties of the theory are discussed.

TABLE 1

DRIVING FORCE			TYPE OF STAR
1. thermal			Sun
2. radiative forces	a)	resonance lines	OB supergiants
	b)	dust	M supergiants
3. Alfvén waves			K giants
4. pulsation	a)	radial	Mira's
	b)	non-radial	Be stars

2. THERMAL WINDS

Thermal stellar winds result from a hot corona round the star. The wind is driven by the conversion of the internal energy of the hot gas in the corona by expansion which provides the work necessary to accelerate the flow against the attraction of gravity.

   An important part of this process is the very high thermal conductivity of a hot plasma. This maintains a high coronal temperature out to many stellar radii. Chapman (1957) showed that such an extended corona cannot remain hydrostatic because the pressure at infinity produced by the corona is order of magnitudes greater than the pressure of interstellar matter. This problem was resolved by Parker who developed the solar wind theory.

2.1. Assumptions of the theory and results

In his original work Parker (1958) made a number of simplifying assumptions.

(i)   heating of the corona occurs at its base
(ii)   no viscosity
(iii) no magnetic field
(iv)   one fluid theory
(v)   spherically symmetry

   By combining the equation of motion with the equations of continuity and state, Parker derived the solar wind equation which specifies the velocity distribution for a given temperature distribution. This equation possesses a singularity, the critical point, and he showed that only the solution passing through the critical point from subsonic velocities at the star to supersonic velocities at greater distances is able to satisfy the physical boundary conditions. In this way a uniquely defined solution is produced.
   Parker predicted that the solar corona would be accelerated to supersonic velocities, a prediction that was later confirmed when satellites were launched with instruments to measure the velocity of the solar wind.

The simple theory of the solar wind produces a slow thermal wind. The final theoretical velocity of the wind is in the range of 100 to 200 km s^{-1}. This should be compared with the escape velocity from the surface of the Sun which is 620 km s^{-1}.

2.2. Later work

Since the original article of Parker, much work has been published developing the theory. All the restrictions made in the original work have been removed, in some cases at the expense of great complexity. A review of all this work is outside the scope of the present article and the reader is referred to reviews made by other authors e.g. Holzer and Axford (1970), Hundhausen (1972) and Kopp (1981).

With the observations made with Skylab (Munro and Jackson 1977) it became obvious that the mass loss per unit surface area from the polar coronal holes was not only significantly higher than for the average solar wind but also that the acceleration of the wind between 2 and 5 solar radii could only be explained by the extra addition of energy or momentum to the wind in that region. Energy can be added by a heating mechanism with an extended dissipation scale length. The fully self consistent calculations of Hammer (1982) and Hearn (1982) show that increasing the dissipation length of the heating mechanism does result in a higher mass loss and in a higher final wind velocity. Momentum may be added to the wind by Alfvén waves propagating in the corona. This was first pointed out by Belcher (1971) and Alazraki and Couturier (1971). A numerical study of Alfvén waves in conduction models of the solar wind by Jacques (1978) shows that Alfvén waves do accelerate the wind to much higher velocities. Since the calculations of Jacques (1978) are not fully self consistent calculations including the transition region and radiative losses, it is not possible to deduce the effect of Alfvén waves on the mass loss.

2.3. The relation to other stationary mass loss theories

The theory of stellar winds driven by thermal pressure resulting from a hot corona is frequently called the Parker theory. Parker studied the topology of the equation resulting from the combination of the equations of motion, continuity and state and showed that the physical solution always goes through the singularity of the equation giving a stellar wind which is subsonic near the star and finishes supersonic. The original Parker theory included the gradient of the thermal pressure as the only driving term in the equation of motion. For that reason the Parker theory is commonly referred to a wind driven only by the gradient of the gas pressure, that is the wind resulting from a hot corona. Nearly all the stationary theories of mass loss reviewed here have the same properties as the Parker theory; the existence of a singularity which defines a unique solution for the mass loss. The only difference from the Parker theory is that the equation of motion contains extra driving terms such as wave pressure or radiation pressure. For this reason it seems better to regard these theories as other special cases of the Parker theory and not to restrict the name

Parker theory to that theory in which the gradient of the thermal pressure is the only driving term.

3. RADIATIVE DRIVEN WINDS - RESONANCE LINES

Stellar winds driven by the radiative forces resulting from the absorption of photospheric radiation by the resonance lines of ions such as CIII, CIV etc. were proposed by Lucy and Solomon (1970) to explain the mass loss from OB supergiants first observed by Morton (1967). The presence of the resonance lines of ions such as CIII in the wind is not consistent with the high coronal temperatures needed to drive mass loss by thermal pressure.

A photon carries momentum as well as energy. If radiation streaming out from a star is absorbed and subsequently re-radiated isotropically, then the momentum of the radiation is transferred to the absorbing ions and then by collisions to the wind. This process can yield a substantial force. A typical unsaturated resonance line can give an accelerating force which is 300 to 1000 times greater than that of gravity inwards.

A difficulty was raised by Marlborough and Roy (1970) who showed in a study of the Parker wind equation modified by a radiative force that if the radiative force exceeds the inwards force due to gravity everywhere the Parker subsonic – supersonic transition disappears, together with the critical point. The subsonic wind is then compressed instead of being accelerated.

Castor, Abbott and Klein (1975) pointed out that the resonance lines are strong and easily saturated and developed a theory of mass loss from hot stars in which the resonance lines in the subsonic region are sufficiently saturated to reduce the radiative force to less than that due to gravity.

3.1. Assumptions of the theory and results

Castor, Abbott and Klein (1975) developed a theory of mass loss from hot stars with the following assumptions.
(i)   The Sobolev approximation.
      This is a simplification in radiative transfer which is valid for rapidly and monotonically accelerating flows. It reduces the radiation transfer equation from an integral equation involving all space to a purely local equation. The result of applying the Sobolev equation to an optically thick resonance line is that the accelerating radiative force is proportional to the velocity gradient of the flow. A saturated optically thick line from a given element of gas will absorb all the radiation from a band of frequencies. If the element of gas is accelerated more rapidly, then the band of frequencies that it can absorb is increased. The radiative force is therefore proportional to the velocity gradient.

(ii) The absorption of radiation takes place in a mixture of optically thick and optically thin lines.
The radiative force coming from an optically thick resonance line calculated with the Sobolev approximation is proportional to the velocity gradient. The radiative force resulting from an optically thin line is independent of the velocity gradient. Castor, Abbott and Klein (1975) assumed that the radiative force coming from a mixture of optically thick and optically thin lines is proportional to the velocity gradient to a power α.

(iii) The radiative force was calculated from a representative set of lines, 900 lines from CIII. With these lines the power α for the velocity gradient is 0.7.

(iv) The star is a point source, and the radiation streams radially.

(v) The continuum opacity of the wind is negligible, the core - halo approximation.

(vi) The photosphere is not modified by the wind.

(vii) Photons are used only once - no multi-line transfer.

The Castor, Abbott and Klein theory gives a modified Parker stellar wind equation with a singularity which defines a unique solution. The radiative forces define both the mass loss and the velocity distribution. The simple properties of the theory can be obtained from the study of the critical point (see for example Cassinelli 1979). This gives the following results

a) $$\frac{V_\infty^2}{V_{esc}^2} = \frac{\alpha}{1 - \alpha}$$

where V_∞ is the final velocity of the wind, V_{esc} is the escape velocity from the surface of the star and α is the power of the velocity gradient in the equation of motion. With an α of 0.7, the final velocity is 1.5 times the escape velocity from the star, so the wind is always fast

b) $$V^2 = V_o^2 + (V_\infty^2 - V_o^2)\left(1 - \frac{r_o}{r}\right)$$

where V is the wind velocity at distance r, V_o is the wind velocity at the surface of the star radius r_o. V_o is negligible compared with the final velocity V_∞ of the wind. This gives a velocity distribution that rises very rapidly close by the star.

c) $$\dot{M} \propto L\left(\frac{\Gamma}{1 - \Gamma}\right)^{\frac{1 - \alpha}{\alpha}} \simeq L^{1.5}$$

where $\dot{M}$ is the mass loss, L is the luminosity of the star and Γ is the ratio of the acceleration due to the scattering of radiation by electrons and the acceleration due to gravity. Γ is a function of the mass luminosity relation. For unevolved OB supergiants the theory predicts a mass loss proportional to $L^{1.5}$ although the exact value of the power depends on the details of α and the mass luminosity relation.

3.2. Later work

Abbott (1982) extended the list of lines used to calculate the radiative forces. He included lines from the 1st to 6th ionization stages for elements from Hydrogen up to Zinc. This involved more than 200000 lines.

The effect of including the star as a source of light of finite size has been studied by Friend and Abbott (1986) and Pauldrach, Puls and Kudritzki (1986).

Both these changes have substantial effects on the final velocity of the wind and on the mass loss. Table 2 shows a comparison of observed values and theoretical values calculated by Pauldrach, Puls and Kudritzki (1986) including both these improvements.

TABLE 2

Comparison of the theory MCAK with observations

Star	Spectral Type	$\dot{M}$ $10^{-6} M_\odot$ yr^{-1} obs	MCAK	V_∞ km s^{-1} obs	MCAK	β
P Cyg	B1 Ia$^+$	20–30	29	400	395	0.98
ε Ori	B0 Ia	3.1	3.3	2010	1950	0.72
ζ Ori A	O9.5 I	2.3	1.9	2290	2274	0.72
9 Sgr	O4(f)V	4.0	4.0	3440	3480	0.81
HD 48099	O6.5 V	0.63	0.64	3500	3540	0.81
HD 42088	O6.5 V	0.13	0.20	2600	2600	0.79
λ Cep	O6 ef	4.0	5.1	2500	2500	0.79

The agreement between the modified Castor, Abbott and Klein theory and the observations is very good. Some adjustment has been made to the stellar parameters to give the best fit but this has been restricted to expected errors in the determinations. β is the power obtained for the theoretical velocity distributions when written in the form

$$V = V_\infty \left(1 - \frac{r_*}{r}\right)^\beta$$

where r_* is the radius of the star. The simple Castor, Abbott and Klein theory predicts a β of ½. It is interesting to note that the modified Castor, Abbott and Klein theory gives a satisfactory prediction of the mass loss and final wind velocity of the hypergiant P Cygni. The simple Castor Abbott and Klein theory gives a mass loss that is far too low and a final wind velocity that is far too high. But the modified Castor, Abbott and Klein theory still gives a β close to 1, that is a wind that is rapidly accelerating. This does not agree with the interpretation of infrared observations obtained with IRAS by Waters and Wesselius (1986) who find a very slow acceleration up to 100 km s^{-1} at 5 stellar radii. The theoretical result implies a velocity of 200 km s^{-1} at 2 stellar radii.

The multiple use of photons in a multi-line theory was suggested by Panagia and Macchetto (1982) as a way of increasing the radiation driven mass loss from a given star. Abbott and Lucy (1985) concluded from a Monte Carlo study of multi line scattering in ζ Puppis that the mass loss is a factor 3 higher than for the single scattering theory. The velocity distribution of the wind in their calculations was assumed and not calculated self-consistently. On the other hand a study by Puls (1987) with a self-consistently calculated velocity distribution concludes that the inclusion of multi-line scattering reduces the actual line force by 10% to 30%. This reduces both the mass loss and the final wind velocity.

3.3 Problems

The large mass loss from Wolf Rayet stars determined from radio, infrared and ultraviolet observations has long been regarded as a problem for radiative driven theories of mass loss (Barlow, Smith and Willis 1981).

Cherapashchuk, Eaton and Khaliullin (1984) deduced from a photometric study of V444 Cygni, A WN5 + O6 double star, that the effective temperature of the WN5 star is 90 000 K instead of the more usually adopted temperature of 30 000 K. Pauldrach, Puls, Hummer and Kudritzki (1985) have shown that with an effective temperature of 90 000 K the modified Castor, Abbott and Klein theory gives good agreement with the observed density and velocity distribution deduced by Cherapashchuk et al. (1984)

But such a high effective temperature is disputed by other workers. Schmutz and Hamann (1986) conclude from non-LTE calculations that the effective temperature is less than 60 000 K. Underhill and Fahey (1987) conclude from IUE measurements that the effective temperature of the WN5 star in V444 Cygni is 30 000 K. Then the explanation of the observed mass loss from Wolf Rayet stars by a radiation driven theory would seem very difficult.

For some hot stars such as the hypergiant P Cygni the mass loss is not constant. It has been known for decades that P Cygni emits shells of matter which are observed in the Balmer lines. A more recent analysis of the data by van Gent and Lamers (1986) concluded that P Cygni emits a shell irregularly every one or two months. No explanation of this behaviour has been given.

Be stars appear to undergo two different types of mass loss. In the ultraviolet a high speed wind is observed which can probably be explained by the Castor, Abbott and Klein theory. The infrared observations obtained with IRAS have been interpreted by Waters (1986) in terms of a disc in the equatorial plane of the Be star. The density distribution obtained from the interpretation implies a very slowly accelerating wind in the disc. In addition the mass loss appears to be 50 to 100 times larger than the mass loss derived from the ultraviolet resonance lines. It is difficult to see how this massive slow wind can be explained by a Castor, Abbott and Klein theory which always gives a rapidly accelerating wind. An attempt has been made by Poe and Friend (1986) by combining the Castor, Abbott and Klein theory with a magnetic field on a rotating star. They find that rapid rotation increases the mass loss easily by a factor of 5, and that for magnetic fields less than 5 gauss the terminal velocity decreases with rotation. But the shallow velocity law that is observed requires a rotation near the breakup speed and this is not observed.

4. RADIATIVE DRIVEN WINDS - DUST

Stellar winds driven by the radiative forces resulting from the absorption of photospheric radiation by dust were proposed to explain the observed mass loss from M giants and supergiants. They have IR excesses which are attributed to dust around the star. Kwok (1975) was the first to work out a hydrodynamic description of the mass loss. A more recent study is that of Tielens (1983).

The driving mechanism is essentially the same as for the hot stars. Photospheric radiation streaming out is absorbed by dust. The energy absorbed heats the dust which then radiates isotropically in the infra red. The momentum of the radiation is transferred to the dust which is then accelerated. The wind is driven by collisions of the dust with the gas.

4.1 Assumptions of the theory and results

The following assumptions are made in the simple theory of dust driven winds.

(i) Dustgrains form at a specified distance from the star. The fraction of dust present and its properties are free parameters in the calculation
(ii) Dustgrain-dustgrain collisions are negligible
(iii) All momentum given to the dust grains is transferred by collisions to the gas.
(iv) Dustgrain-gas collisions are very frequent.
(v) A rather empirical collision cross section is used for the dust grain-gas collisions.
(vi) The mass loss rate remains a free parameter of the calculation.

The dust driven wind theory gives a modified Parker equation with a singularity which defines a unique solution.

The radiative force on the dust is considerable and the dust is accelerated to a velocity in equilibrium with the gas in a distance that is very small compared with the radius of the star. The force on the gas resulting from collisions with the dust grains is assumed to be proportional to $V_d^{3/2}$ where V_d is the drift velocity, the difference between the velocity of the dust and the velocity of the gas. Since the dust, when it is formed at some specified distance from the star, is accelerated rapidly by the radiative forces, the force on the gas also increases rapidly. The result of Marlborough and Roy (1970) is a general result valid for any extra outward force. Since the net force on the gas as a result of collisions with dust grains increases rapidly to become greater than that due to gravity, the critical point of the gas flow must lie very close to the point for formation of the dust. In the work of Tielens this is at 3.5 stellar radii. This means in fact that the property of the whole solution is effectively determined by the distance at which the dust is assumed to form.

The simple theory gives final velocities of 12 km s^{-1} which lie in the observed range of 5 to 20 km s^{-1} typically observed for OH masers associated with Oxygen rich Miras.

4.2 Later work

The simple theories of dust driven winds are essentially one fluid models, achieved by making the assumption that all the momentum gained by the dust grains from the radiation is transferred to the gas. This is equivalent to assuming that the mass of the dust grain is negligible. The restriction has been removed by Berruyer and Frisch (1983) in a two fluid model for dust driven winds. The main effect of removing this restriction is that the critical point moves further outward and the final velocity of the gas is reduced by about 30%.

Gail and Sedlmayr (1985, 1987a) have studied both the formation of dust and its growth, and the radiative transfer involved in calculating the radiative force on the dust grains. In this case the mass loss is no longer a free parameter of the calculation but is determined self-consistently for a given star and element abundances. Their calculations for a carbon star give mass loss rates and terminal velocities in good agreement with the observations. In a following article Gail and Sedelmayr (1987b) conclude that the rapid acceleration of the gas after the beginning of dust formation and the subsequent lowering of the gas density inhibits the growth of dust in the wind. For this reason they conclude that there is a lower limit for the mass loss from M stars of 5 10^{-6} M$_\odot$/year for which dust formation and subsequent driving of the wind by radiative forces on the dust is possible.

4.3 Problems

For the early M and K supergiants, dust cannot form close by the stars. The dust driven wind can only work if some other mechanism brings the gas up to the region where dust can form. This distance appears to be several stellar radii. The current favourite for the other mechanism at present is radial pulsation. Further the work of Gail and Sedelmayr (1987b) suggests that for many types of cool stars the wind must already be substantial before a dust driven wind can take over.

Given that another mechanism is necessary to bring matter up to the dust forming region, it is likely that that mechanism will accelerate the wind to supersonic velocities and then the critical point solution in the dust driven is irrelevant because the wind is already supersonic.

It is further possible that a second mechanism combined with dust forming and radiative driving of the wind will give something that is quite different from them both.

5. ALVEN WAVE DRIVEN WINDS

Winds driven by Alfvén waves were proposed by Hartmann and MacGregor (1980) to explain mass loss from G and K giants on the right hand side of the Linsky Haisch dividing line (Linsky and Haisch 1979). These stars are known to undergo substantial mass loss but the IUE observations in the ultraviolet and X-ray observations from Einstein show no evidence of a transition region or corona. On the left hand side of the Linsky Haisch dividing line the stars have coronae and presumably a modest mass loss.

The winds are driven by the transfer of energy and momentum from the Alfvén wave to the wind. Integrating the time dependent equation of motion over many wave periods gives an extra term in the time averaged equation of motion. This term has the same character as the gradient of the gas pressure and is known as the gradient of the wave pressure. A wave travelling in the direction of the wind will lose energy to the wind. A wave travelling against the wind will gain energy from the wind. For this reason the flux of wave energy is no longer conserved and the conserved quantity is wave action (Lighthill 1978). A wave will accelerate the wind even though there is no dissipation of the wave. The acceleration of the solar wind by Alfvén waves had been suggested by Belcher (1971) and Alazraki and Couturier (1971), but in that application it is modifying an existing thermal wind.

5.1 Assumptions of the theory and results

Hartmann and MacGregor (1980) made the following assumptions in their work.
(i)   Linearized theory - the wave amplitude is small
(ii) The magnetic field is spherically symmetric
(iii) The flux of Alfvén waves is an input parameter
(iv) The dissipation length of the Alfvén waves is treated as a free parameter
(v)   The wind is isothermal with a temperature equal to the photospheric temperature.

The Alfvén wave driven wind theory gives a modified Parker equation. Unlike the simple solar wind theory the equation can have more than one critical point. Usually the solution through only one critical point is a physically valid solution satisfying the physical boundary conditions at the surface of the star and at infinity.

Some of the results given by Hartmann and MacGregor are given in Table 3 for a star of 16 solar masses, 400 solar radii and an effective temperature of 10^4 K. L is the dissipation length of the Alfvén waves, B is the magnetic field at the stellar surface, M is the mass loss and V_∞ is the final velocity of the wind. For each of these results the amplitude of the wave at the stellar surface is $\frac{\delta B}{B}$ equal to 0.1.

TABLE 3

Wave flux erg cm^{-2} s^{-1}	L	B gauss	$\dot{M}$ $M_\odot$/year	V_∞ km s^{-1}
3.4 10^3	∞	1	1.1 10^{-10}	823
4.2 10^5	∞	5	4.3 10^{-8}	508
3.4 10^6	∞	10	5.5 10^{-7}	407
3.4 10^3	1 R_*	1	5.8 10^{-11}	31.6
4.3 10^5	1 R_*	5	3.2 10^{-8}	47.6
3.4 10^6	1 R_*	10	4.5 10^{-7}	51.0

With a given wave amplitude, increasing the magnetic field increases the Alfvén wave flux. The gives an increasing mass loss. For the results with the dissipation scale length equal to infinity the final velocity of the wind is high and decreases slowly with increasing wave flux. When the dissipation length is equal to one stellar radius, the mass loss becomes slightly less, but the final velocity is substantially reduced.

The mass loss deduced from observations is uncertain. For example for α Ori (M2 Iab) the mass loss estimates range from 10^{-5} to 10^{-7} $M_\odot$/year depending on the author. Wind velocities vary from 5 to 100 km s^{-1}. The final wind velocities obtained with an infinite dissipation length are far too fast to be in agreement with the observations. With a dissipation length equal to one stellar radius the agreement is reasonable.

5.2 Later work

Holzer, Flå and Leer (1983) made a more detailed study of the Alfvén wave driven models, with a greater range of parameters.

They find that when the dissipation length is in the region of one stellar radius, the final velocity is exceedingly sensitive to the assumed value of the dissipation length. The observed value of the final wind velocity is for a large number of stars about 0.5 times the escape velocity. To obtain this the wave dissipation length must be close to one stellar radius. They argue that it is rather unlikely that the actual dissipation length in the winds of many different types of stars would be so precisely defined.

To obtain a low final wind velocity, a small but closely specified fraction of the wave momentum must be deposited in the supersonic flow. It is for this reason that the final velocity is so sensitive to the dissipation length. The constant dissipation length is an assumption in the theory and perhaps the inclusion of a more physically realistic dissipation length which will not be constant but dependent on local quantities such as the wave amplitude, would remove this difficulty.

5.3 Problems

The physics of the dissipation of Alfvén waves is uncertain. Holzer, Flå and Leer (1983) included the dissipation of the Alfvén waves resulting from collisions with neutrals. If an Alfvén wave propagates through a partially ionized gas the ionized component is frozen in to the field lines and the neutral component is not. The collisions between the two dissipates the wave energy. Presumably the amplitude of the Alfvén waves propagating outwards in a decreasing magnetic field increases with distance. At some stage non-linear effects must become important as they do with sound waves propagating outwards in a decreasing density with the formation of shocks.

Alfvén wave driven winds obtained with little or no dissipation of the waves are very fast winds, with a final velocity many times the observed final velocity. To obtain agreement with the observations the dissipation scale length must be specified precisely because of the sensitivity of the final velocity to the dissipation length.

6. WINDS DRIVEN BY RADIAL PULSATIONS

Winds driven by radial pulsations of the star were proposed to explain the observed mass loss from long period variables such as Miras. The violet displaced resonance line absorption in the spectra of luminous M giants and supergiants was interpreted by Deutsch (1960) and Weyman (1963) as evidence of matter flowing outward from these stars.

The winds are driven by the shocks that are formed in the outer layers of the star by the radial pulsation. The shocks greatly extend the atmosphere of the star and the gas eventually acquires an average velocity which is supersonic and greater than the local escape velocity from the star

Such winds were studied by Wood (1979) and Willson and Hill (1979). These theories are not linearized wave theories but computer solutions.

6.1 Assumptions of the theory and results

In the work of Wood the following assumptions were made.
(i) The radial temperature distribution of the starting model is given
(ii) The equation of motion contains only the gravity as an external force.
(iii) The energy equation is replaced by the assumption that the waves are isothermal or adiabatic.
(iv) The equations of state allows for the formation of H_2 molecules, but the ionization of hydrogen and helium is not included.
(v) The Lagrangian equations were solved numerically with an explicit method including artificial viscosity.
(vi) The physical boundary condition was the pressure varying sinusoidally at the base of the calculation with a period of 373 days, the pulsation period of Mira.

After an initial transient the solution settles down to a steady periodic solution. The results depend strongly on whether the shocks are isothermal or adiabatic.

If the shock is adiabatic the temperature and density distributions of the atmosphere are very greatly extended and the calculation gives a mass loss of 10^{-2} $M_\odot$/year with a final velocity of 7 km s^{-1}. This mass loss is considerably greater than the observed mass loss of 10^{-6} $M_\odot$/year.

If on the other hand the shocks are assumed to be isothermal the density distribution is much less extended and the mass loss is only 10^{-12} $M_\odot$/year.

6.2 Later work

Further numerical work by Bowen (1987) has included a more realistic energy equation with radiation losses from neutral hydrogen. Bowen finds a solution which has predominantly isothermal shocks except in the outer low density region of the atmosphere where they become adiabatic. This raises the temperature in that region substantially and drives a slow outflow of matter. The mass loss is rather small, only 10^{-8} to 10^{-10} $M_\odot$/year. Bowen has repeated the calculation with dust formed at about 2 stellar radii. This increases the mass loss substantially to 10^{-6} to 10^{-8} $M_\odot$/year with a final velocity of 10 km s^{-1}. These calculations are giving results in reasonable agreement with the observations.

6.3 Problems

It is clear from the work of Wood (1979) and Bowen (1987) that the details of the calculations of the energy balance and the dust formation are critical in determining the results and further work is needed in both these problems.

Willson and Hill (1979) have shown that for effective driving of mass loss by radial pulsation the period of the pulsation must be less than the ballistic period of the atmosphere. If a particle is ejected with a given velocity from the atmosphere and moves under the influence of gravity alone, then it will return to its original place in a time called the ballistic period. Since the amplitude of the shocks observed in Miras are less than the escape velocity, a pulsation period greater than the ballistic period will give a periodic solution with no mass loss. If on the other hand the pulsation period is less than the ballistic period, the matter ejected by one shock will meet the next one coming up and so the solution is no longer periodic and mass loss will result.

There has been one study of mass loss caused by a single shock, which was done by Hillendahl (1976). He studied shocks formed in the pulsation of Cepheids. The mechanical analogy of this type of mass loss is a long bar of metal tapering slowly to a very small diameter. If the broad end of the bar is hit with a hammer a shock propagates down the bar. The amplitude of the shock grows because of the reducing diameter of the bar. Eventually the shock has such a large amplitude that the thin end of the bar is disrupted and flies off.

7. WINDS DRIVEN BY NON-RADIAL PULSATION

As already mentioned in the section on radiation driven winds, infrared observations of Be stars obtained with IRAS have been interpreted by Waters (1986) in terms of a disc in the equatorial plane of the star. The interpretation indicates a very slowly accelerating wind which is inconsistent with the rapid acceleration characteristic of a radiation driven wind. Although it is clear that there is some relationship between the Be phenomenon and rotation, the old idea that the stars are rotating almost at break up speed is not consistent with the obervations

A very interesting development in the last few years has been the conclusion that deformations in absorption lines of Be stars can be interpreted in terms of non-radial pulsations (Vogt and Penrod, 1983, Baade 1984). Vogt and Penrod (1983) found an $\ell = 8$ m = 8 mode provides a good fit for the observations from ζ Ophiuchi (O9.5 Ve). This is a high order non-radial pulsation which is confined to the equatorial region of the star. There is some evidence that Be stars and normal B stars have different types of non-radial oscillations. The B stars have non-radial pulsations with one or two short periods and high ℓ (4 to 10) whereas the Be stars have a long period $\ell = 2$ non-radial pulsation with a short period high ℓ oscillation. There is further some evidence that changes in the mode of the non-radial pulsation coincide with a Be mass loss episode.

All this work provides evidence that the Be phenomenon is connected causally with the non-radial pulsations and the link with rotation is perhaps that only in fast rotating stars are the non-radial pulsations excited to sufficient amplitude.

What the physics is of this connection with non-radial pulsations, is not clear. There is no theory of mass loss driven by non-radial pulsations. There are no results. But it is clear that there are a lot of interesting problems.

REFERENCES

Abbott, D.C. (1982): Astrophys. J. 259, 282
Abbott, D.C., Lucy, L.B. (1985): Astrophys. J. 288, 679
Alazraki, G., Couturier, P. (1971): Astron. Astrophys. 13, 380
Baade, D. (1984): Astron. Astrophys. 135, 101
Barlow. M.J., Smith, L.J., Willis, A.J. (1981): Mon. Not. R. Astr. Soc. 196, 101
Belcher, J.W. (1971): Astrophys. J. 168, 509
Berruyer, N., Frisch, H. (1983): Astron. Astrophys. 126, 269
Bowen, G.H. (1987): Astrophys. J. in press
Cassinelli, J.P. (1979): Ann. Rev. Astron. Astrophys. 17, 275
Castor, J.I., Abbott, D.C., Klein, R.I. (1975): Astrophys. J. 195 157
Chapman, S. (1957): Smithsonian Contrib. Astrophys. 2, 1

Cherepashchuk, A.M., Eaton, J.A., Khaliullin, Kh.F. (1984): Astrophys. J. 281, 774
Deutsch, A.J. (1960): Stellar Atmospheres, ed. J.L. Greenstein, University of Chicago Press, p. 543
Friend, D.B., Abbott, D.C. (1986): Astrophys. J. 311, 701
Gail, H.P., Sedlmayr, E. (1985) Astron. Astrophys. 148, 183
Gail, H.P., Sedlmayr, E. (1987a): Astron. Astrophys. 171 197
Gail, H.P., Sedlmayr, E. (1987b): Astron. Astrophys. 177, 186
Hammer, R. (1982): Astrophys. J. 259, 779
Hartman. L., MacGregor, K.B. (1980): Astrophys. J. 242, 260
Hearn, A.G. (1982): Astron. Astrophys. 116, 296
Hillendahl, R.W. (1967): Proc. I.A.U. colloquium no. 2, N.B.S. Spec. Publication 332, p. 300
Holzer, T.E., Axford, W.I. (1970): Ann. Rev. Astron. Astrophys. 8, 31
Holzer, T.E., Flå, T., Leer, E. (1983): Astrophys. J. 275, 808
Hundhausen, A.J. (1972): Solar wind and coronal expansion, Springer-Verlag, Heidelberg
Jacques, S.A. (1978): Astrophys. J. 226, 632
Kopp, R.A. (1981): The Sun as a star, NASA-SP450, p. 373
Kwok, S. (1975): Astrophys. J. 198 583
Lighthill, J. (1978): Waves in fluids, Cambridge University Press
Linsky, J.L., Haisch, B.M. (1979): Astrophys. J. 229, L27
Lucy, L.B., Solomon, P.M. (1970): Astrophys. J. 159, 879
Marlborough, J.M., Roy, J.R. (1970): Astrophys. J. 160, 221
Morton, D.C. (1967): Astrophys. J. 150, 535
Munro, R.H., Jackson, B.V. (1977): Astrophys. J. 213, 874
Panagia, N., Macchetto, F. (1982): Astron. Astrophys. 106 266
Parker, E.N. (1958): Astrophys. J. 128, 664
Pauldrach, A., Puls, J., Hummer, D.G., Kudritzki, R.P. (1958): Astron. Astrophys. 148, L1
Pauldrach, A., Puls, J., Kudritzki, R.P. (1986): Astron. Astrophys. 164, 86
Poe, C.H., Friend, D.B. (1986): Astrophys. J. 311, 317
Puls, J. (1987): Astron. Astrophys. in press
Schmutz, W., Hamann, W.R. (1986): Astron. Astrophys. 166, L11
Tielens, A.G.G.M. (1983): Astrophys. J. 271, 702
Underhill, A.B., Fahey, R.P. (1987): Astrophys. J. 313, 358
Van Gent, R.H., Lamers, H.J.G.L.M. (1986): Astron. Astrophys. 158, 335
Vogt, S.S., Penrod, G.D. (1983) Astrophys. J. 275, 661
Waters, L.B.F.M. (1986): Astron. Astrophys. 162, 121
Waters, L.B.F.M., Wesselius, P.R. (1986): Astron. Astrophys. 155, 104
Weyman, R.J., (1963): Ann. Rev. Astron. Astrophys. 1, 97
Willson, L.A., Hill, S.J. (1979): Astrophys. J. 228, 854
Wood, P.R. (1979): Astrophys. J. 227, 220

INTERACTING BINARIES AND MASS LOSS IN HOT STARS

F. C. Bruhweiler
Department of Physics
Catholic University of America
Washington, D.C. 20064
United States of America

ABSTRACT. Studies of the ultraviolet spectra of hot interacting binaries reveal methods that may represent important new tools for probing the physical conditions in the winds emanating from hot stars. Two systems, Mu Sagittarii and UW CMa are taken as examples of how these methods can be utilized to investigate the physics of mass loss in such objects.

1. INTRODUCTION.

Rather than providing you a review of our current understanding of mass loss in early-type binary systems, and since one of the goals of this meeting is to determine some of the future requirements for space missions, I wish to present to you my view of the potentials that space observations of binary stars have for giving us a more detailed picture of the basic characteristics of mass loss in hot stars. My talk will cover two basic aspects.

First, I wish to show that interacting binaries containing cool B stars can produce the so-called "superionized" species O VI, N V, C IV, and Si IV. As is well known and pointed out by Lamers and Pauldrach in these proceedings, these species represent the tracers of mass loss in hot O and B stars. The important point is that in interacting binaries the origin of these species may be different from that in normal single hot stars.

Second, I would like to demonstrate that hot interacting binaries represent a largely untapped potentential for probing the actual physics of mass loss. As discussed by Reimers in these proceedings, we have seen how the hot companions of cool late-type supergiants have been extensively used to probe the winds of the cool primaries. Even though it is more difficult, the observed interactions of two hot stellar companions at ultraviolet wavelengths can also be utilized to probe the physical conditions in the winds emanating from hot stars. To illustrate the potentials of this observational tool, I will show how the preliminary UV studies of the two interacting binaries Mu Sagittarii (B8

Ia + B1-2 V) and UW CMa (O7 Ia + O) are beginning to give us a more detailed picture of mass loss in interacting binaries.

Before we begin, we must mention many of the excellent reviews of the results of interacting binaries that have been presented in the last few years. One must remember that to prepare a review of the current knowledge of interacting binaries is an overwhelming task. The subject is much too broad and any such comprehensive review should include topics as X-ray binaries (which we will hear more about later in this meeting), cataclysmic variables, Wolf-Rayet binaries, Algol systems, Type I supernovae, and lengthy discussions of intriguing objects like SS 433.

Much of what we know about mass loss and mass transfer in early-type binaries has come from observations at ultraviolet wavelengths, especially using instrumentation aboard the Copernicus and IUE satellites. For a review of the work accomplished with the IUE Observatory one should see McCluskey and Sahade(1987).

Other important reviews on interacting binaries also include the comprehensive work of Greve (1986) on semi-detached systems. In addition, the recent proceedings of the NATO Advanced Study Institute on Interacting Binaries (Eggleton and Pringle 1983) and the monograph, "Interacting Binaries", by Sahade and Wood (1978) represent important source material for anyone wishing to study interacting stellar systems.

2. THE EQUIPOTENTIAL SURFACES - THE TERRAIN FOR STELLAR MASS LOSS.

In order to fully understand the process of mass loss and mass transfer in binary systems, one must first consider the "terrain" or potential surfaces resulting from the net gravitational and radiative forces in the binary system.

If we neglect radiative forces, as might be the case in cooler binaries without significant mass loss, we have the resticted three-body problem which leads to the "classical" Roche or Jacobian equipotential surfaces. These are illustrated in Figure 1. In this figure, each equipotential surface corresponds to a zero-velocity surface beyond which particles of greater energy cannot penetrate. Of great importance to mass exchange and mass transfer are the three Lagrangian points L_1, L_2, and L_3. These points represent local minima or saddle points in the potential through which matter can more easily flow. It is easy to see that, if one of the stars expands to fill its Roche lobe, the inner Lagrangian point, L_1, is the pathway for mass transfer between the two stars. While L_1 is the main conduit for mass transfer, the outer Lagrangian points, L_3 and especially L_2, are the principal pathways for mass loss in binaries.

Of course, in hot stars the effects of radiation pressure cannot be ignored. Modified equipotential surfaces accounting for radiation pressure have been calculated by Schuerman (1972) and Vanbeveren (1977). In Figure 2, examples are presented where both stellar components have significant radiation pressure. The examples in Figure 2 are simplified examples and the ratio of radiative to gravitational force e_i for each

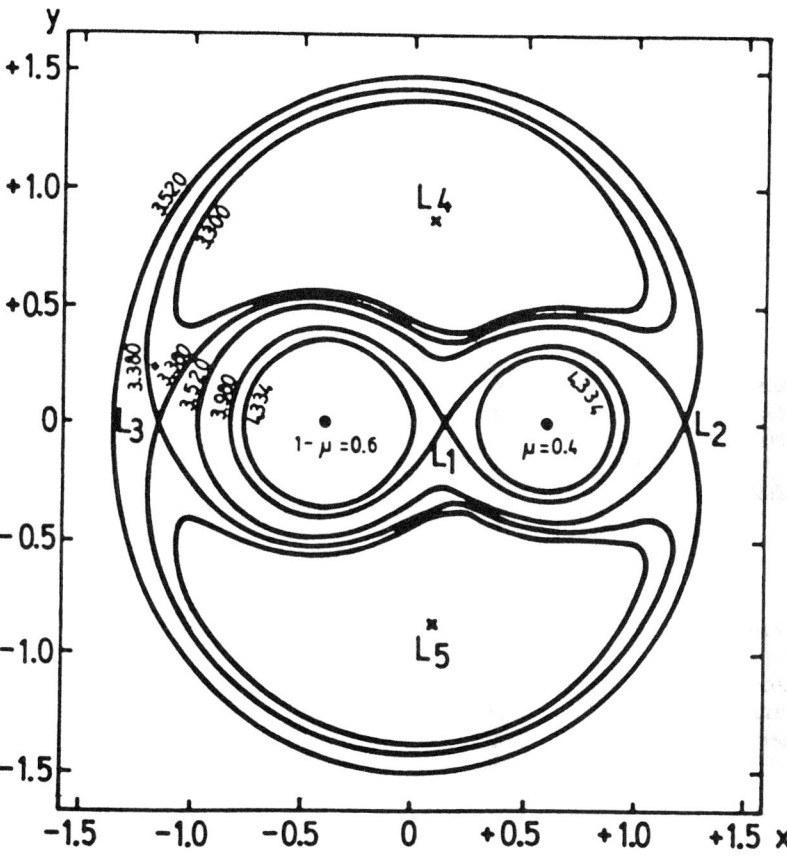

Figure 1. The Jacobian or Roche Lobe Equipotential Surfaces for a Binary Star System. The equipotential gravitational surfaces are shown for a binary system in which the effects of radiative forces are negligible. The fractional mass of the stellar components are indicated. The x and y axes are in units of the separation distance between the stars. The Lagrangian points (L_1 through L_5) are denoted. (from Sahade and Wood 1978)

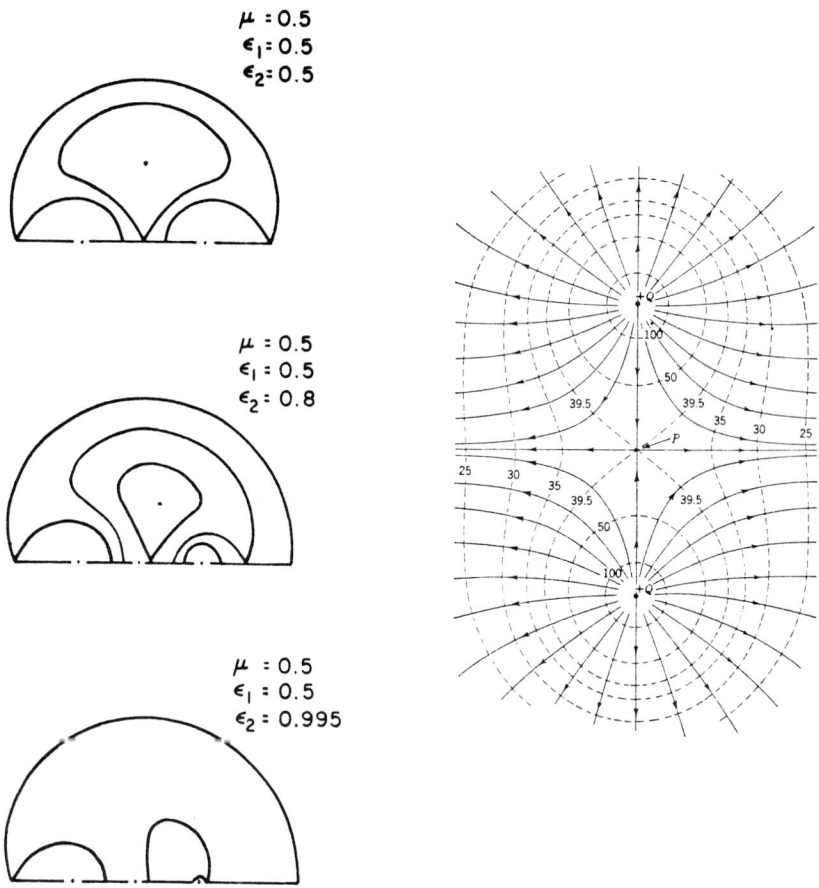

Figure 2. The Jacobian Equipotential Surfaces with Radiative Forces Included. The parameters are as defined in Figure 1. The values e_i are the ratios of radiative to gravitational acceleration, which is assumed to be constant with stellar radius. The three examples on the left are for $e_i < 1$, while the single example on the right is for $e_1 = e_2 > 1$. The mass loss rates are the same for both stars. (Figure is after Sahade and Wood 1978.)

stellar component is given. In actuality, one would expect e_i to be a function of stellar radius, but here it is assumed to be constant.

For cases in which $e > 1$, it easy to see that the solution is analagous to the electrostatic potential produced by particles of like charge, a subject which is discussed in any textbook on electromagnetic theory. An important aspect of these solutions is that in systems in which both stars have prominent stellar winds, an interaction region or shocked interface should be produced between the stars. This region is denoted in the example given on the right in Figure 2. This region, formed by the colliding winds, might produce enhanced X-ray emission in some hot binaries or its presence might be observed in other ways. Unfortunately, this is a subject that has only been treated in a very preliminary way, theoretically (Cooke, Fabian, and Pringle 1979; Prilutskii and Usov 1976). This is a subject we will discuss further in Section 6.

If the effects of radiation are included in early-type binaries, one finds that conservative mass transfer cannot be assumed. Even though this has been known for some time, many theoretical modelling of interacting binaries still use this assumption. In fact, most of the mass transferring through the inner Lagrangian point will not be accreted, but will be lost from the system in early-type binaries (cf. Vanbeveren 1977; Kondo 1979).

3. THE ORIGINS OF THE HIGHLY IONIZED SPECIES IN INTERACTING BINARIES.

As mentioned earlier, the so-called superionized species, O VI, N V, C IV, and Si IV, represent the tracers of the stellar winds in O and B stars. The recent work of Kudritski and his colleagues has been quite successful in expanding the earlier work of Castor, Abbott and Klein (1975) and shows that the radiation field alone is sufficient, with the help of non-LTE effects, to explain the presence of these ions and to drive the winds in luminous O and B stars.

Yet, IUE spectra do reveal N V and C IV in B stars of later spectral type with insufficient radiation fields to produce these ions. For example, variable mass loss features due to these ions are seen in binaries containing late-type B stars. Also, narrow components of C IV and S IV have been seen in nearby ($d < 200$ pc) late-type B stars, and even in early main-sequence stars (Ferlet et al. 1986; Bruhweiler, Grady, and Chiu 1987). Although Ferlet et al. propose a possible interstellar origin for these features, work by Bruhweiler, Grady, and Chiu show that an interstellar origin can be completely ruled-out in all the cases where these narrow features are seen in these nearby objects. Of course interstellar C IV and Si IV are seen in spectra of distant luminous O and B stars, but the presence of these ions in these nearby objects seems to be more closely linked to rapid rotation or the Be phenomenon.

The ions, N V, C IV, and Si IV, have been observed in several Algol binaries. Observations by Peters and Polidan (Fig. 3) find the resonance lines of these ions are not only present but are variable with orbital phase in the systems AU MON (B5e + F-G), CX Dra (B2.5e + F:), and U CrB

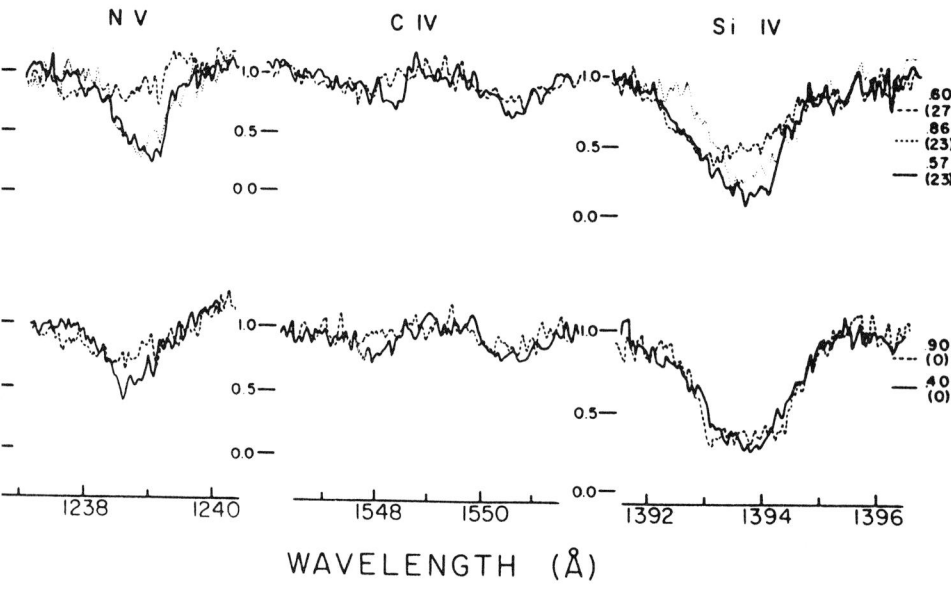

Figure 3. Variable Features of High Ionization in the Algol System CX Dra. The data clearly show that C IV and Si IV are present and vary with phase. These features probably reflect collisional heating. Also, the N V and Si IV features are asymmetric at phase 0.56, indicating a stellar wind. (From Peters and Polidan 1984)

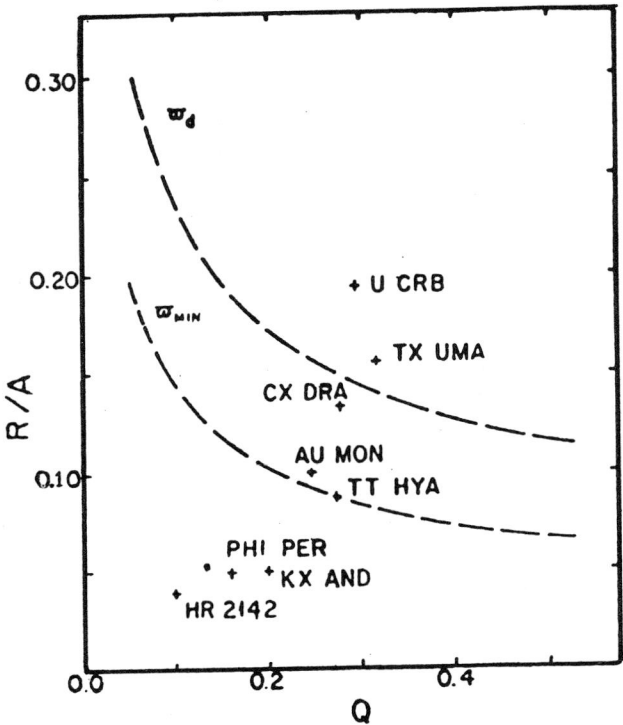

Figure 4. Locations of Algol Systems and Similar Binaries in r-q plane. The locations of the objects observed by Peters and Polidan are plotted as a function of R/A (fractional radius of the primary) and Q (the mass ratio of contact component to the detached star). The two dashed lines represent, ϖ_{min}, the distance of closest approach of a test particle released from L_1 to the center of the detached component, and ϖ_d, the radius of the disk about the detached component. (Figure from Peters and Polidan 1984)

(B6 V + F8 III-IV). Peters and Polidan explain these results by expanding upon the simple model of Lubow and Shu (1975). In this model, mass transfer occurs through the inner Lagrangian point L_1. The accreting material, in the form of a narrow mass stream, either produces a disk (or gaseous halo) about the accreting star and may also impact the surface of the star. Whether the accreting material actually impacts the stellar surface depends upon the accreting star's radius. If the stellar radius is larger than w_{min}, where w_{min} is the minimum distance that the accreting mass stream can get to the center of mass of the accreting star, then the stream will impact the surface. This process will therefore result in collisional heating of the plasma and give rise to the observed highly ionized species. The minimum distance w_D is determined by the angular momentum of the gas in the stream (See Lubow and Shu 1975).

As is shown in Fig. 4, the inferred radii of the stars studied by Peters and Polidan show a clear correlation of the presence of N V with stellar radii graeater than w_{min}.

The Lubow and Shu model does give us a qualitative picture of interacting binaries, but the model cannot give us an accurate detailed picture. For example, the Lubow and Shu model assumes that the stellar orbits are completely circular and that the stars are in synchronous rotation. this may be a workable assumption for some cases, but not all. Furthermore, radiative forces, magnetic fields and other effects are not treated in their simple model. Although Lubow and Shu assume a narrow stream, there are many effects that yield a much wider width than they envision. Some of the problems associated with the Lubow and Shu model have been detailed in Kondo (1979 and references cited therin).

The applicability of the Lubow and Shu model is not our primary concern. The point to be emphasized is that there is more than one way to produce highly ionized species. In addition, The mass loss characteristics are somewhat modified in binaries and in Be stars from what in inferred from the radiatively driven wind picture for more luminous early-type stars.

4. MU SAGITTARII: THE CASE FOR LARGESCALE PHASE DEPENDENT P CYGNI PROFILE VARIABILITY.

Ultraviolet observations of this system, using the IUE (Bruhweiler, Kondo, and McCluskey 1987), indicate very large profile variations in the mass loss features of C IV and Si IV. In Figure 5a, b, and c, we present data for the P Cygni features of the C IV resonance doublet at three phases, 0.05, 0.56, and 0.77. The data, illustrated in heliocentric velocity units, show strong P Cygni emission lobes at all phases. Near secondary eclipse (0.56) and quadrature (0.77), the C IV absorption components are narrow, and displaced in velocity to about -150 km/s. However, at primary eclipse (0.05), these profiles are quite different. We see very enhanced C IV absorption, extending up to -700 km/s. This behaviour is also mirrored in the Si IV resonace doublet as well.

These largescale profile variations suggest two possible explanations. In the first possibility, the B8 supergiant is the source

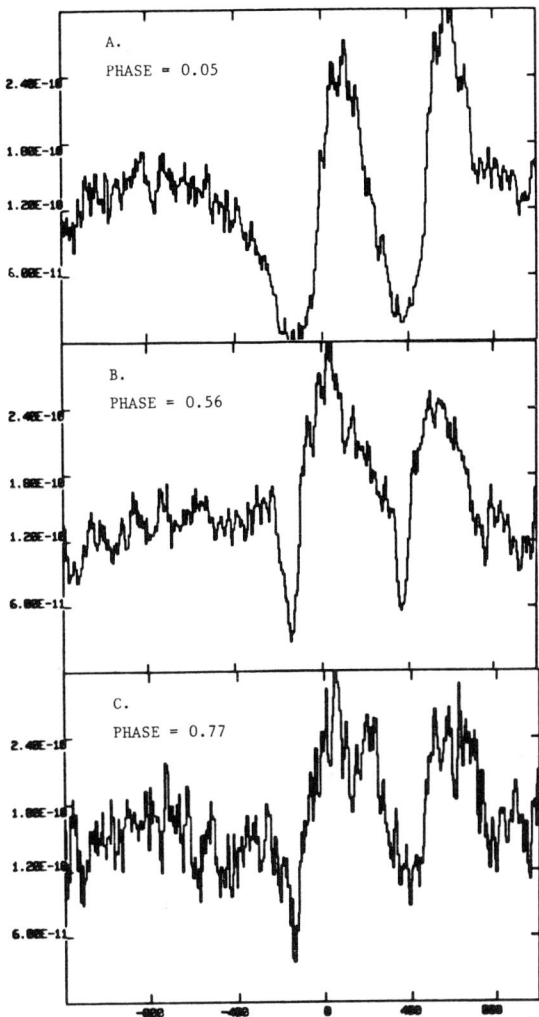

Figure 5. Phase Dependent Variability of C IV P Cygni Profiles in Mu Sgr. Data are presented for the three phases. 0.05, 0.56, and 0.77. The horizonal axis is given in units of heliocentric velocity (km/s) based upon the wavelength of the stronger member of the C IV resonance doublet at 1548.185 A. Note the strong enhancement in C IV absorption at phase 0.05, when the B1 secondary is in front of the B8 supergiant. (From Bruhweiler, Kondo, and McCluskey 1987)

of all the mass loss. Mass that is lost from the system is
preferentially passed from the B8 Ia primary through the inner
Lagrangian point L_1 to the B1 main-sequence secondary. Because of the
strong radiation field of the B1 secondary, little of this mass is
accreted. Instead, this mass flows out of the system through the outer
Lagrangian point at L_2. In the meantime, this material is accelerated to
the observed high velocities by the radiation field of the B1 secondary.
this picture follows the basic picture presented earlier of how mass is
lost in interacting binaries.

A second possibility exists if the B1 secondary has significant
mass loss. Then, we can talk about colliding stellar winds and our
picture is quite similar to the case $e > 1$ in Fig. 2. Even though there
is no doubt that the mass loss rate of the B8 Ia primary dominates. An
estimate (Lamers 1981) yields $M = 2 \times 10^{-6}$ M_0/yr. The mass loss rate of
the B1 secondary is unknown, but it could be as high as 10^{-7} M_0/yr. If
the mass loss from the B1 is significant, we can expect that a shocked
interface, as in the $e > 1$ case in Fig. 2, separates the winds of the
respective stars. The profile variations seen Mu Sgr would represent two
distinct divided domains or "sectored" mass loss. One sector, by far the
largest, would be the region of mass flow from the B8 supergiant. The
second, much smaller region would show mass outflow from the secondary.
The much larger mass loss of the supergiant would focus the secondary
mass loss such that high velocity absorption from the B1 could only be
observed near primary eclipse. Thus, the mass flow from the secondary
would be decidedly non-spherical and the unusually strong P Cygni
profiles observed at phase 0.05 would be a consequence of this
focussing.

We cannot, yet, differentiate between these two scenarios in Mu
Sgr. However, I point out that these pictures do not represent two
different physical models, but only different aspects of the same
description presented in Figure 2. The second possibility discussed
above only comes about if we have significant mass loss from the
secondary.

There is no question that there is much to be learned from
continued study of these largescale profile variations. It also shows
the inherent uncertainties in attempting to deduce mass loss estimates
from interacting binary systems. It is quite evident that the mass flow
is decidedly different from uniform purely radial flow, which is usually
assumed in deriving mass loss estimates. Thus, it is probably still
premature to conclude anything about whether stellar mass loss rates are
enhanced in early-type binaries compared to those for single stars.

5. UW CMa - USING HOT SECONDARIES AS PROBES OF STELLAR WINDS OF HOT PRIMARIES.

In these proceedings, Reimers has discussed how hot companions in
binaries with late-type supergiants can be used as probes of winds of
late-type stars. In essence, the early-type star is used as a UV
continuum source against which the absorption of the wind from the cool
star can be seen.

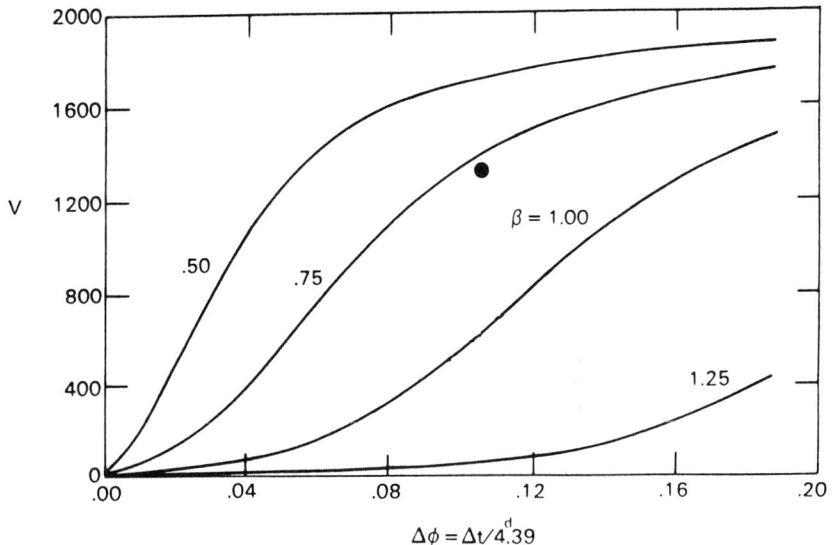

Figure 6. Actual Versus Predicted Phase Lag for Different Velocity Laws of the Form $v(r) = v_t(1-r/R_*)^B$. The value deduced by Drechsel et al. (1981) from C III 1175 Å feature is indicated. (Curves for predicted phase lag are courtesy of S.R. Heap.)

Generally, one does not think about using hot secondaries as probes of winds of hot primaries, but it can be done. So far, its potential has been greatly under-utilized.

The binary system UW CMa consists of two O stars. The more luminous primary is a O7f Ia, while the less luminous secondary is an O star of uncertain sub-class. This system is a short period binary with a period of 4.4 days. The optical spectral (Struve et al. 1958; Hutchings 1977) and photometric variability (Leung and Schneider 1978) of this system have been investigated. Ultraviolet investigations using Copernicus (McCluskey, Kondo and Morton 1975; Drechsel et al. 1981) and IUE data (Heap; in preparation) reveal that the O7 primary dominates the observable mass loss from the system. Moreover, both ultraviolet and optical spectral features indicate a large semi-amplitude for the radial velocity variations, namely K = 240 km/s.

The presence of the secondary and the large K amplitude produces a modulation and phase lag in the measured terminal velocity of the wind of the primary. If we assume a velocity law in the wind of the form

$$V(r) = V_T(1-r/R_*)^B,$$

then we can hopefully determine the value of B. The data of Drechsel et al. clearly show a phase lag of approximately 0.10 period in the variation of the C III 1175 A profile as seen in the Copernicus data. As seen in Figure 6 for a terminal velocity of the wind of V_T = 1750 km/s, this phase lag corresponds to a B of near 0.75 to 0.8. This value is in close agreement with recently derived theoretical values from the work of Kudritzki and his colleagues. Results from other UV lines in UW CMa by Heap (private communications) may suggest some variations from this value, but this is still not clear. However, this method of probing the velocity law shows great potential and may be one of the few ways of obtaining direct information about the velocity law in winds of O stars without theoretical modeling (with all the implied uncertainties), of the observed P Cygni profiles.

It has been well-known for some time that the strength of the Si IV P Cygni profiles correlate with luminosity in O stars (Henize et al. 1975; Walborn and Panek 1984). Likewise, in UW CMA, the Si IV is associated with the luminous primary. The results by Heap show clear occultations of the Si IV emission region about the supergiant by the less luminous O star secondary. These observations suggest that eventual detailed modeling of this system and other systems like it could be an excellent means of probing the ionization structure in O star winds.

Other profile variations contain much information about physical conditions in the winds of these stars. Certainly, more detailed UV studies of UW CMa and other short period binaries such as AO Cas and Plaskett's star (HD 47129), and in X-ray binaries and W-R stars (See also the reviews by Willis, 1983; White, 1983) would provide important probes of the physical conditions in the winds of O stars.

6. A COMMENT ON X-RAY EMISSION FROM COLLIDING WINDS.

Previous published results indicate that binaries do not show any enhancement in X-ray emission in normal binaries over that of single stars. This was in spite of the fact that simplified calculations suggested that an X-ray enhancement should be seen. The lack of any observed increase was thought to imply that the large column density of overlying wind effectively attenuated any X-ray flux or that the simplified treatment overestimated the X-ray emission and was far from adequate in describing the actual physical condition at the shock interface of the winds in binaries. Recent re-analysis of Einstein data (Chlebowski 1987; also see Pollack in these proceedings) now show that O star binaries do have enhanced X-ray emission over that from single stars.

How the properties of this X-ray emission varies with mass loss, and orbital phase must be explored. It remains to be seen if this X-ray emission, too, will be an important new tool for studying the characteristic of mass loss in early-type stars.

REFERENCES.

Bruhweiler, F.C., Kondo, Y., and McCluskey, G. E. 1987, in Preparation.
Bruhweiler, F.C., Grady, C., and Chiu, W. 1987,in Preparation.
Castor, J.I., Abbott, D.C., and Klein, R.I. 1975, Ap.J., 195, 157.
Chlebowski, T. 1987, Bull. American. Astron. Soc., 19, 703.
Cooke, B.A., Fabian, A.C., and Pringle, J.E. 1979, Nature, 273, 645.
Drechsel, H. Rahe, J. Kondo, Y., and McCluskey, G.E. 1981, Astr. Ap., 94, 285.
Eggleton, P.P. and Pringle, J.E. 1983, Proceedings of NATO ASI on Interacting Binaries, Reidel Publ.: Dordrecht, Holland.
Greve, J.P. 1986, Space Science Rev., 43, 139.
Heap, S.R. 1987, in Preparation.
Henize, K. et al. 1975, Ap.J. (Letters), 199, L119.
Hutchings, J.B. 1977, Publ. A.S.P., 89, 668.
Kondo, Y. 1982, Ast. Sp. Sci., 32, 605.
Lamers, H.J.G.L.M. 1981, Ap. J., 276, 677.
Leung, K.C. and Schneider, D.P. 1978, Ap.J., 222, 924.
Lubow, S.H. and Shu, F.H. 1975, Ap.J., 198, 383.
McCluskey, G.E., Kondo, Y., and Morton, D.C. 1975, Ap.J., 201, 607.
McCluskey, G.E. and Sahade, J. 1987, in "Exploring the Universe with the IUE Satellite", ed. Y. Kondo et al., Reidel Publ.: Dordrecht, Holland.
Peters, G.J. and Polidan, R.S. 1984, Ap.J., 283, 745.
Prilutskii, O. and Usov, V.V. 1976, Sov. Astr., 20, 2.
Sahade, J. and Wood, F.B. 1978, "Interacting Binaries", Pergamon Press: Oxford.
Schuerman, D.W. 1972, Astrophys. Sp. Sci., 19, 351.
Struve, O., Sahade, J., Huang, S.S., and Zebergs, V. 1958, Ap.J., 128, 328.
Vanbeveren, D. 1977, Astr. Ap., 54, 877.
Walborn, N.R. and Panek, R.J. 1984, Ap.J., 286, 718.

White, N.E. 1983, in NATO ASI Proceedings on Interacting Binaries, ed. Eggleton, P.P. and Pringle, Reidel. Publ.: Dordrecht, Holland.
Willis, A.J. 1983, in NATO ASI Proceedings on Interacting Binaries, ed. Eggleton, P.P. and Pringle, Reidel. Publ.: Dordrecht, Holland.

MASSES OF PLANETARY NEBULAE AND THEIR CENTRAL STARS

S.R. Pottasch
Kapteyn Astronomical Institute
Postbus 800
9700 AV Groningen
The Netherlands

ABSTRACT: The nebular masses are dependent on the distance. Therefore a discussion is given of nebulae for which the distance is reliable. Especially nebulae in the Magellanic Clouds or the galactic bulge are used, together with a sample of galactic nebulea whose distance is known from independent consideration. It is found that there is a general relationship between the nebular mass and its size. There is considerable real scatter in this relationship. This scatter indicates that an additional parameter is important in determining the nebular mass. This parameter may be the central star mass (or luminosity).

The central star mass can only be determined at present from its luminosity and evolutionary considerations. The present status is summarized. A range of masses between 0.5 and $1M_\odot$ are found. Some inconsistancies between theory and observations are discussed.

1. INTRODUCTION

The mass of the nebula emitted by a planetary nebula is not a directly measureable quantity. What can be measured is that part of the mass which has been ionized by the central star. In addition to this ionized component there may be a substantial amount of material surrounding it. This material could be in the form of neutral hydrogen or it could be mainly molecular. It is not easily observed so that one must be content with measuring the ionized mass and deducing the total mass from this.

The most important difficulty in measuring the ionized mass is the uncertainty in the distance, which enters essentially as the 5/2 power. Therefore great care must be taken in the selection of nebulae to those whose distance is reasonably well known, or is determined by some independent method. Therefore in the discussion a heavy releance is placed on the planetary nebulae in the Magellanic Clouds and near the galactic center. There are disadvantages connected with both groups however. The extinction in the direction of the galactic center is very high and spotty as well, making measurements difficult. The clouds have a rather low extinction (Hodge, 1974, Jacoby, 1980) but are comparatively far away. This has two separate difficulties. The first is the faintness of the nebulae and the large selection effect that follows. Fainter nebulae are strongly selected against and even when they are discovered it is very difficult to obtain an accurate spectrum. But as long as it is clear that selection effects are very important they can at least partially be taken into account. The second difficulty is that most nebulae are too small to be spatially resolved in the Clouds, which is necessary if the mass is to be determined. Thus one is limited to the few large nebulae with sizes between one and three arc seconds which are resolved. In addition, Speckle interferometry has recently been used to resolve some of the brightest nebulae (Wood et al., 1986, Barlow et al. 1986) so that some sizes are known for objects as small as 6×10^{16} cm.

The mass of the central star is even more difficult to determine. There are no accurate direct measurements of this quantity for any central stars. The determination must be made indirectly by using the relation between the core mass and the stellar luminosity first discussed by Paczynski (1971) and later improved upon by other authors. This relation assumes a carbon-oxygen core surrounded by an inner helium burning shell and an outer hydrogen burning shell. The problem now become one of determining the luminosity of the central star, for which the distance uncertainty play a similar role as in the determination of the nebular mass. Thus the samples which will be discussed as having well determined (or at least independent) distances are applicable (and necessary) both to finding the nebular mass an the central star mass.

2. SAMPLES OF NEBULAE WITH WELL DETERMINED DISTANCES

Before listing the samples used, one property of the nebulae should be mentioned which must be born in mind. There is an extremely large range of intrinsic nebular flux among nebulae, covering a range of 1000. This can be illustrated by comparing the observed fluxes of the nebulae in the Magellanic Clouds and the galactic bulge, since their distances are reasonably well known. The distance to the LMC is assumed to be 52 kpc, to the SMC 66 kpc and to the galactic bulge 8 kpc (this last taken from the discussion of Feast, 1987).

The comparison is shown in Fig. 1. Before going into detail concerning the histogram, the various scales on the abscissa should be explained. One scale is given in terms of the $H\beta$ flux as would be observed if the object was at a distance of 8 kpc. The second scale is the comparable radio 6cm continuum flux density assuming that the nebula is optically thin at this frequency, that its electron temperature $T_e = 10^4$ K and that it contains 10% helium, all in the singly ionized state. The third scale gives the luminosity of the central star. This is found from the calibration that the total energy emitted by the star is 150 times the $H\beta$ flux by Gathier and Pottasch (1987). Previous discussions (Pottasch, 1984; Gathier, 1984) gave slightly lower values. The calibration is valid only for nebulae which are ionization bounded (the great majority, see section IV).

Consider first the distribution of fluxes in the MC. Only about 40% of the known nebulae are shown in Fig. 1, partly because fluxes aren't available for the rest. The darkened nebulae are those for which the Hβ flux is given by Webster (1975) and they clearly represent the bright nebulae which are typically found by objective prism surveys. Many other of the known nebulae are in this brightness range. The exception to this are the faint nebulae found by Jacoby using a blink technique in a limited region of each of the Clouds. His results are shown as the open boxes. As can be seen the fluxes of the faint objects are sometimes 1000 times fainter. In the area Jacoby surveyed, there are 7 times as many faint nebulae as bright nebulae. This can only be construed as a lower limit

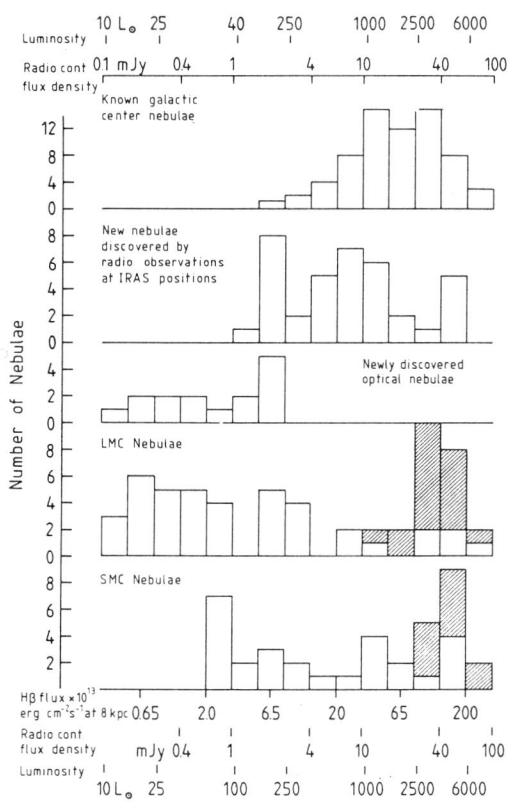

Fig. 1. Histogram of the number of known nebulae as a function of the intrinsic Hβ flux (or radio continuum flux density). The galactic bulge nebulae have been divided into 3 groups emphasizing the selection effects in each group. In the MC distribution, the darkened boxes are the nebulae listed by Webster (1975) illustrating the rather limited range of brightness found in most surveys. The other MC nebulae are taken from the deep (but spatially limited) survey of Jacoby (1980). The exciting star luminosity scale is explained in the text.

since the problems of crowding, etc. become greater as fainter objects are considered. Similarly, it is by no means certain that still fainter nebulae do not exist. On the other hand, no extinction corrections have been applied so that the nebulae may be systematically slightly brighter.

Consider now the distributions of galactic bulge nebulae. The upper diagram shows the 6-cm continuum flux density for all known bulge planetaries which have been measured with the VLA. About half are published (Gathier et al. 1983) and the rest are more recent measurements (Pottasch, Bignell and Zijlstra). Since they are selected from optical surveys they are strongly weighted toward the brighter objects, but they are on the average a factor of 2 or 3 intrinsically fainter than the brighter Cloud nebulae.

The second diagram shows the distribution of new nebulae discovered by measuring (extended) radio continuum radiation at the position of an IRAS source which has infrared colors typical of a planetary nebula (Pottasch, Bignell and Zijlstra). These measurements were made both by the VLA and Westerbork within 10° of the galactic center. Sources which are clearly foreground objects have been removed. This sample can be seen to be clearly fainter than the previous one. There is a sharp cutoff, however, on the faint side of 2 mJy. This is caused by the fact that these observations were short ('snapshot') and the 2 mJy represents the sensitivity limit of the telescope.

The third diagram refers to new nebulae found optically by Kinman, Feast and Lasker (unpublished) in one of the 'Baade windows' near the galactic center. These are the faintest nebulae for which well calibrated spectra are available. The Hβ flux plotted in the diagram has been corrected for extinction by making the observed Balmer decrement agree with the theoretical one. It is clear from this diagram that very faint nebulae exist in the galactic bulge as well as in the MC. They are difficult to find and study, both in the Clouds and in the bulge, but are of fundamental importance for studying the evolution of PN.
The samples which will be considered further may now be listed. They are:

A) nebulae in the Magellanic Clouds
B) nebulae in the galactic bulge
C) nearby nebulae with independent distances
   i) nebulae discussed by Gathier et al. 1986 a,b and Gathier and Pottasch, 1987
   ii) nebulae discussed by Mendez, Kudritzki et al. 1987.

This last sample is divided into two groups because the distances are found by entirely different methods are are therefore subject to much different selection effects. The first sample by Gathier et al. depends on distances found from nebular extinction or 21 cm hydogen line absorption which requires resonably bright nebulae located near the galactic plane. There are also a few nebulae included for which distances are found from the expansion method or as companions of normal stars. The second sample has distances determined form an analysis of the hydrogen and helium stellar atmospheric line profiles, which yields T_{eff} and the gravity. The gravity contains two unknown, the stellar mass and radius. Both can be determined by comparison with evolutionary models. From the radius, temperature and the observed magnitude the distance can be found. Use is made of NLTE model atmospheres in this analysis. With this sample the masses (nebular and central star) may now be discussed.

3. NEBULAR MASSES

As already discussed the subject is complicated by uncertainties of the distances and the fact that the nebulae may only be partially ionization bounded or optically thick). This is subject where there is not yet a general agreement.

The problem arises because of the difficulty to determine the distance to the nearby nebulae. Shklovsky (1956) suggested that the distance could be determined from the observed Hβ flux and the assumption that all nebulae have the same ionized mass. Only one independently determined distance is then necessary to calibrate this relationship. Notice that it is implicitly assumed that the nebulae are optically thin (density bounded) to hydrogen ionizing radiation.

The widespread use of this suggestion for several decades made any reasonable discussion of nebular masses impossible and only a few attempts were made until quite recently. Pottasch (1980) using all the independent distances then available was able to determine reasonably reliable nebular masses for a selection of 25 nebulae. He was able to show that the nebular ionized masses vary over at least 3 orders of magnitude and the assumption of a uniform, common nebular mass is wrong. He also showed that the larger the dimension of the nebula, the higher the nebular mass. This can be explained if the nebulae are really optically thick, since their expansion, which is accompanied by a decrease in density, allows a greater mass to be ionized by the same number of photons. Thus the implicit assumption that the nebulae are optically thin is wrong as well.

This was confirmed by Gathier et al. (1983) who measured a selection of about 30 nebulae in the galactic bulge, thus with a common distance. They found the same effect: the nebulae showed a large range of ionized mass, and the higher masses were found for the larger nebulae. But in spite of this confirmation the original ideas are still discussed in the literature; in particular the idea of using the nebulae masses found in the MC to determine a distance scale for galactic planetaries.

Jacoby (1980) was able to spatially resolve 9 faint PN in the MC and obtain diameters of the first time. Since the distance is known he was able to obtain nebular masses for which he found a spread of a factor of 40. Since then more diameters have been measured. Wood et al. (1986)

using Speckle techniques, were able to measure diameters for 10 very bright nebulae (only the brightest nebulae can be studied with this technique). Wood et al. (1987, quoted by Wood, 1987) have spatially resolved 18 other nebular using high time resolution imaging. The resultant masses are plotted in Fig. 2 as a function of the nebular radius. Several things are apparent from the figure. First, there is a general trend for increasing mass with increasing radius. Second, the very bright nebulae whose diameters were measured by the speckle technique lie in a different part of the diagram and show remarkably little scatter in the mass-radius relation. In contrast the other nebulae have considerably more scatter in this relation.

Before interpreting this diagram, consider the galactic nebulae. These are shown in Fig. 3, which also plots nebular mass against radius. The nebulae are divided into several groups. The galactic bulge nebulae measured by Gathier et al. (1983) are shown as open circles, while those discovered by Kinman et al. (unpublished) are indicated by crosses. The filled triangles are the selection of nearby nebulae for which the dis-

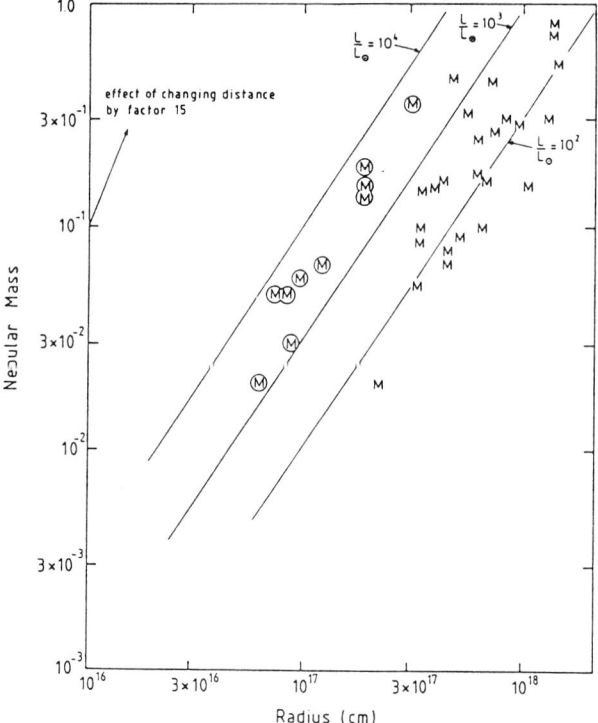

Fig. 2. Masses of MC planetary nebulae are plotted against the radius. The dimensions of the circled nebulae are found by speckle techniques (Wood et al. 1986). The other nebulae are the only resolved nebulae yet measured (Jacoby, 1980; Wood et al. 1987). The solid lines are the theoretical predictions for optically thick nebulae with various central star luminosities as discussed in the text.

tance is known from some independent method (Gathier et al., 1986a,b; Gathier and Pottasch, 1987). Two things are immediately apparent from this diagram:

1) a relationship exists between mass and radius for all the nebulae, and
2) for each of the two galactic bulge subgroups a tighter relationship exists, but these two individual relationships do not overlap eachother. The nearby nebulae have a broader distribution on this diagram, overlapping with both these two subgroups.

What can one expect from optically thick nebulae? This depends on the number of ionizing photons emitted by the exciting star. We will express this in terms of the total luminosity of the star, which in turn

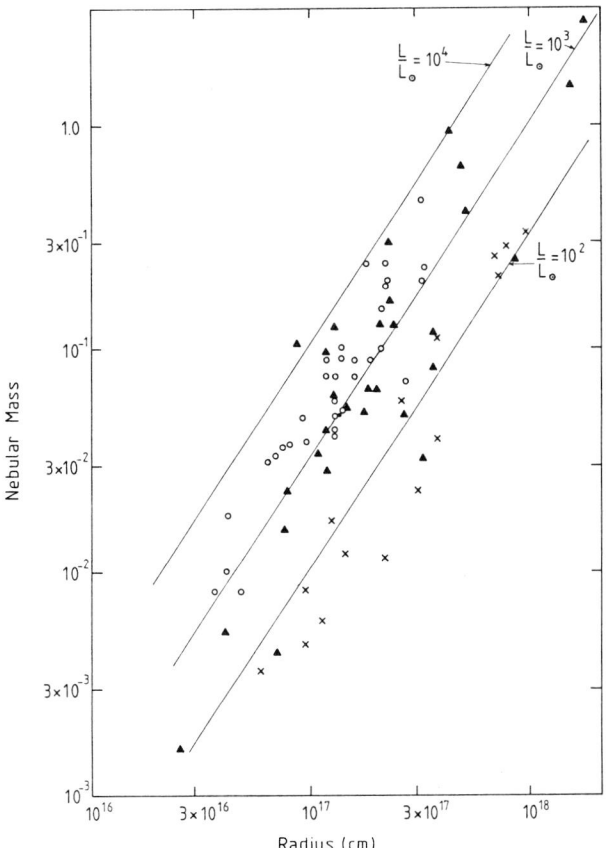

Fig. 3. Masses of galactic planetaries are plotted against the radius. Three groups are defined: the open circles represent the galactic bulge nebulae measured by Gathier et al., while the crosses represent the low luminosity nebulae in the galactic bulge discussed by Kinman et al. (1987, private communication). The nearby nebulae with independent distance determinations (Gathier and Pottasch, 1987) are shown as solid triangles. The solid line are the same as in Fig. 3.

is related to the Hβ flux by the empirical relation given in section 2 (L = 150 d² Hβ). The resultant relation between mass and radius is shown by the solid lines in Fig. 3, for 3 different values of exciting star luminosity. These predicted lines have the same slope as the observed points. Furthermore the range of luminosity is roughly what is expected from nuclei of PN. Thus the assumptions used, and especially that the nebulae are in general optically deep to ionizing radiation, are good approximations.

Comparing the expected relations with the observed points in Figs. 2 and 3, it is now clear how the spread of masses at a given radius, and the formation of various subgroups, can be interpreted. For example, the group of nebulae in the MC whose radius have been determined with Speckle techniques, and the galactic bulge group measured by Gathier et al. (1983) overlap. This is because the same selection effect is working in both cases: the brightest (and therefore most luminous) nebulae have been selected. The subgroup found by Kinman et al. is displaced to the left in

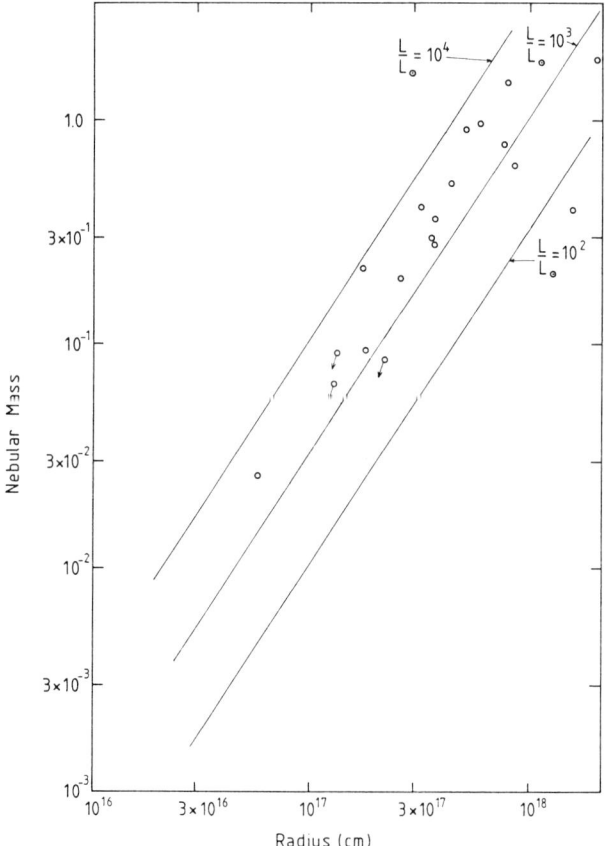

Fig. 4. Mass of galactic planetaries from the Mendez, Kudritzki et al. sample, plotted against radius.

the diagram because it is a group containing the intrinsically faintest known nebulae (see Fig. 1). It appears that parallel evolutionary tracks occur, some taking place with exciting stars of 10^3 to 10^4 $L_\odot$, others occurring with central stars having luminosities as low as 10 to 10^2 $L_\odot$. This is confirmed by the fact that the low luminosity objects contain roughly the same distribution of excitation classes (i.e. central star temperatures) as the high luminosity objects.

In Fig. 4 the subgroup of nearby nebulae studied by Mendez, Kudritzki et al have been plotted. The same trend of increasing ionized mass with radius is seen. The nebulae fall systematically on the high luminosity part of the diagram. This in understandable because they were selected on the basis of having very bright central start so that high resolution spectroscopy could be done.

Another interesting conclusion that can be drawn from Figs. 3 and 4 is that there is no evidence for any optically thin nebulae. These nebulae would be expected to form a horizontal line or band (constant mass) in these diagrams. This is simply not seen. This may be is suppressed for a number of reasons. First, the masses have been calculated assuming a constant filling factor close to unity, which is certainly justified for small nebulae. Large nebulae appear less uniform, however, and could have smaller filling factors. The mass is only proportional to the square root of the filling factor, so that a very large effect is not expected. Second, very large nebulae are apparently very difficult to find, even in the MC, presumably because of their low surface brightness. But at present we can only conclude that the nebulae we see are either optically deep or not very far removed from this stage, up to the largest mass nebulae observed, which is about 1 $M_\odot$. Furthermore, there is no evidence that those nebulae which have faint central stars are less massive than their brighter counterparts.

This has important consequences for the use of the Shklovsky method to determine distances to individual nebulae. Since it is based on the assumption of a constant mass, it requires the presence of nebulae forming a horizontal band in a diagram such as Figs. 3 or 4. Since no such band exists, the assumption of a nebula being optically thin becomes extremely arbitrary. Likewise the search for a 'distance scale' for nearby nebulae from observations in the MC is proving fruitless.

4. CENTRAL STAR MASSES

As discussed above, central star masses may be derived from the predicted relationship between core mass and luminosity for hydrogen and helium shell burning star. This relationship no longer applies after the shell burning has ceased. Therefore the position on the HR diagram must be determined to see if the relationship may be applied. This requires that the stellar temperature be known as well as its luminosity.

The temperature determination for the nearby sample is usually readily available: they are well known nebulae with temperatures found from the Zanstra and energy balance methods (e.g. Preite-Martinez and Pottasch, 1983). In addition the stellar atmosphere analysis (Mendez et al.) yield the effective temperature, which are in good agreement with the hydrogen Zanstra tempertures when these are available.

The temperature determination for the Magellanic Cloud sample, and to a lesser extent the galactic bulge sample, is harder. The reason that measuring stellar magnitudes for small, faint nebulae is very difficult because of the confusion of nebular light. Zanstra temperatures are therefore unreliable at best. Energy balance temperatures are possible, especially if ultraviolet spectra are available. The excitation class will also give an approximate temperature determination. This class depends on the nebular density as well as the stellar temperature, but the temperature is most important factor. A calibration of the excitation class for well studied nebulae, with the Zanstra and energy balance temperature shows a good correlation for excitation classes between 1 and 7,

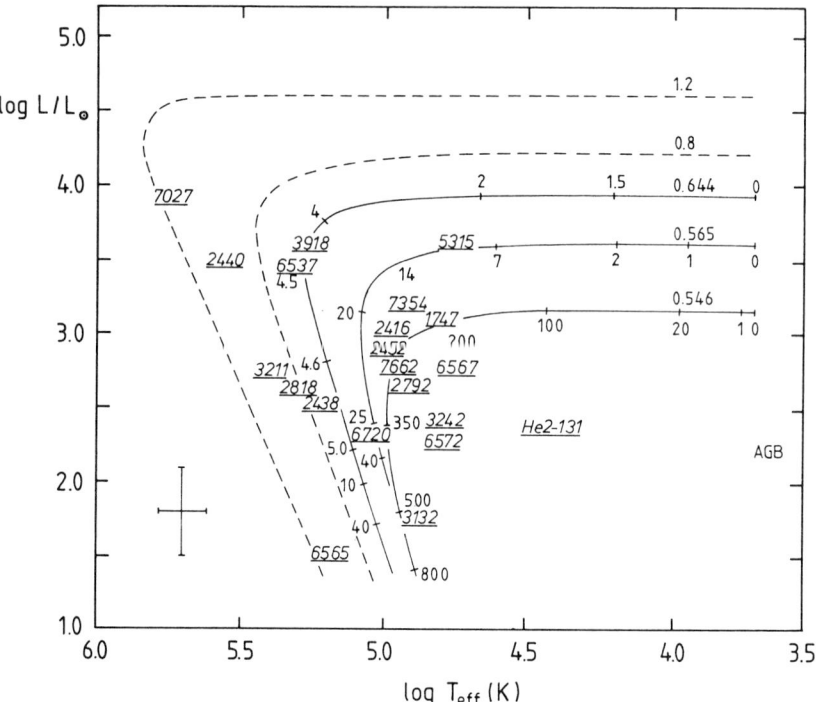

Fig. 5. HR diagram for the sample of central stars nebulae studied by Gathier and Pottasch (references in text). The stars are indicated by the NGC (or other) number of the nebulae. The solid lines are theoretical evolutionary tracks discussed in the text. The theoretical prodiction of the age of the central star is indicated at various prositions on the tracks, in units of 10^3 years.

while classes 8, 9 and 10 have central star temperatures above 10^5 K (Preite-Martinez and Pottasch, 1983, Pottasch, 1987). The excitation classes used have been given by Feast (1968), Webster (1975), Morgan (1984) and Kinman et al. (1987).

The resultant masses range from 0.5 $M_\odot$ to 1.1 $M_\odot$ Examples of HR diagrams indicating the range found are shown in Figs. 5 and 6. In both diagrams the lines indicate the theorical evolutionary tracks calculated by Schonberner (1981) for the lowest three masses and by Paczynski (1971) for 0.8 and 1.0 $M_\odot$. In Fig. 5 the nebulae from the sample of Gathier et al. are shown, while in Fig. 6 the sample of Mendez et al. and Kinman et al. are shown. The very hot stars are only seen in the sample of Gathier et al. because they are very small and faint in the visual making high resolution spectra of the central star impossible. The highest masses of about 1 $M_\odot$ for NGC 7027 and NGC 2440 are form this sample. A group of rather high mass (0.8 $M_\odot$) lower temperature (30 to 50.000 K) central starts are found in the sample of Mendez et al.

It is impossible to say anything about the distribution of the central star mass for nebulae from this data. This is because the selection

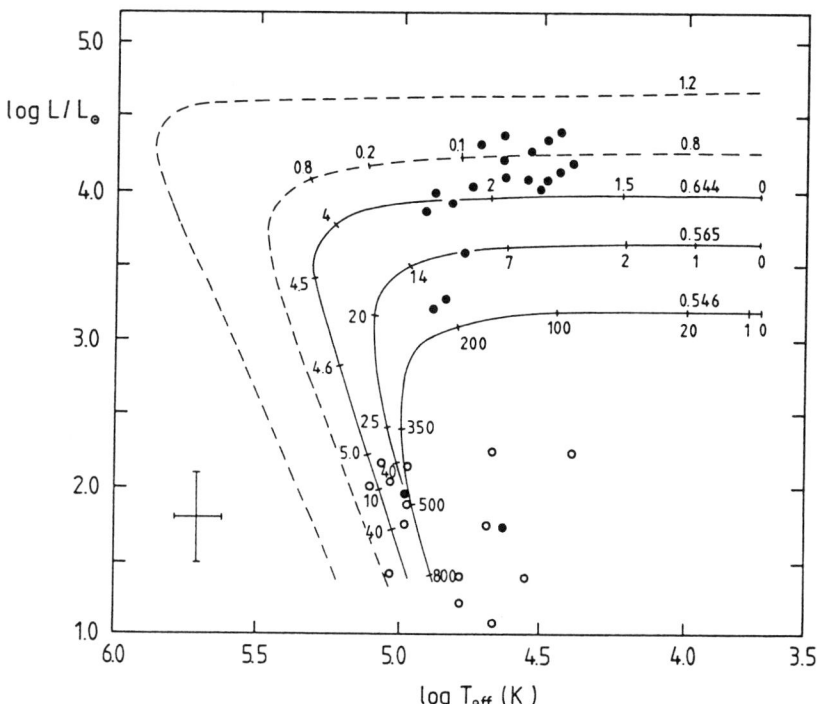

Fig. 6. HR diagram for the sample of central stars studied by Mendez, Kudritzki et al. (filled circles). The open circles are approximate positions for the central stars of the faint nebulae discoverd by Kinman et al.

effects are quantatively not well understood. It is clear that the higher mass stars are brighter and easier to measure.

The expected time for the central star to evolve along a given track is also indicated on four of the tracks. The time is given in units of 10^3 years, beginning when the star has reached 5000K. The time of evolution depends on the mass of the hydrogen rich envelope remaining after the nebulae has been ejected and how quickly it is lost. The mass of the evelope depends critically on the ejection process and may therefore be uncertain. The rate of loss of the evelope depends not only on the luminosity (hydrogen and helium burning) but on the rate of mass loss. This later is a poorly known quantity, either from observations or from theory. It may therefore be expected that these times are uncertain.

TABLE 1

Comparison of predicted lifetimes with observed nebular ages
(in 10^3 years)

NEBULA	SAMPLE	PREDICTION	OBSERVATION
NGC 246	GP	100	8
NGC 2392	MK	0.05	5
NGC 2440	GP	0.5	8
NGC 2452	GP	250	6
NGC 2818	GP	3	22
NGC 3132	GP	600	6
NGC 3242	MK	3	10
NGC 3918	GP	4	4
NGC 4361	MK	50	7
NGC 6567	GP	300	2
NGC 6627	MK	0.5	15
NGC 6891	MK	0.1	20
NGC 7027	GP	0.05	1
He2-131	GP	> 1000	0.5
EGB 5	MK	> 1000	17

These theoretical lifetimes may be compared with the observed nebular ages. These are defined by dividing the observed nebular radius by the observed expansion velocity. The comparison has some uncertainty because the zero points of these ages may not be the same, i.e. the nebula may have been ejected before the central star reached a temperature of 5000 K. The results of this comparison for 15 nebulae are given in Table 1. Six of the nebulae listed are from the Mendez, Kudritzki et al. (MK) sample nine are from the Gathier and Pottasch (GP) sample. As can be seen from the table, the agreement between prediction and observation is poor.

Those nebulae with high mass central stars are found to be considerably older than expected, while those of lower mass (or luminosity) are found to be much younger than predicted. This is true of nebulae in both samples. In addition the sample of Kinman et al. contains at least four and possibly ten nebulae which are considerably younger than expected.

Because of the uncertainties in the expected ages discussed above, it is not expected that the poor agreement brings the core mass - luminosity relation into question. Thus the central star masses found are expected to be reasonably reliable. It would still be interesting to find a single case for which the mass could be found from an independent method, however.

References:

Barlow, M.J. Morgans, B.L. Standley, C., Vine, H. 1986, Mon. Not. Roy. Astron. Soc.
Dopita, M.A., Ford, H.C., Webster, B.L. 1985, Astrophys. J. 297, 593
Feast, M.W. 1968, Mon. Not. Roy. Astron. Soc. 140, 345
Feast, M.W. 1987,. Cambridge Summer School
Gathier, R., Pottasch, S.R., Goss, W.M., Van Gorkom, J.H. 1983, Astron. Astrophys. 128, 325
Gathier, R., Pottasch, S.R., Goss, W.M., 1986a, Astron. Astrophys. 157, 191
Gathier, R., Pottasch, S.R., Pel, J.W. 1986b, Astron. Astrophys. 157, 171
Gathier, R., Pottasch, S.R., 1987, Astron. Astrophys. (to be submitted)
Hodge, P.W. 1974, Astrophys. J. 192, 21
Iben, I. 1984, Astrophys. J. 277, 333
Jacoby, G.H. 1980, Astrophys. J. Suppl. 42, 1
Jacoby, G.H. 1983, IAU Symp. 103, p. 427, ed. D.R. Flower, (Reidel, Dordrecht)
Kenman, T. Feast, M.W., Lasker, B. 1987 (preprint)
Mendez, R.H. Kudritzki, R.P. et al. 1981, Astron. Astrophys. 10, 323 and 1987 (unpublished)
Morgan, D.H. 1984, Mon. Not. Roy. Astron. Soc. 208, 633
Paczynski, B. 1971, Acta Astron. 21 417
Pottasch, S.R. 1980, Astron. Astrophys. 89, 336
Pottasch, S.R. 1984, Planetary Nebulae (Reidel, Dordrecht)
Pottasch, S.R. 1987, ESO Workshop 'Stellar Evolution"
Preite-Martinez, A., Pottasch, S.R., 1983, Astron. Astrophys. 126, 31
Sanduleak, N., MacConnell, D.J., Davis Philip, A.G. 1978, Publ. Astron Soc. Pacific 90. 621
Schonberner, D. 1981, Astron. Astrophys. 103, 119
Shklovskii, I.S. 1956, Sov. Astron. J. 33, 315
Webster, B.L. 1975, Mon. Not. Roy. Astron. soc. 173, 437
Webster, B.L. 1978, IAU Symp. 76, p. 11 ed. Y. Terzian (Reidel, Dordrecht)
Wood P.R. Bessel, M.S., Dopita, M.A. 1986, Astrophys. J. 311, 632
Wood, P.R. 1987, Late Stages of Stellar Evolution, ed. S. Kwok, S.R. Pottasch (Reidel, Dordrecht)

THE FORMATION OF PLANETARY NEBULAE

Sun Kwok
Department of Physics
University of Calgary
Calgary, Alberta
Canada T2N 1N4

ABSTRACT. Although red giants have been suggested as progenitors of planetary nebulae since the 1950's, the exact mechanism of the transition from red giants to planetary nebulae has never been settled. In this review, we identify mass loss on the asymptotic giant branch as the most important physical process that leads to the formation of planetary nebulae. The dynamics of planetary nebulae is discussed in terms of the Interacting Stellar Winds model of Kwok, Purton and FitzGerald. We also propose the existence of two intermediate evolutionary stages: the Late Asymptotic Giant Branch and the Proto-Planetary Nebula phases that separate the Asymptotic Giant Branch and the planetary nebulae phase. Transition objects which belong to these two short, invisible phases are also discussed.

1. INTRODUCTION

Planetary nebulae have been extensively studied since their discovery by William and John Herschel in the last century but new questions on their origin have only arisen during the last decade. In general terms, planetary nebulae are outerlayers of stellar atmospheres which have been ejected in an earlier phase of evolution and are presently ionized by the parent star. In 1956, Shklovskii (1956) identified the progenitor stars of planetary nebulae to be red giants and this is supported by dynamical arguments of Abell and Goldreich (1966). We now know that planetary nebulae are likely to be descendants of intermediate mass stars which have high enough mass to evolve beyond the first giant branch and through the asymptotic giant branch (AGB) but not high enough in mass to ignite carbon. More recently, we have come to believe that two more evolutionary stages - the late AGB and proto-planetary nebulae stages - separate the AGB and the planetary nebulae phases (Kwok, Hrivnak and Boreiko 1987a,b). These two phases are relatively short in duration and consist of mostly stars which are invisible in the optical part of the spectrum. In this review, we will discuss the progress in observational and theoretical understanding of these two phases and the subsequent evolution that leads to the formation of planetary nebulae.

2. ASCEND OF THE ASYMPTOTIC GIANT BRANCH

The major physical process which characterizes AGB evolution is mass loss. As a star ascends the AGB, its luminosity (L_*) and radius (R_*) both increase dramatically, resulting in a steady decrease in photospheric temperature (T_*). As the star evolves beyond the spectral type of approximately M3, the rate of mass loss from the surface becomes so large that it is comparable to the nuclear burning rate in the core. At the same time, dust begin to form. The dust ejected in the mass loss process gradually obscure the photosphere of the star. When the star reaches the spectral type of approximately M10, the star is completely surrounded by an optically thick circumstellar envelope and the conventional methods of classifying the spectral types by molecular bands become impossible. The study of the stellar evolution beyond M10 (which we will call the Late AGB phase) therefore has to rely on the analysis of circumstellar envelope.

During the decade of the 1970's several observational techniques have been developed as diagnostic tools of the circumstellar envelope. The 9.7 μm feature due to silicate grains were discovered in 1969 by Woolf and Ney (1969). This feature was later found to be common among oxygen-rich AGB stars and can be considered as a defining characteristic of such stars (Gehrz and Woolf 1971). Using the 1.5m UM-UCSD infrared telescope at Mt. Lemmon, Merrill and Stein (1976a,b,c) were able to show that the silicate feature in AGB stars range from emission to absorption. Many stars with the silicate feature, in particular those with lower color temperatures, are identified with OH/IR stars discovered by Wilson and Barrett (1968).

For carbon-rich stars, an infrared signature at 11.2 μm was identified by Treffers and Cohen (1974) as due to SiC grains. In 1973, the rotational transition of CO was discovered in the carbon star IRC+10216 by Solomon et al. (1971). This circumstellar molecular emission can be considered as the counterpart of OH maser emission in oxygen rich stars. The spectral shapes of both OH and CO emissions were soon recognized as the result of the a steady expansion of the circumstellar envelope and are manifestations of the mass loss process (OH: Kwok 1976; Elitzur et al. 1976; Olnon 1977; CO: Kuiper et al. 1976).

These infrared and radio observations of the circumstellar envelopes not only suggest the presence of mass loss in AGB stars but more importantly, allow the quantitative determination of the mass loss rates involved. While early estimates of mass loss rates based on optical circumstellar lines suggest mass loss rates of $\sim 10^{-8}$ to 10^{-7} $M_\odot$ yr^{-1} (Weymann 1963; Reimers 1975), the mass loss rates of the "infrared" stars are often found to be much larger (Gehrz and Woolf 1971). It was recognized in the early 1970's, as least at the University of Minnesota where many of the pioneering observations were made, that mass loss must be a very important element in the evolution of AGB stars.

3. DETERMINATION OF MASS LOSS PARAMETERS ON THE AGB

The radio observations of circumstellar molecular emissions allow the accurate determination of the expansion velocity (V) of the circumstellar envelope. Since the gas quickly settles on a constant terminal expansion velocity (Kwok 1975), V can be simply measured as half of the total width of the OH or CO lines.

The determination of the mass loss rate ($\dot{M}$) is not as easy. For late AGB stars, the use of the optical circumstellar lines is neither practicable nor desirable. Gehrz and Woolf (1971) suggest that the strength of the silicate feature (when it is in emission) can be used as an indicator of the optical depth (τ_d) of the circumstellar envelope and the mass loss rate can be estimated by:

$$\dot{M} \sim L\tau_d/Vc \sim 2\times10^{-8}\ \tau_d\ (L_*/L_\odot)(V/km\ s^{-1})^{-1}\ M_\odot\ yr^{-1} \tag{1}$$

Alternatively, the brightness temperature of the CO line can be used. Comparison with radiative transfer model of the CO molecule leads to the following formula by Knapp and Morris (1985):

$$\dot{M} = \frac{(T_A^*/K)\ (V/km\ s^{-1})^2 (D/pc)^2}{const\ \times\ f^{0.85}}\ M_\odot\ yr^{-1} \tag{2}$$

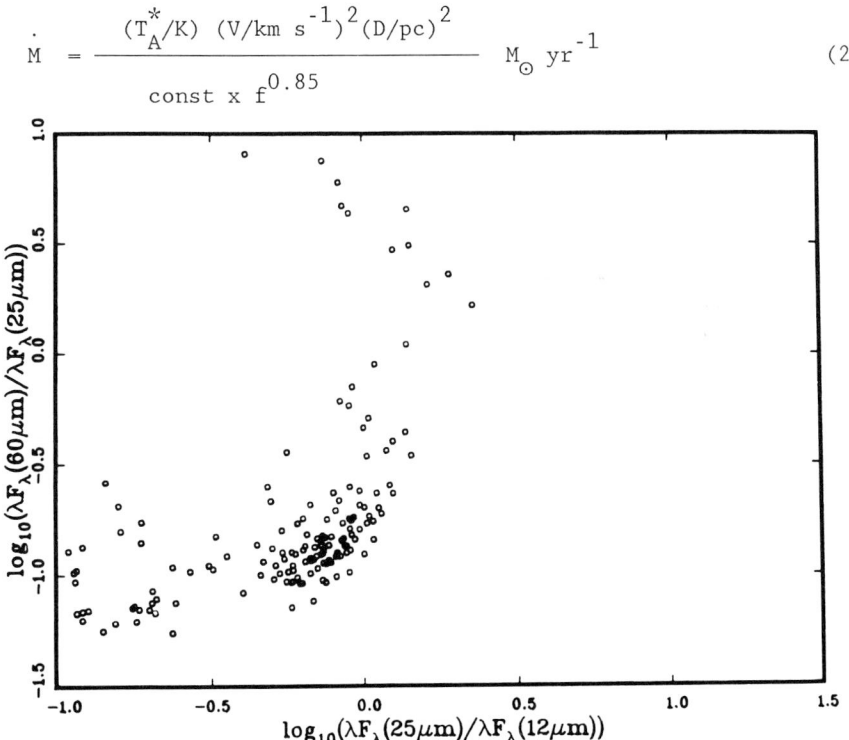

Figure 1. Colour-colour diagram of IRAS LRS Class 30 objects.

where T_A^* is the antenna temperature of the CO line, D is the distance and f is a constant depending on the telescope used. Both the infrared and radio methods suggest that the mass loss rate of AGB stars ranges from 10^{-7} to 10^{-4} $M_\odot$ yr^{-1}, depending mainly on the spectral type (or degree of "redness") of the star.

Such a large variation in the mass loss rate is in fact evident from the behaviour of the silicate feature, which ranges from strong emission to strong absorption. This implies a large variety of optical depths of the circumstellar envelope and correspondingly, a large range of mass loss rates. While the photospheric temperatures of extreme "infrared" stars cannot be directly measured, it is now generally believed that the mass loss rates of stars increase as they evolve up the AGB. Early AGB stars, such as optical Mira Variables, have $\dot{M}$ of $\sim 10^{-7}$ $M_\odot$ yr^{-1} whereas an extreme OH/IR stars will have $\dot{M}$ as high as 10^{-4} $M_\odot$ yr^{-1}.

It has been suggested that the onset of a high rate of mass loss on the AGB is an abrupt one (the "superwind" phase, cf Renzini 1983), possibly due to a switch of the pulsation mode from first overtone to fundamental (Jones et al. 1983). However, observational evidence is consistent with a continuous and gradual change from low to high $\dot{M}$. It is therefore preferable to classify such "infrared" stars with no optical counterparts as members of a late AGB phase. The beginning of the Late AGB is approximately coincident with the 10 μm silicate feature going from emission to absorption.

The existence of a Late AGB is clearly illustrated in Figure 1 which shows the distribution in a colour-colour diagram of stars observed to have silicate absorption feature by the Infrared Astronomical Satellite (*IRAS*). The stars form a well-defined band in the diagram which is consistent with an extension of the sequence defined by the Mira Variables (Kwok, Hrivnak, and Boreiko 1987b). We believe that these stars are more evolved than Mira Variables and represent the progenitors of planetary nebulae.

4. EFFECTS OF AGB MASS LOSS ON THE FORMATION OF PLANETARY NEBULAE

Conventional theories of planetary nebula formation rely on various instabilities of the hydrogen atmosphere of AGB stars for a sudden ejection at the tip of the AGB. These include thermal instabilities (Rose 1966; Smith and Rose 1971), dynamical instabilities (Lucy 1967; Roxburgh 1967; Paczynski and Ziolkowski 1968; Tuchman, Sack and Barkat 1979), and radiation pressure (Faulkner 1970; Sparkes and Kutter 1972). Also popular is the idea that planetary nebula ejection is associated with the onset of helium flash (Trimble and Sackman 1978). It has also been argued that a large fraction of planetary nebulae descends from binary stars and the ejection is induced by instabilities in the mass transfer process (Livio, Salzman and Shaviv 1979). While none of these mechanisms have convincingly simulated envelope ejection of the appropriate mass (Wood 1981), the basic premise of sudden ejection was never questioned until the mid-1970's when the importance of mass loss was recognized.

If the planetary nebular shell is ejected at the tip of the AGB, then the core of the progenitor star must evolve from the right to the left side of the H-R diagram very quickly in order to allow the nebula to be ionized and observed as planetary nebulae. Since we now know that the ejection process is probably preceded by a prolonged phase of steady mass loss, the amount of material in the remnant circumstellar envelope is expected to be

$$M_{CSE} \sim \dot{M}(R_{PN}/V) \qquad (3)$$

For typical parameters of $\dot{M} \sim 10^{-5}$ $M_\odot$ yr^{-1}, $V \sim 10$ km s^{-1}, $R_{PN} \sim 0.1$ pc, we have $M_{CSE} \sim 0.1$ $M_\odot$, which is similar to the masses of planetary nebulae (M_S). Since the transition time from the tip of the AGB to the formation of planetary nebula must be much shorter than the dynamical lifetime of planetary nebulae ($\sim R_{PN}/V_S$, where V_S is the expansion velocity of the nebular shell), the remnant of the AGB circumstellar envelope would not have yet dispersed into the interstellar medium. The fact that M_{CSE} and M_S have similar values implies that mass loss on the AGB must play a significant role in the formation of planetary nebulae.

The implication of the large mass loss rates from infrared stars on the formation of planetary nebulae was recognized by Paczyński (1971). However, planetary nebulae have well-defined shell-like morphologies and cannot be just remnants of circumstellar envelopes of AGB stars. Furthermore, the gas densities and expansion velocities are also higher in planetary nebulae than in circumstellar envelopes. In the mid-1970's, it became clear that while the mass ejected during the AGB has to be taken into consideration (cf Field 1978), it is not sufficient by itself to create the planetary nebulae as we know them.

5. THE INTERACTING STELLAR WINDS MODEL

The Interacting Stellar Winds model of planetary nebulae formation was first proposed by Kwok, Purton and FitzGerald (1978) and later expanded by Kwok (1982, 1983), Chevalier and Imamura (1983), Kahn (1983), and Bedogni and D'Ercole (1986). The ionization structure in the Interacting Stellar Winds Model is studied by Giuliani (1981) and the coupling of the dust and gas components is examined by Okorokov et al. (1985). This model assumes that mass loss on the AGB can continue until the exposure of the core and the nebular shell is formed by remnants of the AGB wind being swept up by a later-developed high-speed wind from the central star. The postulation of a fast wind is necessary in order to compress the remnant circumstellar material into a high-density, shell-like structure as well as to accelerate the swept-up shell from the AGB wind velocity of 3-20 km s^{-1} to the typical planetary nebula expansion velocity of 20-50 km s^{-1}.

Figure 2 shows a schematic diagram of the model. Region (a) is the fast wind from the central star; region (b) is the shocked central-star wind; and region (e) is the remnant of the AGB wind. If the shell is ionization bounded (see Pottasch, this volume), then the ionized and

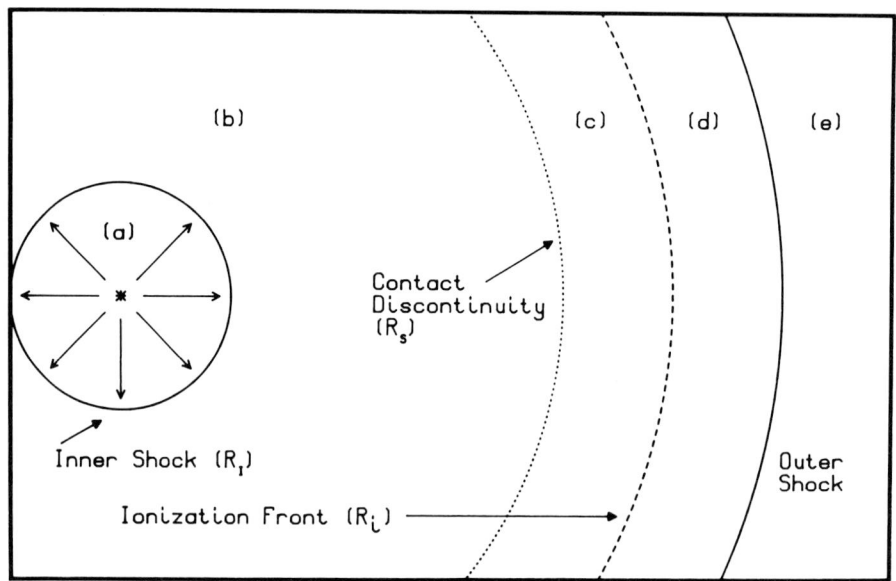

Figure 2. Schematic diagram of the Interacting Stellar Winds Model

neutral zones are represented by regions (c) and (d) respectively. Assuming that most of the PN shell is made up of swept-up mass and the interaction between central-star wind and the PN shell is adiabatic, then the conservation of mass, momentum and energy give the following three equations:

$$dM_s/dt = 4\pi R_s^2 \rho(r) (dR_s/dt - V) \qquad (4)$$

$$M_s d^2R_s/dt^2 + dM_s/dt (dR_s/dt - V) = 4\pi R_s^2 P \qquad (5)$$

$$\frac{d}{dt}(2\pi R_s^2 P) = \frac{1}{2}\dot{m}v^2 - 4\pi R_s^2 dR_s/dt \qquad (6)$$

where M_s and R_s are the mass and velocity of the shell, V and $\rho(r)$ are the velocity and density of the AGB wind, $\dot{m}$ and v are the mass loss rate and velocity of the central-star wind, and P is the pressure in the shocked region which is responsible for pushing the shell. If we assume a steady mass loss from the central star ($\frac{1}{2}\dot{m}v^2$=constant) then equations (4)-(6) can be solved by similarity analysis yielding the following solutions:

$$R_s = V_s t \qquad (7)$$

$$M_s = \dot{M} (V_s/V - 1)t \qquad (8)$$

$$P = \frac{\frac{1}{2}\dot{m}v^2}{6\pi V_s^3} t^{-2} \qquad (9)$$

where V_s is given by

$$(\dot{M}/V)V_s^3 - 2\dot{M}V_s^2 + \dot{M}VV_s = \frac{1}{3}\dot{m}v^2 \qquad (10)$$

Assuming that the shocked gas in region (b) is isobaric and applying the jump condition for a strong adiabatic shock, the inner shock S_i is found to be located at

$$R_I = [(9/4)(V_s/v)]^2 R_s \qquad (11)$$

For $V_s \sim 50$ km/s and $v \sim 2000$ km/s, one finds that $R_I \sim 0.24\, R_s$, or approximately 99% of the volume inside R_s is shocked (Kahn 1983). Using the ideal gas law and assuming that the shocked region is uniform in density, we have

$$T = \frac{\mu m_H v^2 \epsilon}{9 k} \qquad (12)$$

where μ (~0.6) is the mean atomic weight, k is Boltzmann constant, and ϵ is the filling factor. With the above parameters the shocked region is found to have a temperature of $\sim 10^7$K.

6. TIME-DEPENDENT DYNAMICAL EVOLUTION OF PLANETARY NEBULAE

The existence of planetary nebulae requires that the central star evolves in step with the expansion of the nebula. If the central star evolves too slowly, the nebula would not be ionized before it is dispersed into the interstellar medium. On the other hand, a rapidly evolving central star would greatly shorten the observable lifetime of planetary nebulae. A dynamical lifetime of $\sim 10^4$ yr therefore limits the central star mass to within a narrow range of 0.6 to 1 $M_\odot$ (Renzini 1983, Schönberner 1983). Since the central stars of planetary nebulae evolve from the right to the left of the H-R diagram over this short period, any dynamical study of nebula expansion must take into account the rapidly evolving nature of the central star.

A time-dependent dynamical model for planetary nebulae was developed by Volk and Kwok (1985) who use the central star evolution model of Schönberner (1983). In order to be consistent with Schönberner, $\dot{m}$ is assumed by Volk and Kwok (1985) to be $\propto L_* M_*^{-1} R_*$ (the Reimers' formula). The central-star wind velocity v is assumed to be 4 times the escape velocity. As it happens, under these two time dependent forms, the kinetic energy of the wind, $\frac{1}{2}\dot{m}v^2$ is directly proportional to L_*, and therefore approximately constant for most of the lifetime of the central star. This implies that $R_s \propto t$ as discussed in §5.

However, the mass loss formula for central star of planetary nebulae is completely unknown and there is no *a priori* reason that it should obey the Reimers' formula. It is possible for the nebula to accelerate with time under a different mass loss formula.

The Schönberner model is also used by Schmidt-Voigt and Köppen (1987a,b) to compute the ionization structure of planetary nebulae. In particular, they compared the observed strength variation of the He line with nebular size with the prediction of the Interacting Stellar Winds model. They suggest that the observation of HeII Zanstra temperatures of $\sim 10^5$ K in small ($R_s <0.1$ pc) nebulae requires a combination of sudden ejection and the interacting winds process in action.

7. OBSERVATIONAL SUPPORT FOR THE INTERACTING STELLAR WINDS MODEL

The Interacting Stellar Winds model postulates the existence of three physical components in a planetary nebulae system - the shell, remnant of the AGB wind, and the wind from the central star. Only one component (the shell) is easily observable through recombination lines and forbidden lines because of its relatively high emission measure. In order to test the theory, it is important to detect the other two components predicted by the model. While broad lines in the spectra of planetary nebulae nuclei were noted by Minkowski as early as 1945 (cf Aller 1982), it was not until the launch of the IUE satellite in January of 1978 that fast winds from central stars of planetary nebulae were definitely established (Heap et al. 1978). Many central stars of planetary nebulae are now found to have P Cygni profiles in UV resonance lines. The wind velocities range from 1400 to 5000 km s^{-1}, and mass loss rates have been estimated to be 10^{-9} to 10^{-7} $M_\odot$ yr^{-1} (Bianchi, Cerrato and Grewing 1986; Perinotto, Cerruti-Sola and Lamers 1987).

Similarly, haloes around planetary nebulae have been found since 1937 (Duncan 1937), but their common presence was not known until high-dynamic range CCD detectors became available. Faint haloes are now commonly seen outside the main shells of planetary nebulae (Balick 1987) and it is tempting to identify them as remnants of the circumstellar envelopes of the AGB progenitors. Jewitt et al. (1986) estimate that the electron density in the halo is approximately 10% of that in the main shell. In the next few years, improvements in the observational techniques should allow better determination of the masses and velocities of such haloes and therefore lead to critical tests on the dynamics of planetary nebulae.

8. ENERGY AND MOMENTUM EFFICIENCIES OF THE INTERACTING WINDS MODEL

In order to decide whether the wind interaction is energy or momentum conserving, it is useful to define the efficiency parameters ϵ and π which represent the respective energy and momentum ratios of the shell to the central-star wind:

$$\epsilon = \frac{\tfrac{1}{2} M_s V_s^2}{\tfrac{1}{2} \dot{m} v^2 t} \qquad (13)$$

$$\pi = \frac{M_s V_s}{\dot{m}vt} \tag{14}$$

If one adopts the following values as typical observational results: $M_s \sim 0.2 M_\odot$, $V_s \sim 20$ km s^{-1}, $\dot{m} \sim (5-200) \times 10^{-9}$ $M_\odot$ yr^{-1}, $v \sim 2000-4000$ km s^{-1} and a dynamical age of 10^4 yr, then $\epsilon \sim 0.003-0.4$ and $\pi \sim 0.5-40$. While these values may be consistent with a momentum-conserving interaction, they seem to favour the energy-conserving case. Figure 3 shows the value of ϵ as a function of $\dot{m}$ in the energy conserving case, assuming that $\dot{M}=10^{-5}$ $M_\odot$ yr^{-1}, $V=10$ km s^{-1} and $v=2000$ km s^{-1}. It can be shown that for a strong central star wind, ϵ converges to a value of 0.33 independent of other parameters. The observed values of ϵ are therefore consistent with the theoretical efficiency. For small V, $\pi \sim 0.48$ $(\dot{M}v/\dot{m}V)^{1/3}$. If we have $\dot{M}/\dot{m} \sim 10^3$, $v/V \sim 10^2$, then $\pi \sim 22 \gg 1$. This is in agreement with the observed range of π and the interacting winds process seems entirely adequate in driving the expansion of planetary nebulae.

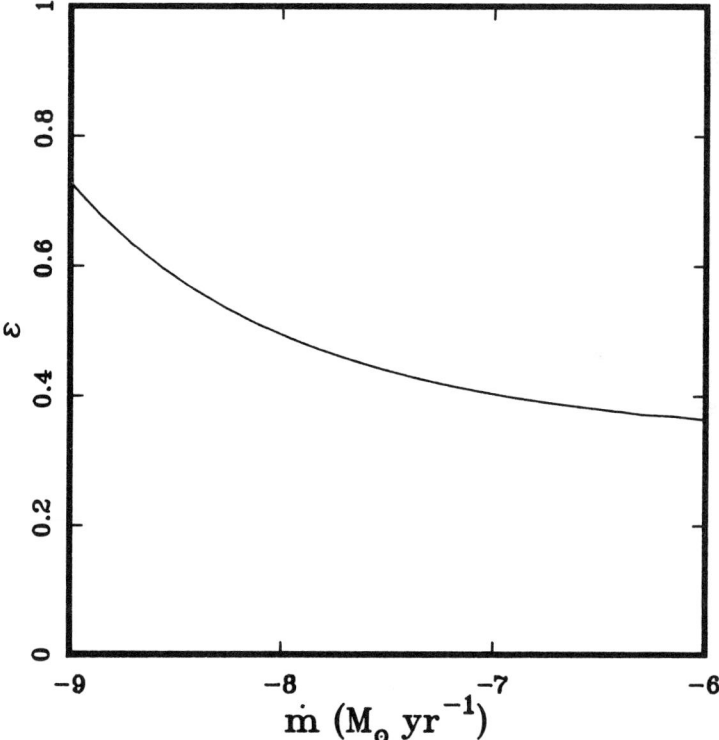

Figure 3. Energy efficiency of the ISW model as a function of $\dot{m}$.

9. TRANSITION FROM AGB TO PLANETARY NEBULAE

Planetary nebulae are bright and easily observable because of recombination lines and forbidden lines in the visible. They are also strong radio sources due to thermal free-free emission. However, the immediate progenitors of planetary nebulae are neither optical nor radio continuum sources. In fact, the only observational link between the AGB and planetary nebulae is in the infrared and such a link is only recently established because of IRAS.

The infrared behaviour of AGB and Late AGB stars have been discussed in §2. The colour temperature of AGB decrease monotonically from >600 K for Mira Variables to ~250 K for extreme OH/IR stars. Since the remnant of the circumstellar envelopes are likely to be present in planetary nebulae, it is quite plausible that planetary nebulae should also be far infrared emitters (Kwok 1980). Recent IRAS observations show that the dust continuum temperature decrease from ~100-200K for young planetary nebulae (Kwok, Hrivnak, and Milone 1986) to 40-100 K for evolved planetary nebulae (Pottasch et al. 1984). An example of the infrared spectrum of a young planetary nebula is shown in Figure 4. We believe that the infrared emissions from AGB and planetary nebulae have a common origin, i.e. dust emission from the AGB circumstellar envelope. However, the monotonic sequence of decreasing colour temperature is in fact due to two factors. In the AGB, the decrease in colour temperature is the result of increasing mass loss rate and optical depth in the circumstellar envelope whereas for planetary nebulae the drop in colour temperature is due to geometric dilution. The optical depth is increasing on the AGB but decreasing beyond the AGB. This turn around in optical depth is used by Kwok (1987) to define the proto-planetary nebula phase which begins at the cessation of the mass loss and ends at the commencement of photo-ionization of the nebula.

Adoption of this definition allows one to search for candidates of proto-planetary nebulae which should have colour temperatures in between those of late AGB stars and young planetary nebulae. The IRAS source 19454+2920 has been suggested as one of such candidates (Kwok, Hrivnak and Boreiko 1987a, see Figure 5). As the star continues to evolve to the left of the H-R diagram, the remnant circumstellar envelope will eventually become optically thin and the central star will re-emerge from the dust envelope and be detected as a near-infrared source (Habing, van der Veen and Geballe 1987). OH17.7-2.0 can be in such an evolutionary phase. By extending the radiative transfer models for AGB stars to beyond the AGB, Volk and Kwok (1987) find a distinct spectral shape for proto-planetary nebulae. A search of the IRAS catalog reveals a number of objects which show such behaviour. One candidate (IRAS 18095+2704) was observed by Hrivnak, Kwok and Volk (1987) and was found to have a bright central star of 11th magnitude and spectral type F0. A number of stars with large infrared excesses and intermediate spectral types have also been suggested as candidates of proto-planetary nebulae (Parthasarathy and Pottasch 1987; Lamers et al. 1987).

The above suggested evolutionary scenario is summarized in Table 1.

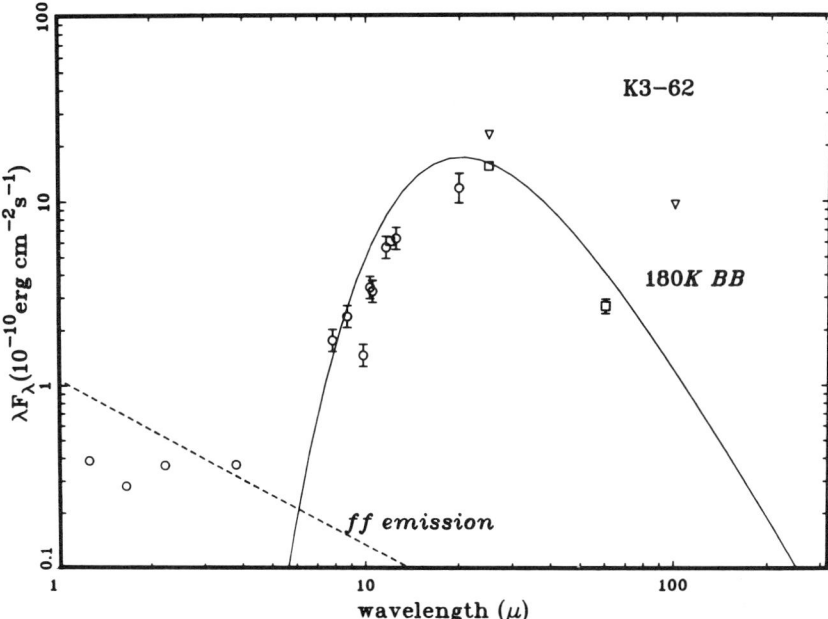

Figure 4. Infrared spectrum of the young planetary nebula K3-62. Circles and triangles are gournd-based (CFHT) and IRAS observations respectively. The dotted line represents the expected level of free-free emission extrapolated from radio measurements. An 180K blackbody curve is also shown for comparison.

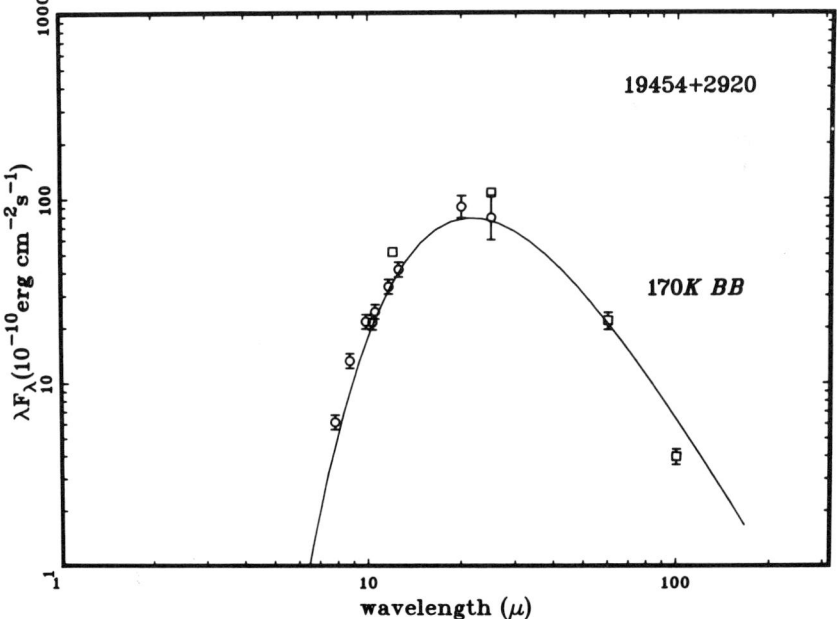

Figure 5. Infrared spectrum of IRAS 19454+2920. An 170 K blackbody curve is also shown for comparison.

TABLE 1
EVOLUTION FROM AGB TO PLANETARY NEBULAE

evolutionary phase	example	optical image	period (days)	colour temperature (K)	silicate dust	OH
AGB	Mira Variables	bright	300-600	>600K	emission	yes
LAGB	OH/IR stars	no optical counterpart	600-2000	250-600K	absorption	strong
post-AGB	19454+2920	no	non-variable	150-250	?	weak
proto-PN	18095+2704	yes	non-variable	150-250	emission	weak
young PN	Vy2-2 Hb 12	bright	non-variable	100-200	emission	single peak
PN	many	bright	non-variable	<100	no	no

9. CONCLUSIONS

Infrared and radio observations in the last decade have clearly established the existence of a late AGB which is beyond the traditional spectroscopic classification limit of M10. Stars on the Late AGB are losing mass at a very high rate and such mass loss must have an effect on the formation of planetary nebulae. It now appears quite probable that most of the masses in a planetary nebula shell are in fact swept-up masses but not due to the result of a sudden ejection. The detections of fast winds from central stars of planetary nebulae and faint haloes outside planetary nebula shells give strong support to the Interacting Stellar Winds model which provides a simple explanation to the morphological, dynamical and spectral properties of planetary nebulae.

Significant progresses have also been made in the study of the transition phases between AGB and planetary nebulae. Observations in the far infrared from the *IRAS* satellite have led to the identification of a continuous infrared sequence connecting the AGB, LAGB, proto-planetary nebulae and planetary nebulae phases. Further identification and observations of proto-planetary nebulae in the next few years will hopefully settle the question of the formation of planetary nebulae.

I would also like to express my thanks to Luciana Bianchi and the Local Organizing Committee for their hospitality. This work is supported by an operating grant from the Natural Sciences and Engineering Research Council of Canada.

REFERENCES

Abell, G. and Goldreich, P. 1966, *Publ. Astron. Soc. Pacific*, **78**, 232.
Aller, L.H. 1982, in *Observational Basis for Velocity Fields in Stellar Atmospheres*, ed. R. Stalio (Osservatorio Trieste), p. 389.
Balick, B. 1987, in *The Late Stages of Stellar Evolution*, eds. S. Kwok and S.R. Pottasch (Reidel:Dordrecht), p. 413.
Bedogni, R., and D'Ercole, A. 1986, *Astron. Astrophys.*, **157**, 101.
Bianchi, L., Cerrato, S., and Grewing, M. 1986, *Astron. Astrophys.*, **169**, 227.
Chevalier, R.A., and Imamura, J.N. 1983, *Astrophys. J.*, **270**, 554.
Duncan, J.C. 1937, *Astrophys. J.*, **86**, 496.
Elitzur, M., Goldreich, P., and Scoville, N. 1976, *Astrophys. J.*, **205**, 384.
Faulkner, D.J. 1970, *Astrophys. J.*, **162**, 513.
Field, G. 1978, in *I.A.U Symp. 76: Planetary Nebulae, Observations and Theory*, ed. Y. Terzian, (Reidel: Dordrecht), p. 367.
Gehrz, R.D., and Woolf, N.J. 1971, *Astrophys. J.*, **165**, 285.
Giuliani, J.L. 1981, *Astrophys. J.*, **245**, 903.
Habing, H.J., van der Veen, W., and Geballe, T. 1987, in *The Late Stages of Stellar Evolution*, eds. S. Kwok and S.R. Pottasch (Reidel:Dordrecht), p. 91.
Heap, S.R. et al. 1978, *Nature*, **275**, 385.
Hrivnak, B.J., Kwok, S., and Volk, K.M. 1987, in preparation.
Jewitt, D.C., Danielson, G.E., and Kupferman, P.N. 1986, *Astrphys. J.*, **302**, 727.
Jones, T.J., Hyland, A.R., Wood, P.R., and Gatley, I. 1983, *Astrophys. J.*, **273**, 669.
Kahn, F.D. 1983, in *I.A.U. Symp. 103: Planetary Nebulae*, ed. D.R. Flower (Reidel:Dordrecht), p. 305.
Knapp, G.R. and Morris, M. 1985, *Astrophys. J.*, **292**, 640.
Kupier, T.B.H., Knapp, G.R., Knapp, S.L., and Brown, R.L. 1976, *Astrophys. J.*, **204**, 408.
Kwok, S. 1975, *Astrophys. J.*, **198**, 583.
Kwok, S. 1976, *J. Roy. Astron. Soc. Canada*, **70**, 49.
Kwok, S. 1980, *Astrophys. J.*, **236**, 592.
Kwok, S. 1982, *Astrophys. J.*, **258**, 280.
Kwok, S. 1983, in *I.A.U. Symp. 103: Planetary Nebulae*, ed. D.R. Flower (Reidel:Dordrecht), p. 293.
Kwok, S. 1987, in *The Late Stages of Stellar Evolution*, eds. S. Kwok and S.R. Pottasch (Reidel:Dordrecht), p. 321.
Kwok, S., Purton, C.R., and FitzGerald, M.P. 1978, *Astrophys. J. (Lett.)*, **219**, L125.
Kwok, S., Hrivnak, B.J., and Milone, E.F. 1986, **303**, 451.

Kwok, S., Hrivnak, B.J., and Boreiko, R.T. 1987a, *Astrophys. J.*, **312**, 303.
Kwok, S., Hrivnak, B.J., and Boreiko, R.T. 1987b, *Astrophys. J.*, in press.
Lamers, H.J.G.L.M., Waters, L.B.F.M., Garmany, C.D., Perez, M.R., and Waelkens, C. 1987, *Astron. Astrophys.*, **154**, L20.
Livio, M., Salzman, J., and Shaviv, G. 1979, *Mon. Not. Roy. Astron. Soc.*, **188**, 1.
Lucy, L.B. 1967, *Astron. J.*, **72**, 813.
Merrill, K.M., and Stein, W.A. 1976a, *Publ. Astron. Soc. Pacific*, **88**, 285.
Merrill, K.M., and Stein, W.A. 1976b, *Publ. Astron. Soc. Pacific*, **88**, 294.
Merrill, K.M., and Stein, W.A. 1976c, *Publ. Astron. Soc. Pacific*, **88**, 874.
Okorokov, V.A., Shustov, B.M., Tutukov, A.V., and Yorke, H.W. 1985, *Astron. Astrophys.*, **142**, 441.
Olnon, F. M. 1977, Ph.D. Thesis, Univ. Leiden.
Paczyński, B. 1971, *Astrophys. Letters*, **9**, 33.
Paczyński, B., and Ziolkowski, J. 1968, *Acta Astronomica*, **18**, 255.
Parthasarathy, M., and Pottasch, S.R. 1987, *Astron. Astrophys.*, **154**, L16.
Perinotto, M., Cerruti-Sola, M., and Lamers, H.J.G.L.M. 1987, in *The Late Stages of Stellar Evolution*, eds. S. Kwok and S.R. Pottasch (Reidel:Dordrecht), p. 387.
Pottasch, S.R. et al. 1984, *Astron. Astrophys.*, **138**, 10.
Reimers, D. 1975, *Mém. Soc. Roy. Sci. Liège*, Ser. 8, p. 369.
Renzini, A. 1983, in *I.A.U. Symp. 103: Planetary Nebulae*, ed. D.R. Flower (Reidel:Dordrecht), p. 267.
Rose, W.K. 1966, *Astrophys. J.*, **146**, 838.
Roxburgh, I.W. 1967, *Nature*, **215**, 838.
Schmidt-Voigt, M. and Köppen, J. 1987a, *Astron. Astrophys.*, **174**, 211.
Schmidt-Voigt, M. and Köppen, J. 1987b, *Astron. Astrophys.*, **174**, 223.
Schönberner, D. 1983, *Astrophys. J.*, **272**, 708.
Shklovskii, I. 1956, *Astr. Zh.*, **33**, 315.
Smith, R.L., and Rose, W.K. 1972, *Astrophys. J.*, **176**, 395.
Solomon, P.M., Jefferts, K.B., Penzias, A.A., and Wilson, R.W. 1971, *Astrophys. J. (Lett.)*, **163**, L53.
Sparkes, M.W., and Kutter, G.S. 1971, *Astrophys. J.*, **221**, 616.
Treffers, R. and Cohen, M. 1974, *Astrophys. J.*, **188**, 545.
Trimble, V., and Sackman, I.J. 1978, *Mon. Not. Roy. Astron. Soc.*, **182**, 97.
Tuchman, Y., Sack, N., and Barkat, Z. 1979, *Astrophys. J.*, **234**, 217.
Volk, K. and Kwok, S. 1985, *Astron. Astrophys.*, **153**, 79.
Volk, K., and Kwok, S. 1987, in *The Late Stages of Stellar Evolution*, eds. S. Kwok and S.R. Pottasch (Reidel:Dordrecht), p. 305.
Weymann, R. 1963, *Ann. Rev. Astron. Astrophys.*, **1**, 97.
Wilson, W.J., and Barrett, A.H. 1968, *Science*, **161**, 778.
Wood, P.R. 1981, in *Physical Processes in Red Giants*, eds. I. Iben and A. Renzini (Reidel:Dordrecht), p. 205.
Woolf, N.J., and Ney, E.P. 1969, *Astrophys. J. (Lett.)*, **155**, L181.

MASS LOSS AND POST-ASYMPTOTIC GIANT-BRANCH EVOLUTION

Detlef Schönberner
Institut für Theoretische Physik und Sternwarte
der Universität
Olshausenstrasse 40, 2300 Kiel, F.R.G.

1. INTRODUCTION

Before low-mass ($M < 2.3\ M_\odot$) or intermediate-mass stars ($2.3 < M/M_\odot < 8$) on the asymptotic giant branch (AGB) die as white dwarfs, they evolve quickly (within ~ 2 or 3×10^4 yr) through the planetary nebulae (PN) phase. Considerable progress in our understanding of this post-asymptotic giant-branch (PAGB) evolution has been made during the last years since consistent computations through the AGB, with thermal pulses and mass loss taken into account, became available (Schönberner, 1979, 1983; Kovetz and Harpaz, 1981; Iben 1984; Wood and Faulkner, 1986). It was found that the evolution of an AGB remnant depends very sensitively on the initial condition prevailing during its formation. By this very reason the somewhat approximative calculations of Paczynski (1971) failed to give a satisfactory explanation of the evolution of the central stars (CPN).

AGB remnants with masses of about $0.6\ M_\odot$ can account for the observed temporal evolution of central stars, provided the following two criteria are met: the formation of planetary nebula must occur by means of a strong mass loss ("superwind") that forces the remnant to make its transit across the HR diagram into the hot region ($T_{eff} > 30000°K$) in a time that is short compared to the expansion time scale of the newly created PN (Renzini, 1981; Schönberner, 1983; Iben and Renzini, 1983); ii) the remnant (i.e., the CPN) must then fade down to about $10^2\ L_\odot$ within the typical lifetime (≈ 30000 yrs) of planetary nebulae.

As already mentioned above the finer details of the evolution of PAGB stars depend sensitively on their thermal structure when they leave the AGB. With other words: the PAGB evolution is a function of the phase of the thermal pulse cycle on the AGB during which the PN formation occurs. In short, two major modes of PAGB evolution to the white dwarf stage are possible, according to the two main phases of a thermally pulsing AGB star: the hydrogen-burning or helium-burning mode. If, for instance, the PN formation, i.e., the removal of the stellar envelope by mass loss, happens during a luminosity peak that follows a thermal pulse of the helium-burning shell, the remnant leaves the AGB while still burning helium as the main energy supplier (Härm and Schwarzschild, 1975). On the other hand, PN formation may also occur during the quiescient hydrogen-burning phase on the AGB, and the remnant continues then to burn mainly hydrogen on its way to becoming a white dwarf. The computations now show that a hydrogen-burning PAGB model evolves about 3 times faster through the region occupied by the central stars as a helium-burning model of the same mass. Using hydrogen-burning models, Schönberner (1981), Schönberner and Weidemann (1981), and Schönberner (1984, 1986) demonstrated that the temporal evolution of central stars can be very well explained by models with masses between 0.55 and 0.64 $M_\odot$. The conclusion then follows that obviously the PN ejection is generally <u>not</u> initiated by a thermal pulse, but occurs during the quiescent hydrogen-burning phase on the AGB.

It must be emphasized that the conclusion about CPN being hydrogen burners was not solely based on the time argument as given above. Also the observed shape of the luminosity function of CPN could only be explained by hydrogen-burning PAGB models (e.g., Fig. 9 of Schönberner, 1981, and discussion in Schönberner and Weidemann, 1983). One special feature of this luminosity function is a deficit ("gap") of CPN with $M_v \approx 4 \ldots 5$. This "gap" can be explained by hydrogen-burning PAGB models of ≈ 0.6 $M_\odot$ because they experience a rapid luminosity drop of ≈ 1 dex within only about 10^3 years when hydrogen burning starts to cease. Conversion into a luminosity function leads to a pronounced dip between $M_v \approx 4.5$ and 6.0, the exact position depending somewhat on the mass of the models. Such a luminosity drop is not found in models that leave the AGB while burning helium (cf. Fig. 1 in Iben, 1984), and this fact clearly indicates that at least the majority of CPN must be hydrogen burners. Additional observational support for a fast luminosity drop during the CPN evolution comes from the variation of the nebular ionization during the later phases of evolution. Schönberner (1986) showed that their exist a correlation between the luminosities of the CPN and the degree of nebular ionization, in that PN with a lower ionization also belong to intrinsically faint CPN, whereas highly ionized PN have also luminous central objects.

Therefore, I will concentrate on hydrogen shell-burning stars that are evolving off the AGB in thermal equilibrium under the influence of mass loss.

2. TERMINATION OF THE AGB EVOLUTION

The structure of an AGB star is rather complicated. It has a hydrogen-exhausted core which contains two burning shells, namely the hydrogen-burning shell at the core's surface and the helium-burning shell further inwards. The helium-exhausted inner part of the core consists of carbon and oxygen and is electron degenerated. The core is actually nothing else than a very hot white dwarf which is surrounded by a huge, nearly fully convective envelope containing the unprocessed stellar matter. The stellar radius exceeds the core radius by factors of about 10^4! In the course of evolution along the AGB, the hydrogen-exhausted core is growing in mass at the expense of the envelope due to nuclear burning in the hydrogen-burning shell, while its radius is shrinking. The core of an AGB star may contain up to more than 99% of the stellar mass! The evolutionary track of an AGB star in the H-R diagram is entirely due to the response of the envelope to the masswise growing core: Expansion of the envelope along the AGB, and finally contraction to white-dwarf dimensions if the envelope mass becomes too small (PAGB evolution).

The luminosity of a giant star at the Hyashi limit depends, as is well known from evolutionary calculations, practically only on the mass M_H of its hydrogen exhausted core, and not on its total mass:

$$L \sim M_H^\alpha,$$

where the exponent α is a function of M_H and decreases from 4 to 2 when M_H grows from 0.5 to 1.4 $M_\odot$ along the AGB (cf. Kippenhahn, 1981). Convenient interpolation formulae for L have been proposed by several authors, the most popular being that of Paczynski (1970):

$$L/L_\odot = 59520 \, M_H/M_\odot - 30950$$

The evolution of the stellar core of mass M_H and white dwarf dimensions proceeds independently from the envelope (of mass M_e). The envelope provides the nuclear fuel, and as long as this is possible, the core evolution is decoupled from that of the envelope.

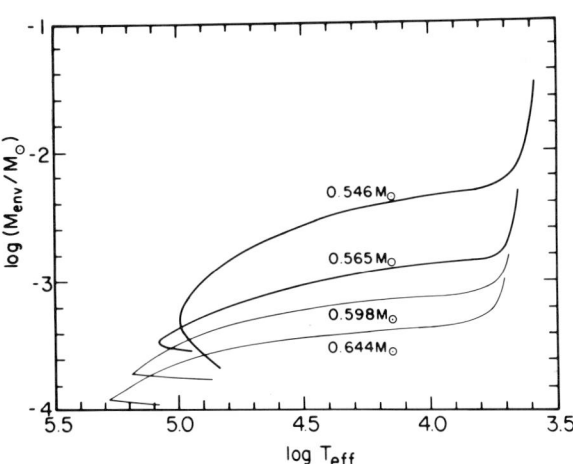

Fig. 1: Envelope mass M_e vs. T_{eff} for PAGB models of different core masses M_H according to Schönberner (1983).

The evolution along the AGB is terminated if either M_e becomes very small by the combined effect of nuclear burning in the hydrogen-burning shell and mass loss from the surface, or M_H approaches 1.4 $M_\odot$. The second possibility leads to a SN explosion and will not be discussed here. The transition from an AGB star to a white dwarf can be split into two steps:

i) If M_e is only of the order of several percent of the stellar mass, the envelope starts to shrink, but is still able to release enough gravitational energy as to maintain the burning temperatures at its base. Consequently, the luminosity stays about constant, and the star evolves horizontally across the HR diagram. The core evolution is still independent from that of the envelope.

ii) If $M_e/M_\odot$ becomes about 10^{-4}, the hydrogen-burning shell starts to cool, the luminosity drops rapidly and the star enters the white-dwarf regime.

Fig. 1 shows the variation of the envelope mass M_e with effective temperature for different PAGB models as given by Schönberner (1983). Note that in these evolutionary phases M_H practically equals the total stellar mass M because of the smallness of M_e ($M = M_e + M_H$). For a given M_H, a unique relation $T_{eff}(M_e)$ for the horizontal evolution from the AGB till the turn-around point at $T_{eff} > 10^5$ K exists. The shapes of the relations $T_{eff}(M_e)$ are similar, but M_e increases with decreasing M_H. Similar relations for a larger range of M_H are given in Paczynski (1971).

The timescale for the crossing of the HR diagram is determined by the total amount of the available fuel M_e and the fuel consumption $\dot{M}_e$ (we neglect in this chapter any mass loss). Let us define a "horizontal speed" by $\dot{T}_{eff}$, which can be written as follows:

$$\dot{T}_{eff} = (dT_{eff}/dM_e)\dot{M}_e = -\dot{M}_H(dT_{eff}/dM_e)$$

$$\dot{M}_c = -\dot{M}_H = -(L/E_H X_e) \; ,$$

where E_H (= $6 \cdot 10^{18}$ erg g^{-1}) is the energy released per gram of hydrogen, and X_e is the hydrogen abundance (mass function) in the envelope. L is practically constant since M_H does not grow appreciably during this relatively short evolutionary phase, and dT_{eff}/dM_e is determined by $T_{eff}(M_e)$ as shown in Fig. 1. The constant term $\dot{M}_H$ determines the general timescale of the envelope contraction, which increases with core mass M_H. The function $T_{eff}(M_e)$ determines the variation of the speed $\dot{T}_{eff}$ during the horizontal evolution:

i) for $T_{eff} \gtrsim 10^4$ K, $|dT_{eff}/dM_e|$ is large, which means a rather fast envelope contraction;

ii) for $T_{eff} \lesssim 10^{3.7}$ K, $|dT_{eff}/dM_e|$ is small, and the envelope contraction is relatively slow.

Table 1 illustrates the evolutionary speeds in different parts of the H-R diagram and for different core masses M_H. The Δt are given by $\Delta M_e/\dot{M}_H$, with $\dot{M}_H$ = const and $T_{eff}(M_e)$ taken from Schönberner (1983) and Paczynski (1970, 1971). For T_{eff} > 10000°K, the evolutionary speed increases very rapidly with mass, and for $T_{eff} \geqslant$ 30000°K, we find approximately $\Delta t \sim M_H^{-10}$. Please note that also the very fast envelope contraction of the 1.2 $M_\odot$ model is completely controlled by nuclear burning. In the cooler region of the H-R diagram, i.e., for T_{eff} < 10000°K, the evolutionary speed does not depend very much on M_H. Moreover, all models evolve extremely slowly in the vicinity of the AGB, as can be understood from the shape of the $T_{eff}(M_e)$-relation. The corresponding timescale exceeds the kinematical lifetime of planetary nebulae by a factor of ten! Thus we are left with the problem that a 0.6 $M_\odot$ PAGB model which burns hydrogen quietly evolves with an appropriate speed through the CPN region, but that its transition from the tip of the AGB to the CPN region is much too long.

Table 1: Transition times Δt in different parts of the H-R diagram for various AGB remnant masses

		$\Delta t/yr$ for $\Delta \log T_{eff}$			
$M_H/M_\odot$	$\dot{M}_H/M_\odot yr^{-1}$	3.5-3.7	3.7-4.0	4.0-4.5	4.5-5.0
1.2	6.0 10^{-7}	4 10^5	6000	3	0.7
0.8	2.4 10^{-7}	4 10^5	5300	100	120
0.6	9.0 10^{-8}	5 10^5	5300	1700	3500
0.57	6.0 10^{-8}	5 10^5	7000	6000	18000

3. INFLUENCE OF MASS LOSS

We shall now show that the problem of the last chapter does not really exist since one basic process has been completely omitted, namely mass loss. In fact, the very existence of nebular shells are a manifestation of some sort of severe mass loss: To create a nebular shell of, say, 0.2 $M_\odot$ within, say 2000 yr, a mass loss of 10^{-4} $M_\odot yr^{-1}$ is necessary (cf. also discussion of Renzini, 1981).

We can easily understand the influence of mass loss on the AGB and PAGB evolution within the framework of the last chapter. Mass loss from the surface by a stellar wind, $\dot{M}_W$, competes with nuclear processing at the base of the envelope, $\dot{M}_H$, and we have now

$$\dot{M}_e = -(\dot{M}_H + \dot{M}_W).$$

If we described the mass loss on the AGB by a Reimers'-type law, we have $\dot{M}_W \sim L^{1.7} M^{-1}$ (Reimers, 1975, Schönberner, 1979), whereas $\dot{M}_H \sim L$. Consequently, mass loss will dominate at large luminosities. Numerical estimates show that $\dot{M}_H \approx \dot{M}_W$ is already reached for $L \approx 10^3 L_\odot$ ($\eta = 1$). We conclude that mass loss reduces the envelope on the luminous AGB ($L > 10^3 L_\odot$) more effectively than nuclear burning. With other words, mass loss terminates the AGB evolution and reduces the maximum possible luminosity for a given initial mass.

The effect on the evolution is as follows: as long as the envelope is massive enough, the evolution along the AGB is not affected since it is determined by the core mass M_H. Of course, the envelope starts to shrink if M_e falls below a certain small value, as discussed in the previous chapter, and the evolutionary speed for the horizontal evolution (i.e., for the PAGB evolution) can now be expressed by

$$\dot{T}_{eff} = -(\dot{M}_H + \dot{M}_W) \, dT_{eff}/dM_e.$$

Mass loss speeds up the horizontal evolution and even determines this speed if $\dot{M}_W > \dot{M}_H$. Please note that in cases where mass loss is important, the evolutionary speed depends only on the actual value of $\dot{M}_W$. This implies that a hydrogen-burning PAGB model evolves independently of its earlier mass loss history, as long as its thermal equilibrium has not been destroyed. This fact allows the modelling of CPN evolution by means of hydrogen-burning PAGB stars without knowing the details of the mass loss phases which lead to the PN formation at the tip of the AGB, provided that the PN-formation time is small compared to the PN lifetime. Another fact also deserves attention: heavy mass loss cannot completely remove the hydrogen-rich envelope as long as hydrogen is still burning. Instead, the star shrinks more quickly, and the time Δt elapsed between two positions along the horizontal track is given by

$$\Delta t = \Delta M_e/(\dot{M}_H + \dot{M}_W) \approx \Delta M_e/\dot{M}_W \quad \text{for} \quad \dot{M}_W > \dot{M}_H.$$

We estimated above that the observations of young PN indicate mass loss rates not much less than $10^{-4} M_\odot \mathrm{yr}^{-1}$ lasting for some 10^3 yr. Such rates are about two orders of magnitudes larger than what one would expect from the application of Reimers' formula (for $\eta = 1$). Such an unusually strong wind was provisionally called a "superwind" by Renzini (1981). A mass loss of about $10^{-4} M_\odot \mathrm{yr}^{-1}$ leads to a somewhat 3 orders-of-magnitudes larger envelope depletion than nuclear burning (cf. Table 1), and the horizontal speed is increased accordingly. The transition from the tip of the AGB to 5000°K (log $T_{eff} = 3.7$) is shortened by the same amount to about 10^3 yr. Thus, we conclude that a very strong stellar wind ("superwind") with rates of the order of $10^{-4} M_\odot \mathrm{yr}^{-1}$ is necessary:

i) to build up a typical PN shell of about 0.1 ... 0.2 $M_\odot$ within some 10^3 yr;
ii) to force the envelope to shrink within the same time until the physical mechanisms that lead to the enhanced mass loss are shut off.

We can, however, safely assume that some sort of mass loss will persist beyond the shut-off point, and its order of magnitude may be estimated by Reimers' law. Such an estimation shows that $\dot{M}_W$ dominates over $\dot{M}_H$ as long as $T_{eff} \lesssim 10^4$ K, and the contraction is speeded up accordingly (cf. Fig. 1 in Schönberner 1983). In the CPN region, i.e., for $T_{eff} \gtrsim 30000°K$ (or log $T_{eff} \gtrsim 4.5$), we have $\dot{M}_W \lesssim 0.1\,\dot{M}_H$, and the evolution is only controlled by the nuclear term.

The occurrence of a "superwind" demands a physical explanation which is presently not available. Most likely, severe pulsations in combination with radiation pressure on grains are responsible. Of importance for the PAGB evolution is the question of when the "superwind" stops, and this instant can then be considered as the beginning of the PAGB phase of evolution. In Schönberner (1983) it has been assumed this happens between $10^{3.7}$ K to $10^{3.75}$ K.

We conclude from Fig. 1 that this is an essential assumption, otherwise too much envelope mass would remain and consequently the model would spend too much time near the AGB ("lazy" remnants according to Renzini, 1981; Iben and Renzini, 1983). Now the hydrodynamical calculations of Tuchman et al. (1979) show that the pulsational instability ceases only when all but $\approx 0.001\,M_\odot$ of the envelope has been ejected. A similar result was obtained earlier by Härm and Schwarzschild (1975). For the core masses involved ($\approx 0.6\,M_\odot$), this residual envelope mass corresponds roughly to $10^{3.7}$ K (Fig. 1). But, as can be also seen from Fig. 1, at this temperature a change in the behavior of PAGB models occurs: below, their effective temperature are only weakly dependent on envelope masses, whereas above this temperature the opposite is true. The location of this transition region does not change for somewhat higher (core) masses (Fig. 1). Therefore, we suggest that any mechanism which leads to enhanced mass loss remains in effect as long as no drastic changes in the envelope structure occur. Because mass loss "pushes" the star towards higher effective temperatures by reducing its envelope mass, it will then continue until the star reaches a configuration which leads to a rapid shrinking of the envelope and correspondingly to a disruption of the physical mechanism responsible for the "superwind". In this scenario there is no room for "lazy" remnants. Even CPN are loosing mass, observations with the IUE satellite revealed. According to the recent compilation of Cerruti-Sola and Perinotto (1985), the rates are generally smaller than $10^{-7}\,M_\odot yr^{-1}$ and hence contribute only little to the evolutionary rates (cf. $\dot{M}_H$ of Table 1). The Reimers mass loss formula, although in principal not valid for hot stars, gives just the same order-of-magnitude rates.

We argue that, although we do not know exactly the moment when a "superwind" stops, it will most likely not be before the envelope is reduced to $10^{-3}\,M_\odot$. Then any residual "normal" wind, together with hydrogen burning, will be efficient enough to provide a sufficiently fast transition such as to convert an AGB remnant into a CPN with an observable PN. We also state that PAGB stars of about 0.6 $M_\odot$ which burn hydrogen quiescently are very useful in modelling CPN without invoking any mass loss. The latter is, however, absolutely necessary to ensure that the formation of a nebular shell and the transition of the remnant into the CPN of the HR diagram occurs within a time which is short compared to the kinematical lifetime of the nebula.

Fig. 2 shows the evolutionary tracks of four hydrogen-burning post-AGB models according to the computations of Schönberner (1979, 1983), generated from Population I stars with initial masses varying from 0.8 to 1.4 $M_\odot$. The two lower-mass remnants were generated by applying a superwind-like mass-loss rate on the AGB, the remaining two are the result of a Reimers'-like mass-loss rate. The transition times are shorter than those given in Table 1 since mass loss accelerates the evolution below 10000°K (cf. discussion above). The horizontal evolution through the CPN region is highly mass sensitive: $\Delta t \sim M^{-10}$. The reason is the larger luminosity (core-mass luminosity relation) and the smaller available amount of fuel (see Fig. 1) if M is increased. A similar mass dependence holds for the luminosity drop when hydrogen burning extinguishes. This luminosity drop is obviously related to the number of thermal pulses which occurred earlier on the AGB: the 0.546 $M_\odot$ model is still below the threshold for the occurrence of thermal pulses, but the 0.565 $M_\odot$ model experienced 4, the 0.598 $M_\odot$ model 10 and the 0.644 $M_\odot$ model 24 thermal pulses on the AGB.

4. THERMAL PULSES AND MASS LOSS

I would like to close this review with a few remarks concerning the role of thermal pulses. It was originally thought that the short-lived luminosity peak during a thermal pulse triggers an (pulsational) instability which leads to the loss of the whole envelope (cf. Härm and Schwarzschild 1975). The computations show that the remaining envelope contracts very rapidly on a thermal timescale because the star is far out of thermal equilibrium. Later on, the evolution slows down when thermal equilibrium is gradually restored, but the whole evolution through the CPN region is about 3 times slower when compared to that of a hydrogen-burning model of the same mass. Several important facts should be noted:

i) The remnant envelope mass depends on the details how the mass is removed (i.e., on the mass loss rate and on its duration), and this envelope mass M_e remains constant during the following helium-burning phase, so that, contrary to the case of hydrogen burning, the location in the HR diagram is not uniquely related to M_e.

ii) Despite the fact that M_e can be smaller than for a corresponding hydrogen-burning remnant, it cannot be removed completely, as one might think at a first glance. It is, however, possible that a modest stellar wind removes this small hydrogen-rich envelope during the helium-burning phase before hydrogen is re-ignited, thereby opening a likely channel for the creation of helium white dwarfs (DB).

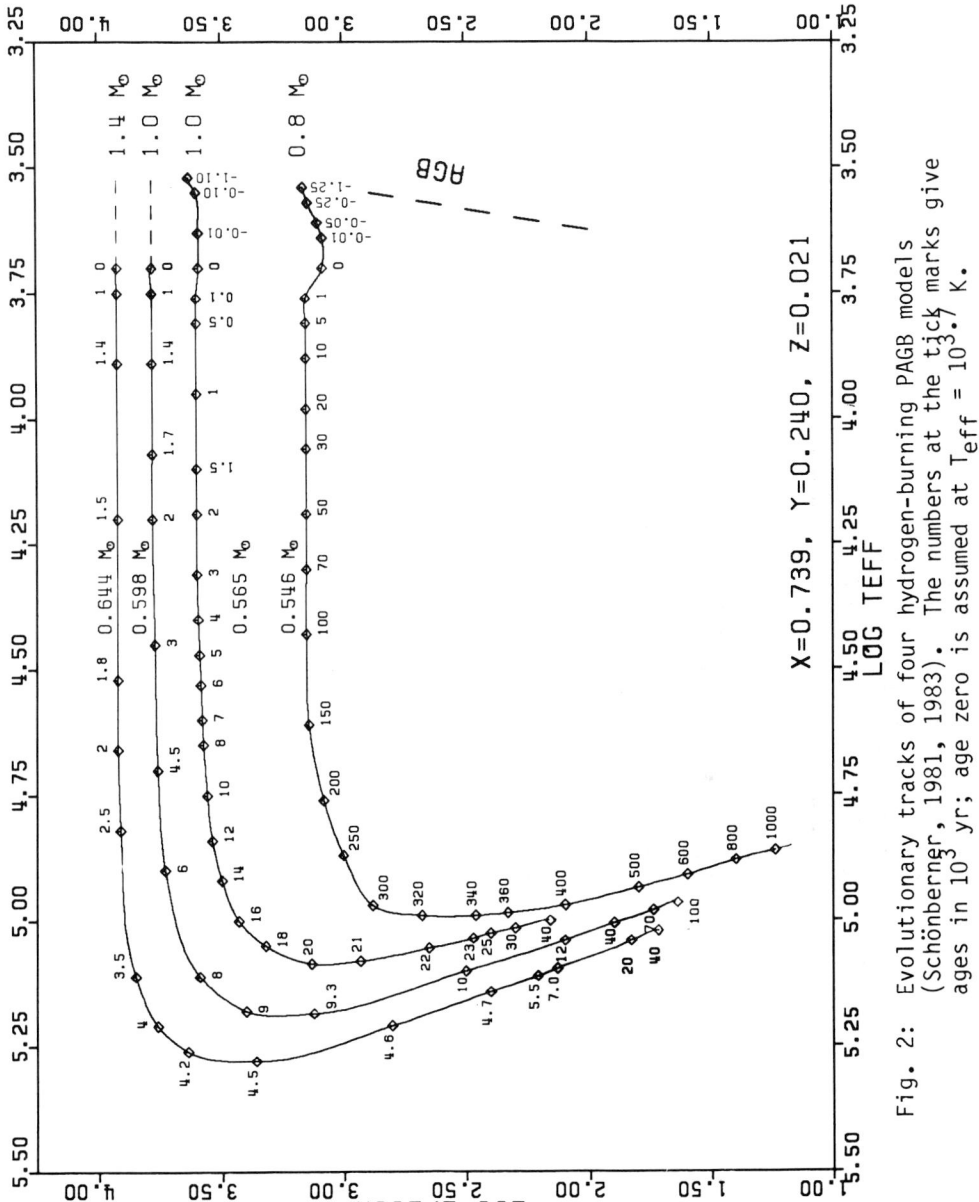

Fig. 2: Evolutionary tracks of four hydrogen-burning PAGB models (Schönberner, 1981, 1983). The numbers at the tick marks give ages in 10^3 yr; age zero is assumed at $T_{eff} = 10^{3.7}$ K.

Thermal pulses lead to a further complication. If the AGB remnant left the AGB shortly before the next thermal pulse, the latter may occur before the remnant becomes a white dwarf (Schönberner, 1979; Iben, 1984). Such a "late" thermal pulse will then convert the hydrogen-burning PAGB star into a "helium burner". For a thorough discussion of the properties of helium-burning PAGB models, the reader is referred to the paper by Iben (1984).

REFERENCES

Härm, R., Schwarzschild, M.: 1975, Astrophys. J. **200**, 324.
Iben, I. Jr.: 1984, Astrophys. J. **277**, 333.
Iben, J. Jr., Renzini, A.: 1983, Ann. Rev. Astron. Astrophys. **21**, 271.
Kippenhahn, R.: 1981, Astron. Astrophys. **102**, 293.
Kovetz, A., Harpaz, A.: 1981, Astron. Astrophys. **95**, 66.
Paczynski, B.: 1970, Acta Astron. **20**, 47.
Paczynski, B.: 1971, Acta Astron. **21**, 417.
Reimers, D.: 1975, "Problems in Stellar Atmospheres and Envelopes," B. Baschek, W. H. Kegel, G. Traving Eds. Springer, Berlin, p. 229.
Renzini, A.: 1981, "Physical Processes in Red Giants", I. Iben, Jr. and A. Renzini Eds., Reidel, Dordrecht, p. 431.
Schönberner, D.: 1979, Astron. Astrophys. **79**, 108.
Schönberner, D.: 1981, Astron. Astrophys. **103**, 119.
Schönberner, D.: 1983, Astrophys. J. **272**, 708.
Schönberner, D.: 1984, IAU Symp. No. 105 "Observational Tests of the Stellar Evolution Theory," A. Maeder and A. Renzini, Eds., Reidel, Dordrecht, p. 209.
Schönberner, D.: 1986, Astron. Astrophys. **169**, 189.
Schönberner, D., Weidemann, V: 1981, "Physical Processes in Red Giants", I. Iben, Jr. and A. Renzini, Eds., Reidel, Dordrecht, p. 463.
Schönberner, D., Weidemann, V.,: 1983, IAU Symp. No. 103, "Planetary Nebulae," R. D. Flower, Ed., D. Reidel, Dordrecht, p. 359.
Tuchman, Y., Sack, N., Barkat, Z.: 1979, Astrophys. J. **234**, 217.
Wood, P. R., Faulkner, D. J: 1986, Astrophys. J. **307**, 659.

COLLIMATED OUTFLOWS FROM YOUNG STARS

R. Mundt
Max-Planck-Institut für Astronomie
Königstuhl 17
D-6900 Heidelberg
Fed. Rep. of Germany

ABSTRACT. It is now generally accepted that energetic, and often bipolar, mass outflows are an important phase in early stellar evolution. Evidence for such outflows is provided by a variety of phenomena. The most important ones are:
(1) the high-velocity molecular gas (HVMG)
(2) Herbig-Haro (HH) objects and optical jets.

In the former case one studies the molecular gas components near young stars with significantly higher velocities than the turbulent velocities of typical molecular clouds ($\sim$ 1 km/s). In the vicinity of stars of low to moderate luminosity (1-100 $L_\odot$) the typical velocity of the HVMG is 5-20 km/s. The HVMG does not represent matter emanating directly from the associated stars, but is molecular cloud material which has been somehow accelerated by the stellar wind. It could for example be located in the shell of a wind-blown cavity. The mass loss rates derived from observations of the HVMG are rather high. Assuming momentum driven flows and wind velocities of about 200 km/s the estimated values range from 10^{-8} $M_\odot$/yr to several 10^{-6} $M_\odot$/yr. About 50% of the molecular outflows are clearly bipolar. However, their degree of collimation is small. Typical length-to-diameter ratios are only 2-3. The bipolar lobes extend over distances of about 0.1-1 pc.

Optical emission-like jets and HH objects associated with young stars trace flow components with a much higher degree of collimation and velocity than the HVMG. The measured radial velocities can reach values of up to 450 km/s. For some jets the opening angle is less than 1 degree and their length-to-diameter ratio can have values of up to 30. The length of these jets reaches from about 0.01 to 1 pc. For most known jets the "driving stars" are T Tauri stars or IR-sources of low to moderate luminosity (1-100 $L_\odot$). Typical velocities, particle densities, mass fluxes, and kinetic luminosities are estimated to be 200-400 km/s, 20-100 cm^{-3}, and 0.05-2x10^{-8} $M_\odot$/yr, 0.01-0.2 $L_\odot$, respectively. Estimated mass fluxes and velocities are consistent with our present knowledge of T Tauri star wind properties. However, estimated mass loss rates are 1 to 2 orders of magnitude lower than those derived from studies of the HVMG; therefore, the jets are probably not driving the HVMG (if it is momentum driven).

HH objects and optical jets are phenomena, which are intimately related. Both objects show the same type of emission line spectrum. These lines are probably formed behind radiative shock waves with shock velocities of 40-100 km/s. Furthermore, several long known HH objects form the brightest knots (hot spots) in these jets. There are a variety of mechanisms, which can in principal excite internal shock waves in these jets. Examples of likely excitation mechanisms are fluid dynamical instabilities or pressure gradients in the ambient medium. These internal shock waves have to be relatively oblique, since the estimated shock velocities ($\sim$ 50 km/s) are much lower than the typical flow velocities.

It is proposed that many HH objects represent the locations of the most strongly radiating shock waves within jets. In some cases they are generated by these jets as they plough into the ambient medium (e.g. through the bow shock). This latter model is supported by the observations which strongly suggest that terminal HH objects trace the working surface of the jet. Not only is this idea consistent with the observed velocitiy field in or near these objects, but also with the typical densities near the edges of their associated molecular clouds, and with the jet parameter given above.

For a more detailed discussion of these jets and the HVMG associated with young stars the reader is refered to the review articles listed below. The two images on the next pages are of the HH34 region. The knotty jet pointing towards HH34S is a striking example of a jet emanating from a young stellar object.

References

Lada, C.J., 1985, Ann. Rev. Astron. Astrophys. 23, 267.
Mundt, R., 1985a in "Protostars and Planets II", eds. D. Black, and M. Matthews, University of Arizona Press, Tucson, p. 414.
Mundt, R. 1985b, in "Nearby Molecular Clouds", ed. G. Serra Lecture Notes in Physics, 217, p. 160, Springer Verlag, Heidelberg.
Mundt, R., 1986, in "Jets from Stars and Galaxies", Can. J. Phys., 64, 407.
Mundt, R., 1987, in "Circumstellar Matter", eds. I. Appenzeller and C. Jordan (Reidel, Dordrecht), p. 147.
Mundt, R., Brugel, E.W., Bührke, T., 1987, Ap.J. 319, 275.
Schwartz, R.D., 1983, Ann. Rev. Astron. Astrophys. 21, 209.
Shu, F.H. Adams, F.C., Lizano, S., 1987, Ann. Rev. Astron. Astrophys., 25.

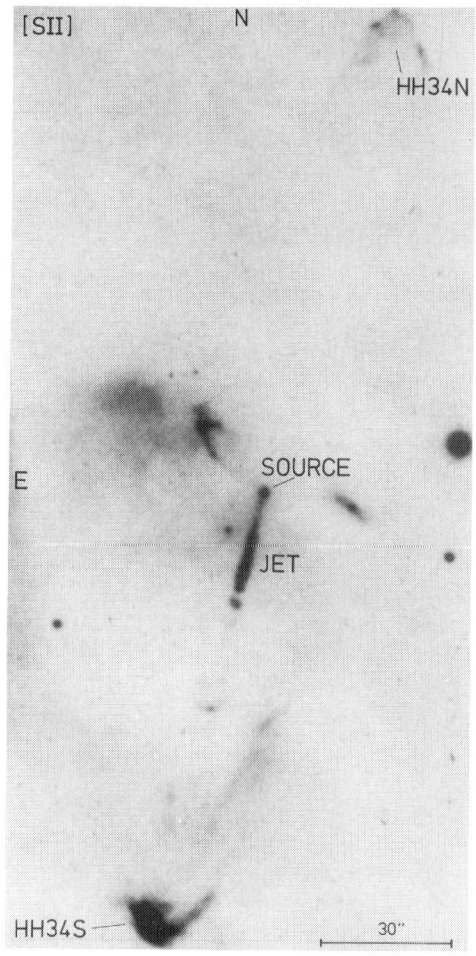

Figure 1: [SII] $\lambda\lambda$ 6716, 6731 image of the HH34 region. The source is located in the centre. A 30 arcsec long jet is pointing from the source towards HH34S. Both, this HH object and the one on the opposite side of the source (HH34N) have the shape of a bow. HH34S and the jet are blueshifted (v_{rad} = -80 - -130 km/s), while HH34N is redshifted (v_{rad} = +150 - +200 km/s). HH34S and HH34N are therefore interpreted as the bow shocks resulting from two oppositely directed jets, which are propagating with high speeds (100-200 km/s) through the ambient medium (for details see Bührke, Mundt and Ray 1987, Astr. Astrophys., submitted). This image is a montage of two CCD frames taken at the prime focus of the 3.5 m telescope on Calar Alto, Spain.

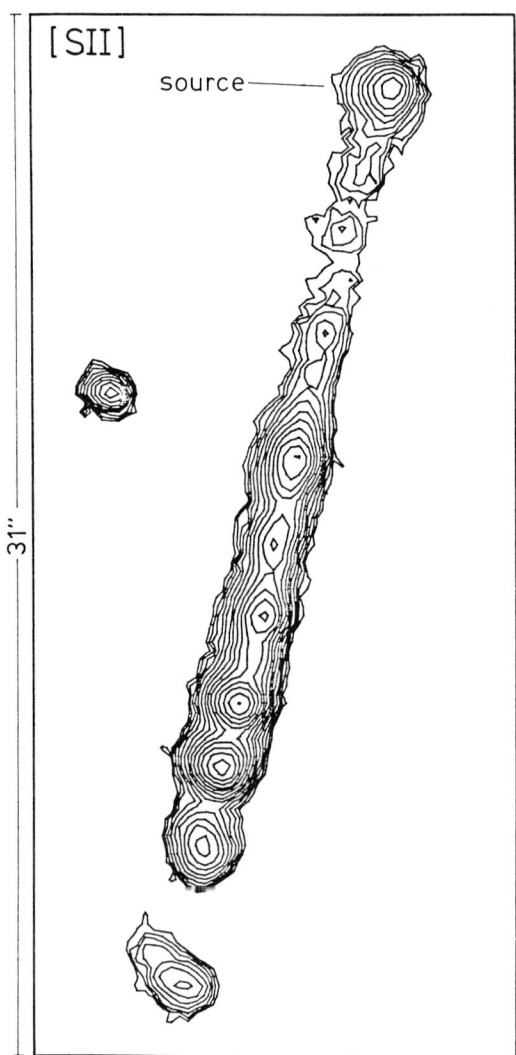

Figure 2: Isointensity contour plot of a [SII] λλ 6716, 6731 CCD frame of the jet associated with HH34. Twelve knots are visible in the section shown with a typical separation of 2 arcsec. This image was also taken at the prime focus of the 3.5 m telescope on Calar Alto (exposure time 1h, seeing 0.8 arcsec).

OBSERVATIONAL EVIDENCE FOR OUTFLOW IN ACTIVE GALACTIC NUCLEI AND IN X-RAY BINARIES

R.T. Schilizzi
Netherlands Foundation for Radio Astronomy
P.O. Box 2, 7990 AA Dwingeloo, The Netherlands

ABSTRACT. Proper motion measurements in superluminal sources and in SS433 and Cyg X-3 provide the strongest evidence for outflow in AGN's and X-ray binaries. Indirect evidence for outflow in AGN's comes from the collimated jet-like structures which are a common feature of extragalactic radio sources. The broad absorption lines in some quasars and possibly the asymmetric narrow emission line profiles in Seyfert and radio galaxies are further indirect evidence of outflow.

1. INTRODUCTION

Mass or energy flow is fundamental to all models of Active Galactic Nuclei and X-ray binaries. In AGN's the enormous luminosity of an active nucleus must create a radiation pressure which exerts an outward force on material in the vicinity of the nucleus. In the case of X-ray binaries the heating up of gas flowing from the companion star to the compact object is held to be responsible for the X-ray emission. In both AGN's and X-ray binaries, accretion disks postulated to exist around a massive black hole in the nucleus (AGN), or a black hole or neutron star (X-ray binary), provide a natural means of channelling mass and energy outwards along the axis of the disk, producing a collimated flow. Direct observational evidence of outflow is in fact to be found only in the measurement of radio proper motions in some twenty superluminal sources and in two X-ray binaries. Indirect evidence can be found in optical line profiles in AGN's and in the morphologies of jet-like features in the radio, optical and X-ray.
    In the following sections, I will briefly review the observational evidence for outflow in both classes of object.

2. ACTIVE GALACTIC NUCLEI

2.1. Superluminal motion in compact radio sources

In more than twenty radio sources with compact core components, VLBI observations of the cores show features which separate from each other at angular rates of the order of 0.5 milliarcsec/year (Porcas, 1987). At the cosmological distances of the objects, these proper motions correspond to speeds several times that of light (Pearson and Zensus, 1987).

The most widely accepted model for superluminal motion assumes that the relativistic electron plasma responsible for the radiation at radio wavelengths is itself moving with a bulk speed close to that of light in a direction towards the observer. This has the effect of apparently compressing the timescale of emission and creating the illusion of "faster than light" motion.

For example, in 3C345, one of the best studied superluminal sources, component C2 (Figure 1) moves with an apparent speed of 10.8c (for H_o =100 kms^{-1}Mpc^{-1}, q_o =0.05), component C3 at 6.8c, while component C4 has been observed at 22 GHz to accelerate from 1.6c to 6.8c (Biretta,

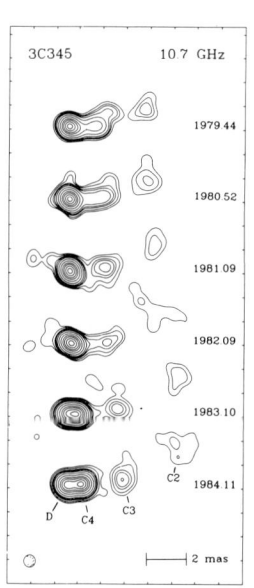
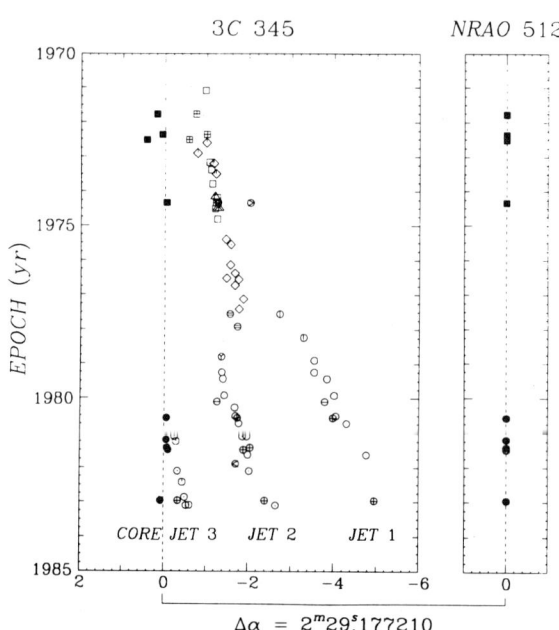

Figure 1 (above left). Maps of 3C345 at 10.7 GHz (from Biretta and Cohen 1987). North is at the top; tick marks are at spacings of 1.2 mas. The FWHM of the restoring beam is 0.6 mas (hatched circle)

Figure 2 (above right). Proper motion of components in 3C345 (from Barthel et al 1986). Jet 1, jet 2, and jet 3 refer to components C2, C3 and C4 in Figure 1.

Moore and Cohen 1986). On the relativistic bulk motion model the bulk relativistic Lorentz factor γ must be ≥ 7, and the angle to the line of sight $\leq 5°$. Assuming a jet consisting of electrons and positrons, the radiating material would accelerate as the external pressure decreases away from the core, so that near the core (e.g. C4) γ is probably ~3, but accelerates to γ~10 further out.

The morphology of superluminal sources can be described in terms of an unresolved core at one end of a "jet" which invariably is composed of individual blobs with, at present levels of sensitivity, no sign of any underlying smooth emission. In one case, 3C345 again, astrometric VLBI measurements by Bartel et al (1986) (Figure 2) have demonstrated that the core is stationary with respect to the inertial frame of reference (represented by the quasar NRAO512) and that the blobs in the jet move away from the core.

Superluminal motion has been detected in radio galaxies (e.g. 3C120), blazars (e.g. BLLac), and in core-dominated quasars (e.g. 3C345) as well as lobe-dominated quasars (e.g. 4C34.47). Descriptions of their properties and of arguments for and against the relativistic bulk motion model can be found in the Proceedings of a Workshop on Superluminal Sources (Eds. J.A. Zensus and T.J. Pearson). For completeness, it should be mentioned that subluminal motion (at 0.19c) has been measured in the nucleus of 3C84 (NGC1275) by Romney et al 1984; this may represent a typical value for an unboosted velocity. In addition there are upper limits of less than c for two nearby galaxies, M87 and NGC6251, and for two quasars (Porcas 1987).

2.2. Jet morphology

The many examples of long thin collimated structures in radio sources (and also in optical and X-ray) provide clear but indirect evidence of outflow in AGN's. This "jet" radio emission is generally thought to arise from dissipation in the energy transfer from the nucleus to the extended lobes, presumably from interaction with the surrounding medium. Examples of radio jets are shown in Figure 3 to 5.

Figure 3 shows the remarkable VLA image of Cygnus A (Perley, Dreher and Cowan 1984), in which one can discern a faint jet directed from the nuclear radio component towards and into the north-west lobe. Good quality reproductions of this image also show what looks to be a faint counter-jet with about one-quarter the surface brightness per beam area in the direction of the south-east lobe. The discovery of jet-like features leading to extended radio lobes well outside the optical confines of the associated parent object in Cygnus A and many other sources like it, leads one naturally to consider outflow of material from the nucleus, even though no proper motions have ever been measured in large scale radio jets. Estimates of velocities of material in the large scale jets vary widely from a few hundred kms^{-1} (e.g. Cornwell and Perley, 1984; Rees, 1982) to relativistic values (e.g. Scheuer, 1987). Detailed reviews of the properties of radio jets can be found in the Proceedings of a Workshop on the Physics of Energy Transport in Extragalactic Radio Sources (Eds. A.H. Bridle and J.A. Eilek) and in Bridle (1986).

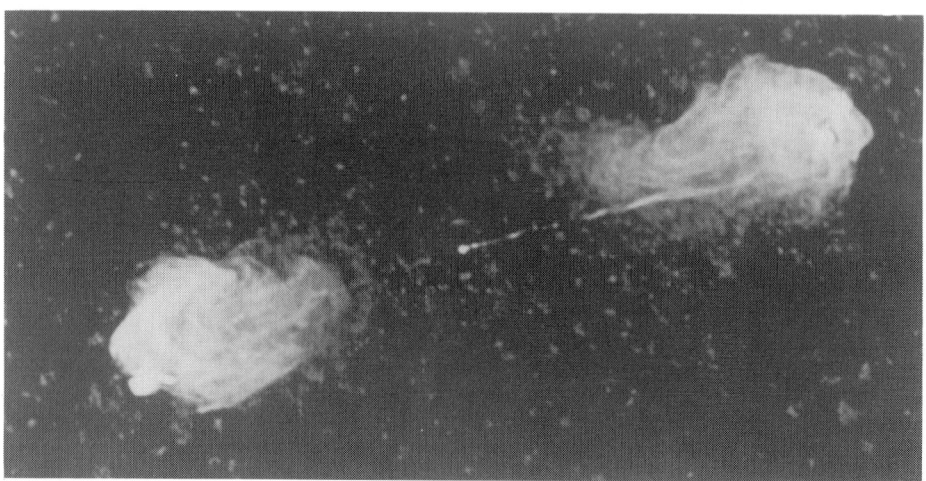

Figure 3. Cygnus A at 5 GHz (from Perley, Dreher and Cowan 1984)

Not all jets are strongly one-sided. In Figure 4, I show a composite of different scale radio structure in 3C236, the largest known radio source in the universe. The top panel shows the large scale structure which is 3.9 Mpc across (Barthel et al 1985), the middle and bottom panels the 2.2 kpc radio structure in the nucleus of the galaxy itself (Schilizzi et al 1987). Careful examination of the bottom panel shows jet emission either side of the nucleus visible at the limit of angular resolution in this image (~15pc). This jet emission is well-aligned with the radio axis defined by the outer lobes. This is a common feature of extended sources with compact cores, and is usually interpreted as meaning that the directional memory of the energy transport process is long ($\geq 10^8$ years) and that it operates over many orders of magnitude in distance ($\leq$10pc to 2Mpc).

This continuity of jet direction can also be seen in Figure 5 where the results on the radio galaxy 3C120 by Walker et al (1987) show a continuous link between the superluminal features and the large scale radio jet. The curvature from the small to the large scale can be explained as a magnification by projection of a small intrinsic curvature if the source axis is oriented close to the line of sight ($\leq$15°).

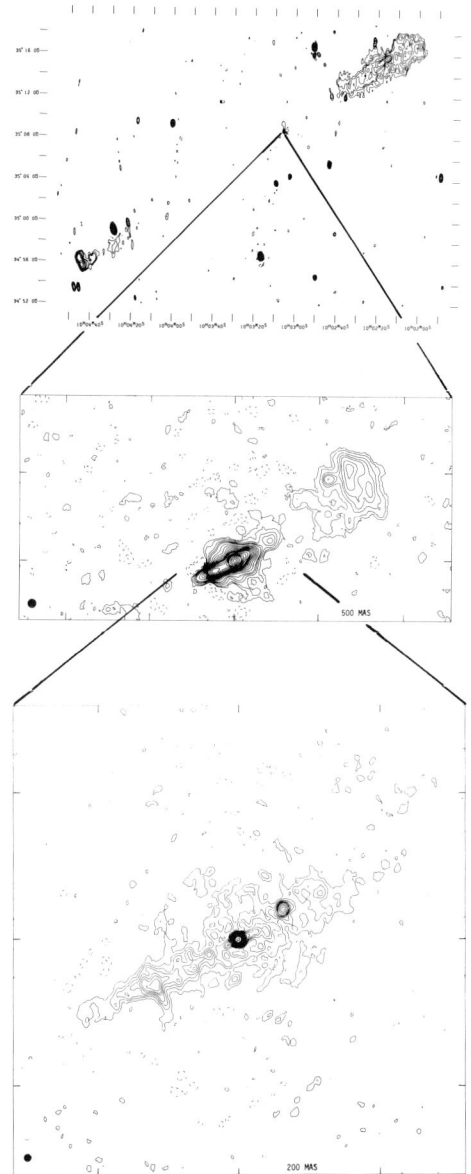

Figure 4. The radio structure of 3C326 on scales from 15 pc to 3.9Mpc. Top panel: WSRT 21 cm map from Barthel et al (1985); restoring beam 13x23 arcsec. Middle panel: VLBI + MERLIN 18 cm map from Schilizzi et al (1987); restoring beam 50 milliarcsec. The linear extent of the nuclear structure is 2.2 kpc. Bottom panel: VLBI + MERLIN 18 cm map from Schilizzi et al (1987) showing a two-sided jet close to the nucleus; restoring beam 10 milliarcsec.

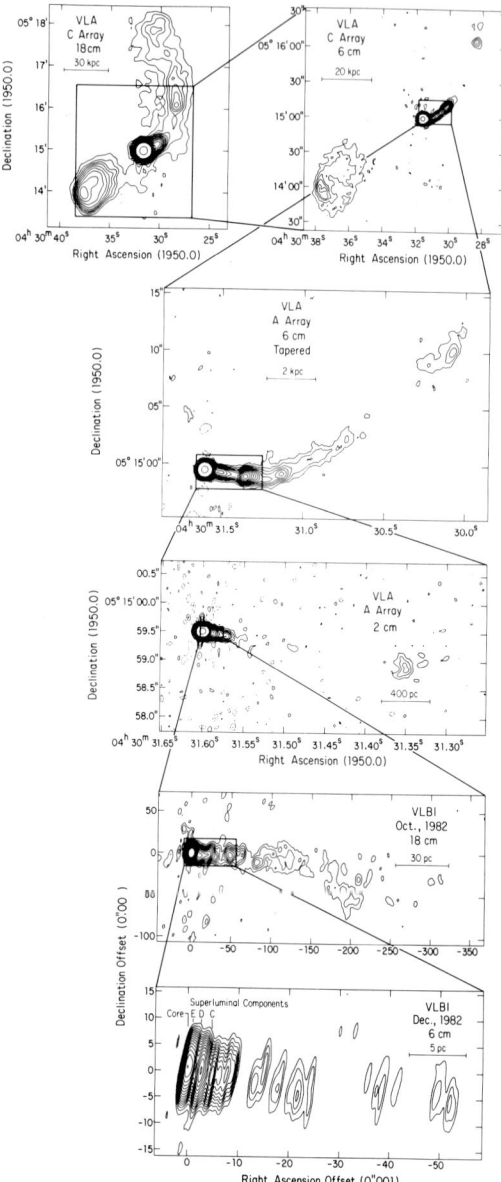

Figure 5. The radio structure of 3C120 on scales from 0.5 pc to >100kpc (from Walker, Benson and Unwin 1984).

2.3. Optical evidence for outflow in AGN's

2.3.1. Broad absorption line quasars

In 5 to 10% of quasars, broad blue shifted absorption lines are found accompanying the strong emission lines (see Figure 6). BAL quasars are defined by Weymann et al (1981) as having (i) broad absorption lines in species other than Lyα, typically CIV, (ii) absorption line widths >2000 kms^{-1}, and (iii) line centres displaced at least 5000 kms^{-1} from the emission line centre. The absorption outflow velocity ranges up to 2 to 3×10^4 kms^{-1}. The BAL gas appears to have higher ionisation, lower density, higher velocity and to be further away from the nucleus than the broad emission line gas (Weymann et al 1981, Turnshek 1986). Various considerations suggest that the BAL region is between 1 pc and 1 kpc from the continuum source in the nucleus (Turnshek 1986).

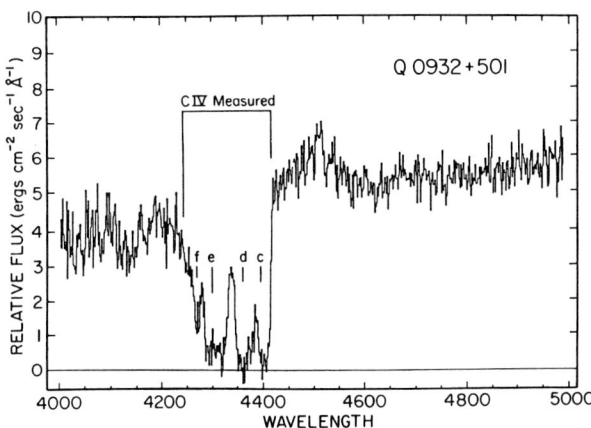

Figure 6. A spectrum of Q0932+501 showing a broad absorption line (from Turnshek et al 1984).

2.3.2. Narrow emission line asymmetries

Spectroscopic studies by Heckman et al (1981), Miley and Heckman (1982) and Whittle (1985 a, b) of forbidden optical emission lines, in particular, [OIII], in Seyferts and radio galaxies show that (i) the majority of the profiles are asymmetric with a sharper fall-off to the red than to the blue. The asymmetry is more pronounced towards the base of the profile. (ii) The degree of asymmetry is correlated with the Hα/Hβ, ratio indicating that dust is a significant constituent of the narrow line region.

These two observations imply radial flow and extinction, but do not unambiguously define the direction of the flow. If the flow is outward, and there is dust distributed within the AGN, the dust will, on average, absorb the radiation from the far side more because of the larger path length and greater optical depth. Alternatively, one can postulate that each cloudlet carries a component of dust with it, so that only the side facing the continuum source will be photoionised and radiate. This would enhance emission from cloudlets in the far hemisphere of the narrow line region which would have to be in infall to account for the profile asymmetry.

3. X-RAY BINARIES

3.1. Cygnus X-3

Cyg X-3 is a well-known variable in the γ, X, infrared and radio regimes. The emission in the γ, X and IR is modulated with a $4^h.8$ period (Lloyd-Evans et al 1983; Brinkman et al, 1972; Becklin et al 1973) that is thought to originate in the orbital motion of the binary. In the radio, Molnar, Reid and Gridlay (1984, 1985) have possibly found a $4^h.95$ period in the quiescent emission. Cyg X-3 also shows spectacular outbursts in the radio which last about a week and occur once or twice a year (Figure 7). Each outburst consists of a small number of flares which rise in a few hours and decay typically in two to three days, reaching the quiescent level after about two weeks. Outbursts are also seen in the IR (Mason, Cordova and White 1986).

The October 1983 radio outburst was followed by the VLA and MERLIN at 5 GHz for several weeks (Spencer and Johnston, 1986; Spencer et al 1986; Johnston et al 1986). The MERLIN observations can be interpreted as indicating that Cyg X-3 expanded at ~12 milliarcsec/day for ~2 weeks after one of the early flares in the outburst, before slowing down when it reached an angular size of ~80 milliarcsec. At the distance generally assumed for Cyg X-3, 10kpc (it is heavily obscured in the optical), the observed proper motion corresponds to a velocity of 0.35c if the two components expand away from a central point (see Figure 9). Figure 8 shows the change in angular separation of radio components in Cyg X-3 as a function of observing date, and Figure 9 shows a contour representation of the source on 13 October 1983.

Spencer and Johnston (1986) point out that the total stored energy in relativistic electrons and magnetic fields in the radio components is $>10^{43}$ ergs which implies a generation rate of $>10^{39}$ ergs/sec during the flare. Possible sources of this energy could be a decrease in the angular momentum of the neutron star, or a favourable increase in accretion rate from the secondary.

3.2 SS433

SS433 has been exceedingly well studied in all wavelength bands since it first became popular in 1978 (e.g. Margon 1984). There is a substantial body of evidence that outflow occurs in SS433, primarily from VLA and VLBI measurements. Interpretation of the moving optical lines in terms of the kinematic model (Abell and Margon 1979) already was strong indirect evidence for outflow. But it was the VLA observations by Hjellming and Johnston (1981a,b) showing outward motion of features on arcsec scales in the polarised maps of SS433 that put the

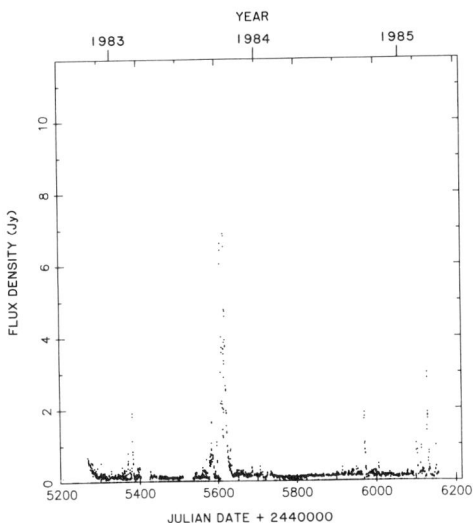

Figure 7. The 11.1 cm flux density of Cyg X-3 during the period October 1982 to March 1985 (from Johnston et al 1986).

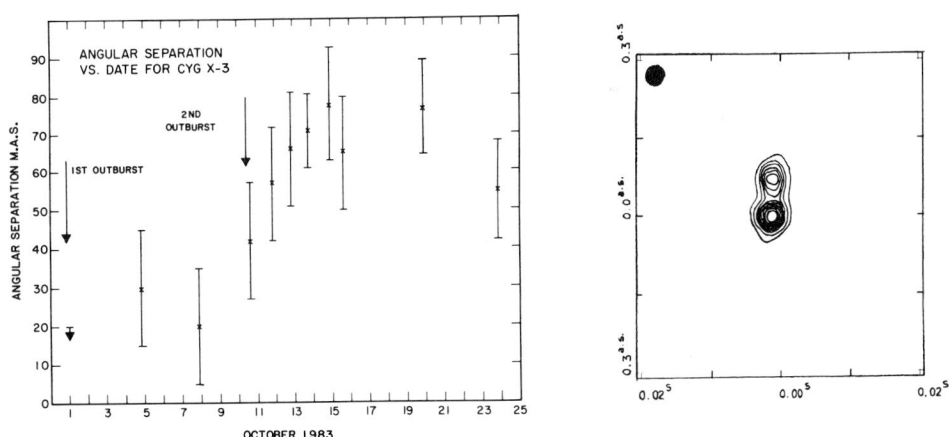

Figure 8 (above left). Angular separation versus date for Cyg X-3 during the October 1983 outburst (from Spencer et al 1986).

Figure 9 (above right). Map of Cyg X-3 at 6 cm made on 13 October 1983, 13 days after a major flare (from Spencer and Johnston 1986).

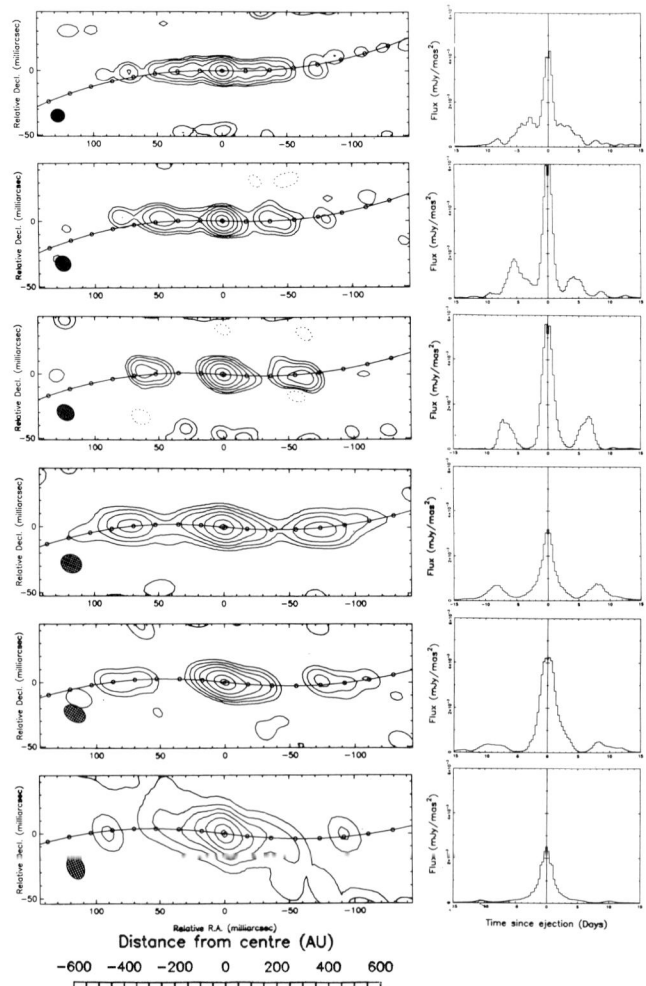

Figure 10a. VLBI maps of SS433 made at intervals of 2 days from 17 May 1985. The ellipse at the lower left indicates the FWHM of the restoring beam. The curve in each map represents the locus of ejecta as predicted by the kinematic model parameters of Margon (1984); open circles mark intervals of two days along the curve. The linear scale in Astronomical Units is also given.

Figure 10b. Crosscuts through the map of Figure 10a along the kinematic model curve shown. The horizontal axis is labelled in time unit (days) since the assumed ejection. All crosscuts have the same flux density scale.

issue of outflow beyond doubt. These observations also resolved the angle ambiguity of the kinematic model as well as establishing the distance to SS433 as being between 5.0 and 5.5 kpc.

Recent VLBI measurements by Vermeulen et al (1987) graphically demonstrate outflow on angular scales 10 to 100 times smaller than available to the VLA. Figure 10a is a sequence of six contour maps of SS433 taken at two day intervals. The maps contain a strong and variable unresolved (<10 milliarcsec) core (see Figure 10b) which is taken to coincide with the stellar source. Connecting the brightest features either side of the core from frame to frame leads to the surprising result that the proper motion is apparently at most three quarters of the speed predicted from the kinematic model (~9 milliarcsec/day or 0.26c). This can however be reconciled to the kinematic model by decomposing the bright features into their component blobs; it is then possible to identify blobs which move through the frames in a way consistent with the kinematic model.

Another interesting feature of these data is the discovery of a brightening zone at about 50 milliarcsec (4×10^{15} cm) from the stellar system. Blobs passing through this zone first brighten and then decay in intensity. To the east this is most obvious between maps 1 and 2, and to the west between maps 2 and 3. In keeping with the evident ejection of gas in discrete clouds, it is possible to explain the brightening as occuring when a new blob overtakes the bowshock of its predecessor. Since the ejection angle precesses, such overtaking manoevres will occur even if the blobs move at the same speed. The brightening itself could be caused by magnetic field compression, density enhancement or in-situ particle acceleration through turbulence.

REFERENCES

Abell, G.O., Margon, B. (1979) Nature 279, 701.
Bartel, N., Herring, T.A., Ratner, M.I., Shapiro, I.I., Corey, B.E. (1986) Nature 319, 733.
Barthel, P.D., Schilizzi, R.T., Miley, G.K., Jägers, W.J., Strom R.G. (1985) Astron. Astrophys. 148, 243.
Becklin, E.E., Neugebauer, G., Hawkins, F.J., Mason, K.O., Sanford, P.W., Matthews, K., Wynn-Williams, C.G. (1973) Nature 245, 302.
Biretta, J.A., Moore, R.L., Cohen, M.H. (1986) Ap.J. 308, 93.
Biretta, J.A., Cohen, M.H. (1987) in Superluminal Sources (Eds. J.A. Zensus, T.J. Pearson, CUP, Cambridge), p. 40.
Bridle, A.H. (1986) Can. J. Phys. 64, 353.
Brinkman, A., Parsignault, D., Giacconi, R., Gursky, H., Kellogg, E., Schreier, E., Tananbaum, H, (1972) IAU Circular No. 2446.
Cornwell, T.J., Perley, R.A. (1984) Proc. NRAO Workshop on Energy Transport in Extragalactic Radio Sources (Eds. A.H. Bridle, J.A. Eilek), p. 39.
Heckman, T.M., Miley, G.K., van Breugel, W.J.M., Butcher, H.R. (1981) Ap.J. 247, 403.
Hjellming, R.M., Johnston, K.J. (1981a) Nature 290, 100.
Hjellming, R.M., Johnston, K.J. (1981b) Ap.J. 246, L141.

Johnston, K.J., Spencer, J.H., Simon, R.S., Waltman, E.B., Pooley, G.G., Spencer, R.E., Swinney, R.W., Angerhofer, P.E., Florkowski, D.R., Josties, F.J., McCarthy, D.D. Matsakis, D.N., Reese, D.E., Hjellming, R.M. (1986) Ap.J. 309, 707.
Lloyd-Evans, J., Coy, R.N., Lambert, A., Lapikens, J., Patel, M., Reid, R.J.O., Watson, A.A. (1983) Nature 305, 784.
Margon, B., (1984) Ann. Rev. Astron. Astrophys. 22, 507.
Mason, K.O., Cordova, F.A., White, N.E. (1986) Ap. J. 309, 700.
Miley, G.K., Heckman, T.M. (1982) Astron. Astrophys. 106, 163.
Molnar, L.A., Reid, M.J., Grindlay, J.E. (1984) Nature 310, 662.
Molnar, L.A., Reid, M.J., Grindlay, J.E. (1985) in Radio Stars, (Eds. R.M. Hjellming, D.M. Gibson, Reidel Dordrecht) p. 329.
Pearson, T.J., Zensus, J.A. (1987) in Superluminal Sources (Eds. J.A. Zensus, T.J. Pearson, CUP, Cambridge), p. 1.
Perley, R.A., Dreher, J.W., Cowan, J. (1984) Ap.J. 285, L35.
Porcas, R.W. (1987) in Superluminal Sources (Eds. J.A. Zensus, T.J. Pearson, CUP, Cambridge), p. 12.
Rees, M.J. (1982) in IAU Symposium 97, Extragalactic Radio Sources (Eds. D.S. Heeschen, C.M. Wade, Reidel Dordrecht), p. 211.
Romney, J.D., Alef, W., Pauliny-Toth, I.I.K., Preuss, E., Kellermann, K.I. (1984) in IAU Symposium 110, VLBI and Compact Radio Sources (Eds. R. Fanti, K.I. Kellermann, G. Setti, Reidel Dordrecht), p. 137.
Schilizzi, R.T., Skillman, E.D., Miley, G.K., Barthel, P.D., Benson, J.M., Muxlow, T.W.B. (1987) IAU Symposium 129, The Impact of VLBI on Astrophysics and Geophysics (Eds. M.J. Reid, J.M. Moran), in the press.
Spencer, R.E., Johnston, K.J. (1986) in RS Ophiuchi (1985) and the Recurrent Nova Phenomenon, (Ed. M.F. Bode, VNU Science Press, Utrecht), p. 215.
Spencer, R.E., Swinney, R.W., Johnston, K.J., Hjellming, R.M., (1986) Ap.J. 309, 694.
Turnshek, D.A., Weymann, R.J., Carswell, R.F., Smith, M.G. (1984) Ap.J. 277, 51.
Turnshek, D.A. (1985) in IAU Symposium 119, Quasars (Eds. G. Swarup and V.K. Kapahi, Reidel Dordrecht), p. 317.
Vermeulen, R.C., Schilizzi, R.T., Icke, V., Fejes, I., Spencer, R.E. (1987) Nature 328, 309.
Walker R.C. (1984) Proc. NRAO Workshop on Energy Transport in Extragalactic Radio Sources (Eds. A.H. Bridle, J.A. Eilek), p. 20.
Walker, R.C., Benson, J.M., Unwin, S.C. (1987) Ap.J. 316, 546.
Weymann, R.J., Carswell, R.F., Smith, M.G. (1981) Ann. Rev. Astr. Astrophys. 19, 41.
Whittle, M. (1985a) M.N.R.A.S. 213, 1.
Whittle, M. (1985b) M.N.R.A.S. 213, 33.

MECHANISMS FOR OUTFLOW IN AGNs

Martin Rees
Institute of Astronomy
Madingley Road
Cambridge CB3 0HA
England

ABSTRACT. In active galactic nuclei (AGNs), radiation pressure is important for the dynamics of the line-emitting clouds. In SS433, the steady speed of the jets (~ 0.26 c) suggests that radiative acceleration may be responsible; but there are difficulties with this interpretation, and it is perhaps unlikely that analogous effects occur in AGNs. In the pair-dominated plasmas expected in AGNs, radiation pressure is competitive with gravity even for luminosities below the normal Eddington limit. For realistic geometries, radiation-driven pair-dominated winds may arise. The most dramatic outflow from AGNs, however, is manifested by the relativistic jets observed predominantly in the radio band. Although the plasma in these jets may be predominantly electron-positron pairs, the outflow is probably driven electromagnetically rather than radiatively: most of the energy may initially be Poynting flux, which is converted into fast particles in the "blobs" whose apparently superluminal motion is revealed by VLBI. Finally, the possibility is raised that sporadic outflow may result from tidal disruption of stars near a massive black hole: such a process may have observable consequences in our Galactic Centre.

1. RADIATION PRESSURE EFFECTS RELATED TO THERMAL EMISSION IN AGNs

A generic feature of AGNs and related objects is the dominance of radiation pressure. From the level of ionization implied by the broad emission lines, one can directly infer the ratio of radiation pressure to gas pressure. Among the well-known consequences of this is the possibility that radiation pressure can accelerate photoionized clouds away from the central source of UV continuum. The actual kinematics are still uncertain: we cannot yet reliably discriminate between inflow, outflow and orbiting models for the clouds, though studies of how the line profiles vary in response to changes in the central continuum can in principle decide this issue.

The radiation energy density in the broad-line region is well below that of a black body at 10^4 K. The clouds, with electron temperature $\gtrsim 10^4$ K, are therefore exposed to <u>dilute</u> radiation — were this not so, approximate LTE would prevail, and the emission lines would not stand out above the continuum intensity. But there may be 3 or 4 powers of ten difference in scale between the emission line region and the central object, so clouds could exist at a range of smaller radii r around a central continuum source. If the density varied as r^{-2}, the ionization parameter would be independent of r; however at higher densities the usual HII region assumptions progressively break down. When the densities are sufficiently high, each cloud behaves like a black body (or, more precisely, like a segment of an irradiated stellar atmosphere). The often-observed UV bump, indicating a thermal black body component, is conventionally attributed to an accretion disc. However, all we really know is that there is a surface area radiating thermally — it could be a lot of small clouds rather than a single surface.

It is not obvious that one can exclude the possibility that clouds exist (with a similar covering factor) at all logarithmic intervals of radius. Theorists who have offered reasons why emission line clouds may form at a special radius $r \simeq 1$ pc (e.g. because a wind from the central object is stopped by a shock at this distance, or because this is where the form of the potential changes from 1/r to a shallower r-dependence determined by the stellar distribution) may be trying harder than they need.

The surface radiation from optically thick clouds and the inner parts of accretion flows (and of course from supermassive stars) would resemble that from ordinary massive O or B stars. There are just two possible differences:

(i) The dominance of electron scattering opacity is even greater, with the result that there may be an even greater disparity between bolometric and colour temperatures.

(ii) The surface gravity is not enormously larger than for ordinary stars, but the specific gravitational binding energy *is* very large, and so therefore is the escape velocity V_{esc}. This means that, if fields were to grow via differential rotation to a strength anywhere near equipartition, the Alfven speed ($\sim V_{esc}$) would be high enough for magnetic fields to induce much stronger flare-like activity than in ordinary stars.

2. THE PRIMARY POWER SUPPLY

There are two quite distinct ways in which massive black holes can generate a high luminosity: straightforwardly by accretion, or via an electromagnetic process, where the power comes from the hole itself (see section 5 below). The latter process tends to give purely non-thermal phenomena, whereas accretion yields an uncertain mixture of thermal and non-thermal power. The properties of an AGN must depend, among other things, on the relative contributions of these two mechanisms.

There are a variety of channels for the power output; the relative importance of thermal and non-thermal processes depends on uncertain physics, but the most purely non-thermal phenomena may be preferentially associated with "starved" holes where the accretion rate is far below the "critical" value and electromagnetic extraction dominates. Figure 1, which summarises these, refers to reprocessing that happens close to the hole. Further reprocessing may occur on all scales from 10 up to as much as 10^{10} gravitational radii, in a manner sensitive to galactic environment.

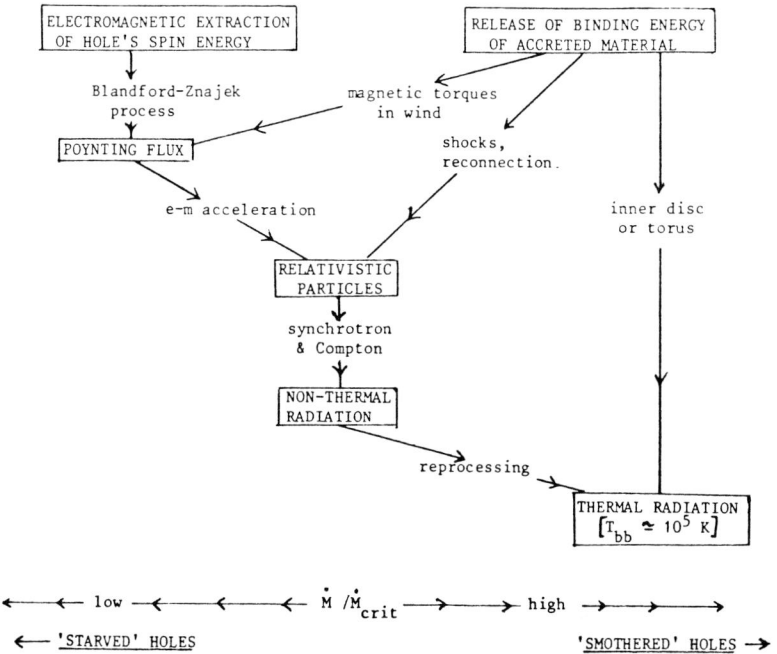

Figure 1. Radiation processes in the central core of an AGN. The relative importance of non-thermal processes is likely to decrease as $\dot{M}/\dot{M}_{crit}$ rises. Further reprocessing occurs on all scales out to $\sim 10^{10}\ r_{schw}$.

Three modes of radiation need to be considered in the central regions of AGNs; depending essentially on the ratio of the energy dissipation rate per unit volume to the plasma density:

(a) When the density is low (or the dissipation rate very high), the only radiative processes with adequate efficiency will be those involving relativistic electrons.

(b) In intermediate cases, subrelativistic electrons may be able to radiate with adequate efficiency via bremsstrahlung and Comptonisation.

(c) For high densities, the radiation approaches a black body; this happens only when the optical depth due to Thomson scattering is $\gtrsim 10^3$.

Regime (c) is not attained in accretion flows unless $\dot{M}$ is large, and/or the effective viscosity is low enough to permit a long residence time for material before it gets swallowed by the central object. In general, we might expect a three-phase structure around massive black holes: dense optically-thick clouds or filaments may be embedded in a hotter (but non-relativistic medium); relativistic particles accelerated by shocks may constitute a third phase, contributing importantly to the radiative output even though the fraction of particles in this phase could be small. The possibility of a pair-dominated plasma is a further complication.

3. TRANSPORT PROCESSES WHEN RADIATION PRESSURE IS OVERWHELMINGLY DOMINANT

Internal conditions in accretion flows may approach thermal equilibrium (case (c) above), especially in thick discs or "donuts" where $\dot{M}$ is high and the effective viscosity low. In the opposite limit of low densities and rapid inflow, the cooling may be so inefficient that gas can remain at the virial temperature. The ratio of radiation and gas pressure at a large optical depth depends on the value of T^3_{eff}/n, where T_{eff} is the effective temperature and n the particle density. Inside stars, the mass-dependence of this ratio is well-known. For typical stars, where radiation pressure support is unimportant, it goes as M^2; but for very massive stars ($M \gtrsim 100\ M_\odot$), radiation pressure is dominant, and the ratio then rises roughly as $M^{\frac{1}{2}}$. This means that for supermassive stars such as might be a stage in AGN evolution, the radiation pressure contributes more than 99 per cent of the total.

The dominance of the radiation pressure is even more extreme in accretion flows. The difference between these and supermassive stars is that the density n can be very much lower, the binding mass being not the gas itself (whose active gravitational effects are generally negligible) but a central collapsed body. The ratio of radiation and gas pressure can now be estimated by the following simple argument. There are two characteristic temperatures: one is

$$T_{virial} \simeq 2 \times 10^{12} (r/r_{Schw})^{-1}\ K \qquad (1)$$

the other is the effective (black body) temperature

$$T_{eff} \simeq 10^6 (M_h/10^8\ M_\odot)^{-1/4} (r/r_{Schw})^{-1/2} [\tau(>r)]^{1/4} \qquad (2)$$

In this approximate expression (for a quantity which is in general geometry-dependent) τ is a measure of the optical depth outside a radius r. If material with temperature T_{eff} is to be supported in a potential well the energy per proton must be of order kT_{virial}; this therefore requires that radiation pressure (or photon density) must exceed gas pressure (or proton density) by (T_{virial}/T_{eff}). Ratios of order 10^6 are now possible. (Such dominance occurs, incidentally, in another very different context: the early universe, where in the fireball before recombination, radiation pressure dominates gas pressure by the photon/baryon ratio, known to be $\gtrsim 10^8$.) The radiative transfer in this situation is discussed in refs [1] and [2]. Because gas pressure is so unimportant, the fractional temperature differences between rising and falling elements are tiny and convection is inefficient.

4. RADIATION PRESSURE ON ELECTRON-POSITRON PAIRS

In compact sources where the primary radiation mechanism is non-thermal, high energy photons interact with each other to create electron-positron pairs. This effect was already recognised in the 1960s (e.g. ref. 3) and several authors have explored it in the last few years [4 - 7]. Electrodynamic processes around black holes can, of course, yield a pure $e^+ - e^-$ plasma. However, even in a conventional accretion flow, where the power derives from the gravitational binding energy of infalling matter, the density of pairs may exceed that of the original electron-ion plasma. The pairs then augment the scattering opacity, and must be included in estimates of the trapping radius and the effects of radiation pressure. This subject was recently addressed by Lightman, Zdziarski and myself [8]. We calculate the factor by which the pairs reduce the effective Eddington limit. More significantly, there is a broad range of values of L and $\dot{M}$ for which the actual luminosity lies only a factor 2 below the modified Eddington luminosity.

These calculations assumed spherical symmetry. However, they suggest that in a realistic geometry, outflow of pair-dominated plasma could be driven by radiation pressure. Two possibilities come to mind:
(i) If the inflow is in a flattened configuration, radiation pressure may exceed gravity at high latitudes.
(ii) The infalling material may be in blobs, the volume between them containing pairs not weighed down by ordinary matter. The pairs, perhaps filling most of the volume, could then flow outwards, in a wind.

5. JETS IN GENERAL

Computer simulations of jet propagation are proceeding apace. It is now possible to include the effects of magnetic stresses, and to do genuinely three-dimensional simulations where axisymmetry is not

artificially imposed. Such studies will help us to understand how
stable the jets are, whether they are confined by magnetic fields or by
the pressure of an external cocoon, and the extent to which they
resemble those mapped by the VLA.

Few entirely uncontroversial statements about jets can be made,
other than simply stating that they somehow transport energy outwards.
However, there is in my view an increasingly compelling case that
in strong sources jets propagate with speeds of order c all the way out to
the hot spots in the extended lobes. The propagation involves a high
Mach number, permitting almost loss-free energy transport. The beams
may consist of electron-positron plasma, and the apparent one-sidedness
may arise from Doppler favouritism, rather than being intrinsic. Lower
powered sources involve jets which are slower moving, having suffered
more entrainment and dissipation.

Only a small fraction of AGNs display these jets. It could be
that they are only generated under special circumstances. Alternatively,
they may in many cases be rapidly stopped, failing to propagate
effectively through the interstellar and intergalactic medium. There
is evidence that large-scale jets require a host galaxy whose potential
well is pervaded by high-pressure hot gas. Big jets are primarily
in ellipticals or in interacting systems (see ref. 9 for a review).

Even though they may be manifest in just a small subset of AGNs,
jets pose a distinct problem: how can the power emerge <u>primarily</u> in
this low-entropy form? The problem is posed especially by strong radio
galaxies such as Cygnus-A, in which the kinetic energy transported by
the jet exceeds the direct radiative luminosity of the galactic
nucleus. The central mass must be at least $\sim 10^8$ $M_\odot$, because of the
huge overall energy content of extended sources. So the nuclear
luminosity is far below the Eddington limit, suggesting a low accretion
rate. How, then, can an intense relativistic outflow be generated,
when M is low, especially in view of the fact that inflowing gas may
then be able to cool in the inflow timescale, so the efficiency of
accretion may be low? The answer is that the jet production mechanism
may be tapping the latent spin energy stored in the hole.

The possibility of tapping the rotational energy of a Kerr black
hole was first discussed by Penrose [10]. Blandford and Znajek [11]
showed how this could happen realistically, by exploiting the analogy
between a Kerr black hole and a spinning conductor, and follow-up work
has led to specific models for strong radio sources [12, 13]. This
process requires a magnetic field threading the hole; a current system
flowing through the hole; and, for maximal power dissipation in the
external medium, a suitable impedance match between the resistivities
of the hole and of its surroundings.

A small amount of plasma around the hole could carry currents
sufficient to maintain a field pervading the ergosphere. Even though a
low density plasma does not radiate very efficiently, it produces a few
gamma-rays by bremsstrahlung and related processes. Collisions between
such photons near the hole could create enough electron-positron pairs
to supply a current flowing into the hole. (Indeed, the charge density
would everywhere be high enough for relativistic MHD to apply, in the

sense that charge neutrality is approximately preserved.)
Electron-positron pairs moving with Lorentz factors of up to 100 would transport kinetic energy outwards, but most of the power outflow would initially be in the form of Poynting flux associated with the magnetic field coiled around the jet axis, and frozen in to the pair plasma. This Poynting flux may be converted into fast particles where the jet encounters ambient material, perhaps on the scale of the VLBI radio components. The expected magnetic field in the jet has just the kind of configuration that could cause magnetic confinement and collimation. The plasma around the hole that supplies the currents and anchors the field is just a catalyst. The power output of a source like Cygnus A could in principle be sustained with zero accretion rate if some of the hole's spin energy were channelled into the surrounding plasma to compensate for its small radiative losses.

Radio galaxies, according to this idea, harbour massive black holes which may, for most of their lifetime, be quiescent because they are not surrounded by plasma. If some event such as interaction with a companion were to trigger renewed infall (maybe at a low rate but sufficient to reactivate the nucleus by applying a magnetic field), the "clutch" is engaged, enabling the hole's spin energy to be converted into non-thermal directed outflow. The ejecta may plough their way out to scales 10^{10} times larger. If this is indeed what happens in Cygnus A and M87, then these very large-scale manifestations of AGN activity may offer the most direct evidence for inherently relativistic effects.

6. SUPERLUMINAL SOURCES

There are three features of the superluminal souces (discussed by Schilizzi in an accompanying paper) that at first sight seem to present difficulties:
   (i) The high apparent transverse velocities, of order 10c, that are characteristically seen, implying (in simple models) bulk motions of 'blobs' with Lorentz factors $\gamma_b \gtrsim 10$.
   (ii) The substantial fraction of sources that display such effects - a fraction far larger than the value γ_b^{-2} expected on the basis of simple unidirectional outflow.
   (iii) The fact that the apparent motions are outwards, rather than inwards, and appear to be, in projection, along the general direction of larger scale jets.

It may be worth emphasising that, of these, it is really only the third that raises real problems for the models. The high velocities require Lorentz factors exceeding 10, but this is in itself no problem, especially if the motion involves a medium where electron-positron pairs are the dominant plasma, and where the Poynting flux may be the dominant energy flow (in the sense that $(B^2/8\pi)c$ exceeds the kinetic energy flux in the plasma). Nor is there any difficulty in understanding how such motions might occur in a large fraction of sources. All that is required is that the compact central source should eject material in a range of directions: Doppler favouritism can then do the rest. Indeed the very first superluminal model [14] had this feature.

More problematic, however, is the interpretation of the motions that are <u>predominantly</u> <u>aligned</u> with the large-scale jet: if the source ejected plasma blobs in random directions, we would see those moving at angles to our line of sight, but these would, in projection, appear to be moving in all directions. If there were a directed outflow along a jet, but the flow were deflected or sprayed into a range of directions by internal shocks or interaction with obstacles, then a high proportion of jets could display superluminal effects [15]. However, although these would emanate from points on the jet, the motion could be directed in any direction unless the probability of ejection changed very sharply over a range of angles $\sim \gamma_b^{-1}$. It would therefore be very important to know whether the motions are indeed closely aligned with the jets. If they are not, and components fade out before having traversed more than a couple of jet widths (see figure 2 and caption) there may be no problem. But it would be very hard to understand the frequent occurrence of motions $\sim 10c$ if they involve blobs moving along the entire length of the jet.

The superluminal components observed at milli-arc-second resolution by VLBI have (deprojected) lengths of $\sim 10^{20}$ cm — 10^{-3} of typical VLA map dimensions — and imply relativistic bulk motion on these scales. There are, however, persuasive reasons for attributing the primary energy production to relativistically deep potential wells on scales $10^{14} - 10^{15}$ cm. However, we have no direct evidence that well-collimated jets exist on scales below those accessible to VLBI. It is important not to forget that many powers of ten difference in scale are involved: if collimation <u>is</u> initiated on scales of 10^{15} cms or less, the jets may face many vicissitudes before they penetrate to the much larger distances where we observe the radio phenomena. They may be destroyed and recollimated; they could even change direction. The small-scale jet may be lined up with the direction of the rotation axis of the central massive object while the large-scale one may lie along the rotation axis of the galaxy.

The stuff ejected from 10^{15} cm may be electron-positron plasma rather than "ordinary" electron-ion plasma. Although there have been several recent investigations of cooling and radiative transfer in such plasmas (e.g. refs [5 - 6]), little attention has yet been given to their dynamics. Energy can also be transported via Poynting flux — either as a large-scale field carried out with the particles (a directed MHD wind) or as low-frequency wave modes — and this can, in principle, swamp the kinetic energy carried by the charged particles themselves.

The flow patterns on the unobservably small scales $10^{15} - 10^{19}$ cm, if we could probe them in the same detail that the VLA provides for scales a million times larger, would no doubt prove just as complex: there would be a multiphase medium (see section 2), entrainment of surrounding gas, bending by transverse pressure gradients, and shocks where the jet impinges on the dense gas clouds that emit the broad emission lines. But one general statement can be made. The flow patterns would not simply be a scaled-down version of those seen on larger scales, because one key number — the ratio of radiative cooling times ($\propto r^2$ for a simple diverging jet) to dynamical times ($\propto r$) — is proportional to r rather than being scale-independent. Consequently, the flows on small scales would tend to be less elastic and more dissipative; they are less likely to maintain a high internal pressure, and would dissipate more energy (as synchrotron radiation at infrared or higher frequencies) if bent through large angles.

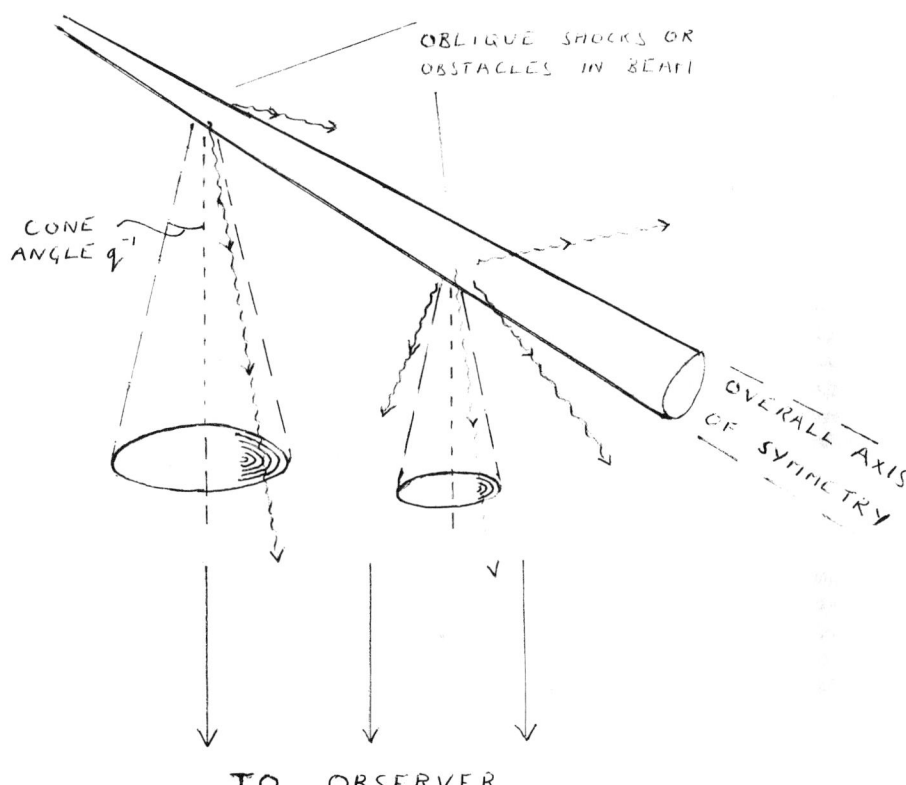

Figure 2. Superluminal motions with apparent speeds $\gtrsim qc$ are produced only by material moving with Lorentz factors $\gamma_b \gtrsim q$ in directions within q^{-1} of the line of sight (i.e. within the cones shown). If the overall source axis is not specially closely aligned with our line of sight, then the material must "spray" in a range of directions, owing to interactions with obstacles, oblique shocks, etc. But it would then be a genuine problem to understand motions that seemed, in projection, to lie along the jet direction: this would require that the range of directions within the cones were not uniformly populated with ejecta, but that there was a strong preference for the right-hand (shaded) sectors.

Progress in this area will surely depend on increasingly sophisticated hydrodynamical codes. Two-dimensional codes have already uncovered some gas dynamical properties of supersonic flows that were unanticipated by analytical models and may have counterparts in radio maps. We must await 3-D codes before we can hope to simulate the non-linear development of instabilities and the bends in jets. The other important computational development is the use of MHD codes. Only in this way can we see if it really is practical to confine jets magnetically, and whether or not the shapes of "hot spots" and the polarization patterns observed in jets can be accounted for.

7. THE JETS IN SS433

Understanding the nature and origin of jets is essential to an understanding of active galaxies. The mildly relativistic jets in SS433 may be a miniature version of some jet-like phenomena in extragalactic radio sources. In some respects, they are more easily studied and more tightly constrained than those in any extragalactic object. Only in SS433, for example, do we directly measure the speed of outflow, by observing the time-varying Doppler shifts of the emission lines. These measurements indicate that this has remained remarkably constant at .26c, as well as being uniform across the jet. This steady speed places a powerful constraint on all models.

A mechanism that naturally explains both the value and the constancy of the jet outflow speed is 'line-locking'; this was first suggested in the context of SS433 by Milgrom [16]. It requires the following: (1) an underlying continuum flux strong enough for radiation pressure to overcome gravity; (2) that the dominant momentum transfer is through Lyman-line absorption by some hydrogenic ion; and (3) that the continuum flux falls off sharply above the Lyman edge for that ion (as occurs, for example, in early-type stars at the H Lyman edge). Gas would then be accelerated up to a terminal velocity such that the relativistic Doppler shift of the accelerated gas with respect to the underlying continuum sources shifted the Lyman-edge wavelength down to the local comoving Lyman α wavelength. This happens for a velocity of 0.28c (actually at a somewhat lower velocity because blanketing by higher Lyman lines would attenuate the continuum at slightly longer wavelengths than the Lyman limit).

The general suggestion of line-locking is not new. There is convincing evidence of 'locking' between closely-spaced lines in some stellar spectra [17]. Studies of radiation-driven acceleration in the quasar context were stimulated by various claims to have found preferred ratios between different redshift systems in quasars displaying multiple absorption lines. The reality of this effect is still controversial, though there is some recent evidence of it [18].

Radiative acceleration via the Lyman lines in SS433 has been more fully explored by Shapiro et al. [19]. An important constraint is that the acceleration cannot start too close to a central compact object because the intensity (and, correspondingly, T_{eff}) would then be too high to permit hydrogen to survive in neutral form. This problem is evaded if the acceleration is attributed to a high-Z hydrogen like ion (Fe, for instance); but there is then a loss of efficiency because the abundances of such ions (and the appropriate cross sections) are lower. Although the line-locking mechanism cannot be ruled out, it is fair to say that fuller investigation has made it seem less likely; it is certainly hard to incorporate the necessary constraints in any astrophysically-plausible model for this unusual system.

SS433 offers the clearest evidence of thermal matter being expelled from an object whose luminosity probably exceeds L_{Edd} - even if the ejection speed did not have the special value suggestive of line-locking, it would still be the most likely object, of all those known, where radiative driving might be important. Begelman and I [20] have considered whether the jets could be accelerated by thermal continuum radiation, Compton scattering then being the relevant opacity. For this to work, some of the space around a compact object must be filled with a mixture of matter and radiation such that $(p/\rho)^{\frac{1}{2}}$ exceeds the escape velocity; if the surrounding material were in a rotationally-flattened distribution, the buoyant material would be expelled in 'twin exhaust' jets along the minor axis rather than in the uncoordinated photon bubbles mentioned in section 3. A volume with high (p/ρ) would naturally arise within the magnetopause of a neutron star: the pressure holding up surrounding material can here be supplied by magnetic stresses, the matter density perhaps even being lower than it is further out.

In this latter mechanism, the acceleration would occur close to the central object, where radiation would be trapped: multiple scattering then permits efficient conversion of radiant energy into bulk kinetic energy (in contrast to the optically thin case, when each photon can be used only once, and its momentum, not energy, is the relevant quantity). The details would be influenced by radiative viscosity, which would tend to entrain matter into the jet. More detailed computations are needed in order to calculate the velocities of these jets, and how sensitive they are to ambient conditions (these points are both very important in the application to SS433).

Radiation pressure acceleration is unlikely to be relevant to the jets in strong radio galaxies. In the nuclei of these galaxies, the photon luminosity is low, and the main theoretical problem is to explain how the energy from the AGN can be channelled predominantly into the low-entropy form of a relativistic outflow and/or Poynting flux; electromagnetic processes around a spinning hole (with low accretion rate $\dot{M}$) offer the most promising possibility (see section 5). However it is worth noting that if there were SS433-like jets in radio-quiet quasars, they might have escaped our attention; the time-scale or precession would be too slow for us yet to have detected it, and the only signature would then be weak emission lines with a redshift discordant from the quasar itself.

8. STELLAR DISRUPTION IN THE GALACTIC CENTRE?

The theme of this conference is outflow and mass loss. I shall therefore conclude by mentioning a dramatic manifestation of this phenomenon in galactic nuclei which can be predicted to occur but which has not yet been conclusively identified: the ejection of debris from tidally disrupted stars.

Whereas the Schwarzschild radius r_{Schw} scales with the hole's mass, the tidal radius r_T (defined as the radius within which a star of given type would be tidally disrupted) grows only as $M_h^{1/3}$. A hole exceeding $10^8\ M_\odot$ would swallow solar-type stars without disrupting them, a fact whose relevance to quasar models was first emphasised by Hills. But a hole within the mass range relevant to our Galactic Centre (below a few million solar masses) would tidally disrupt all stars of density $< 10^4$ gm cm^{-3} before swallowing them. The fate of the debris from such stars depends on the uncertain answers to several interlinked questions: What fraction of the debris actually goes down the hole, rather than being expelled? What is the radiative efficiency for the accretion process? (In other words how many ergs of energy are radiated for each gram that is swallowed?) How long does it take to "digest" one star? In particular, how does the "flare duration" and decay timescale for such a process compare with the interval between one stellar disruption and the next?

If a solar-type star approaching on an almost "parabolic" orbit with small impact parameter is disrupted by passage within a distance r_T ($\sim 10^2\ r_{Schw}$) of a $\sim 10^6\ M_\odot$ black hole, the bits of debris move out along orbits whose mean binding energy (to the hole) is $GM_*/r_* \simeq 10^{-5} c^2$. This is because the energy needed to tear the star apart has come from the incoming orbital energy. These orbits are very eccentric with typical major axes $\sim 10^5\ r_{Schw}$; the "mean orbit" for the debris has a period of $t_{orb} \simeq 30\ M_{h6}$ yrs.

Three mechanisms may be distinguished whereby some of the debris may nonetheless escape:

(a) Debris that ends up bound to the hole, on a range of elliptical orbits with specific binding energy $\sim 10^{-5}\ c^2$, would form an axisymmetric torus after very few orbits. But such a torus would be likely to have such high viscosity that energy would be released at far above the Eddington limit. This would result in much ejection. The fraction of the stellar rest mass energy emitted as radiation depends on the fraction swallowed (rather than expelled) and on the efficiency with which it can radiate before being swallowed. However, all the action would be over within a few times t_{orb}. We cannot empirically rule out $\sim 10^{44}$ erg s^{-1} from the Galactic Centre provided that it is a flare-like event with a short duty cycle (i.e. a duration $\ll 10^4$ yrs). The expulsion driven by radiation pressure would be anisotropic, probably in a twin jet configuration aligned with the angular momentum vector of the original stellar orbit, and with ejection velocities up to $\sim 0.1c$.

(b) Since the original binding energy of the disrupted star must be supplied by its orbital energy, the debris must on average be bound to the hole by an energy $\sim v_*^2$ per unit mass, v_* being the escape velocity from the surface of the star. However as Lacy et al. [21]. emphasised, some fraction of the debris can escape at $\gtrsim 1000$ km s-1. The reason for this is that at peribothron a star undergoing tidal disruption is moving at $V \simeq 3 \times 10^4 \, M_{h6}^{1/3}$ km s^{-1}: it becomes somewhat compressed and elongated into a prolate shape and pressure gradients can impart to material on the leading side of the star an excess velocity δV over the parabolic orbital velocity which is a significant fraction of V. This corresponds to a large excess orbital energy - enough for the debris to escape on a hyperbolic orbit with terminal velocity $\sim \sqrt{(\delta V)v_*}$. Whenever a star is disrupted, some fraction of its "remains" may therefore spray out in a fan or cone.

(c) Stars that penetrate well inside r_T are severely compressed when they pass close to the hole. There is then the possibility of explosive energy release. The p-p reaction is too slow to release much energy on a dynamical timescale: however, proton capture on C, N and O can yield (for solar abundances) enough energy to unbind the star; most of the debris could then escape on hyperbolic orbits. In the rarer cases when the star captured is a C-burning giant, the compression can lead to an explosive increase in reaction rates [22].

Every few thousand years, whenever a star passes close enough to the central hole to be disrupted, the Galactic Centre could expel a 'ballistic' jet with mass 0.1 - 1 $M_\odot$. There are three possibilities [23]:

(i) a double jet, associated with a supercritical torus; the velocity here may be $\sim$ 0.1c.
(ii) a fan-like jet, being the part of a tidally-disrupted star expelled on hyperbolic orbits, with speed $\gtrsim 10^3$ km s^{-1}.
(iii) a nuclear powered fan jet, containing possibly almost the entire mass of a star that has passed close enough to the hole for nuclear reactions to release more than GM_*^2/r_* of energy.

The question of whether this type of outflow might trigger some of the phenomena seen near the Galactic Centre (or have observable effects in nearby external galaxies whose nuclei may contain massive black holes) deserves more study.

REFERENCES

1. Prendergast, K.H. and Spiegel, E.A. Comm. Astrophys. Sp. Phys. **5**, 43 (1973)
2. Rees, M.J. in "Comets, Stars and Galactic Nuclei" ed. W. Hillebrand et al. p166 (Springer 1987)
3. Jelley, J.V. Nature **211**, 472 (1966)
4. Guilbert, P.W., Fabian A.C. and Rees, M.J. MNRAS **205**, 593 (1983)
5. Svensson, R. MNRAS **227**, 403 (1987) and references cited therein.
6. Lightman, A.P. and Zdziarski, A. Astrophys. J. (in press)
7. Fabian, A.C., Blandford R.D., Guilbert, P.W., Phinney, E.S. and Cuellar, L. MNRAS **221**, 931 (1986)
8. Lightman, A.P., Zdziarski, A. and Rees, M.J. Astrophys. J. (Lett.) **315**, L.113 (1987)
9. Begelman, M.C., Blandford, R.D. and Rees, M.J. Rev. Mod. Phys. **56**, 255 (1984)
10. Penrose, R. Rivista Nuovo Cimento **1**, 252 (1969)
11. Blandford, R.D. and Znajek, R.L. MNRAS **179**, 433 (1977)

12. Rees, M.J., Begelman, M.C., Blandford, R.D. and Phinney, E.S. Nature **295**, 17 (1982)
13. Phinney, E.S. Unpublished Cambridge Ph.D Thesis (1983)
14. Rees, M.J. Nature **211**, 468 (1966)
15. Lind, K.R. and Blandford, R.D. Astrophys. J. **295**, 358 (1985)
16. Milgrom, M. Astro. Astrophys. **78**, L.9 (1979)
17. Shapiro, P., Milgrom, M. and Rees M.J. Astrophys. J. Supp. **60**, 393 (1986)
18. Scargle, J.D. Astrophys. J. **179**, 705 (1973)
19. Foltz, C., Chaffee, F., Morris, S. and Weymann, R.J. Astrophys. J. (in press)
20. Begelman, M.C. and Rees, M.J. MNRAS **206**, 209 (1984)
21. Lacy, J.H., Townes, C.H. and Hollenbach, D.J. Astrophys. J. **262**, 120 (1982)
22. Carter, B. and Luminet, J.P. MNRAS **212**, 23 (1985) and references cited therein
23. Rees, M.J. in "The Galactic Center" ed. D. Barker p.71 (A.I.P. 1987)

REQUIREMENTS FOR THEORETICAL MODELS OF OUTFLOWS

Jeffrey L. Linsky[1]
Joint Institute for Laboratory Astrophysics
National Bureau of Standards and University of Colorado
Boulder, Colorado 80309-0440, USA

"A hypothesis or theory is clear, decisive and positive but is believed by no one but the man who created it. Experimental findings on the other hand, are messy, inexact things which are believed by everyone except the man who did the work" - Harlow Shapley

1. INTRODUCTION

The organizers of the Workshop have requested that I summarize the very thorough coverage of mass loss from diverse astrophysical sources presented over the past five days. Clearly this is an impossible task and I cannot pretend to do it justice. During the 31 years since Deutch (1956) first presented indisputable evidence for mass loss from the α Her system, the topic of mass loss has grown to be a major research area in astrophysics. The review papers presented at this Workshop summarize the current status of this field.

I propose instead to articulate what struck me as especially interesting during this Workshop, in particular concerning our theoretical understanding of the mass outflow process. I will attempt to answer the question - what requirements must an acceptable theory of mass outflow fulfill? In my opinion the most important requirement is that such models include the essential physics, even if only crudely. Our difficult task is therefore to identify the essential physics that must be included.

Before proceeding, it is instructive to consider the results that many astronomers are looking for in the literature on mass loss. Like most people they are looking for clear and decisive answers to what is inherently a messy, inexact question. Figure 1

---

[1] Staff member, Quantum Physics Division, National Bureau of Standards.

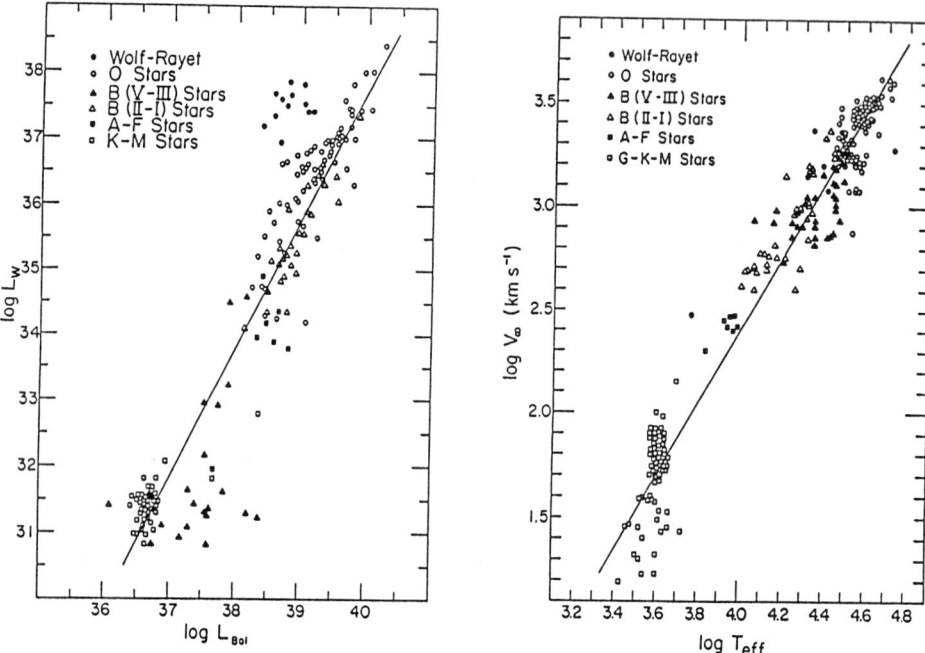

Figure 1. (Left) Empirical correlation of the wind luminosity ($L_w = 1/2 \dot{M} v_\infty^2$) with stellar bolometric luminosity for hot and cool stars. (Right) Empirical correlation of the wind terminal velocity (v_∞) with stellar effective temperature. (From Waldron 1985).

from Waldron (1985) illustrates the point. This figure shows empirical correlations of the wind luminosity ($L_w = 1/2 \dot{M} v_\infty^2$) and wind terminal velocity (v_∞) for both hot and cool stars. This figure gives the impression that the mass loss rate ($\dot{M}$) depends primarily on stellar luminosity and effective temperature. There is no a priori reason, however, why $\dot{M}$ should depend upon only these two quantities. We should probably seek empirical or theoretical relations that include functional dependencies upon chemical composition, rotational velocity, radius, mass, and the magnetic field parameters since various proposed mass loss mechanisms include these parameters explicitly. The reason why many people desire an explicit parameterization for the mass loss rate is that it is useful in understanding other problems like stellar evolution as Chiosi described to us during the Workshop (cf. Iben 1985) and because of an oversimplified understanding of determinism in science.

Even at the risk of being labelled an inveterate pessimist, I remind everyone that the equations of continuity, momentum and energy conservation for fluid flow (see Cassinelli 1979) only relate atmospheric parameters at one location in an atmosphere to those elsewhere in the atmosphere over the range of photon transport. The presumption that these equations require that the mass loss or other global properties of fluid flow are adequately described by the global properties of the star is, however, an enormous leap of faith.

Let me illustrate my point by an example. Can the weather at any point on the Earth's surface be computed (i.e. predicted) knowing only the gross properties of the Earth, the solar luminosity, and the local longitude and latitude? Clearly not. One must also take into account the complex geometry of the Earth's surface (i.e. mountain ranges and bodies of water) and the large amount of energy associated with changes of phase (i.e. condensation and evaporation of water). Inclusion of these and other essential physical processes results in a coupled set of nonlinear equations that do not admit solution a priori. Instead, with great difficulty one can numerically integrate these equations forward in time given initial conditions at all locations. Atmospheric properties on the Earth are characterized by large fluctuations about mean quantities that are themselves not simply predictable. One concludes that this problem is deterministic, but only in the sense of an initial value problem that is characterized by what appears to be individual, nonrepeatable behavior. There are many analogies between stellar mass loss and the terrestrial weather as we shall see. Whether or not globally-averaged stellar mass loss rates may be computed a priori even when localized mass loss is not, is a vital open question. The important lesson of the above example is that one must include the essential physics.

2. POTENTIAL SOURCES OF MISUNDERSTANDING

Since observations generally precede theory, it is important to recognize what the data alone without detailed modelling cannot tell us. During the Workshop I was reminded of a number of limitations of the data that can lead to misunderstandings.

(1) Spectroscopic indicators of mass loss are quantities averaged over the line of sight and across the surface (and off the limb) of a star. They do not provide unique information on the atmospheric parameters at any point on a star or a true mean, since the summation is nonlinear when overlying gas absorbs photons emitted below.

(2) Upper limits of quantities are no more than upper limits. For example, the absence of wind signatures for dwarf stars of spectral types A-M does not imply that these stars have no mass loss but rather upper limits to the mass loss rates that could be appreciable. Similarly, upper limits to the observed x-ray flux indicate only that the emission measure of hot plasma lies below some value, but such stars may nevertheless have hot coronae.

(3) The presence of emission lines in a spectrum does not require outflowing gas or even the presence of gas in a geometrically

extended atmosphere. They may often be simply explained by a large line source function relative to the local continuum.

(4) The presence of a constituent in a wind does not imply that it plays a major role in the acceleration process. For example, when dust is formed beyond the sonic point it plays no role in determining $\dot{M}$, and far from the star where the density is low, the dust can not transfer momentum to the gas (cf. Kwok 1985).

(5) The expansion velocity (v_{exp}) measured with a given diagnostic does not in general measure the asymptotic velocity very far from the star (v_∞). This problem may be especially severe for OH/IR stars for which deJong has presented evidence during this Workshop for wind acceleration at $100R_*$.

(6) Energy and momentum deposition in a stellar atmosphere may be unrelated to the mass loss process. For example, energy and momentum deposited beyond the critical point and energy deposited below the critical point but radiated away locally do not increase $\dot{M}$. Thus dignostics of the energy and momentum transfer to the atmospheric gas may not provide information relevant to the outflow.

(7) Dynamic phenomena even with large kinetic energy relative to the total kinetic and potential energy of the wind will not add to the mass loss rate unless they do not increase the density or change the location of the critical point. Antonucci in her review and Thomas in the discussion pointed out the long known fact that solar spicules have much more kinetic energy than the whole solar wind, but spicules have not been connected directly to the wind expansion.

(8) Important parameters of a theory are often unmeasurable and not readily computed from first principles. A good example is the magnetic field strength, geometry, and input flux of MHD waves which are parameters for MHD wave-driven acceleration theories. This mechanism has been used to explain mass loss for hybrid (Hartmann, Dupree and Raymond 1980; Hartmann and MacGregor 1980), M supergiant (Hartmann and Avrett 1984) and T Tauri star (Hartmann et al. 1985) mass loss but the required field strengths of a few gauss are far beyond present measurement techniques.

(9) The location along the line of sight of circumstellar gas seen as blue shifted absorption features in line profiles is usually unknown, except for certain binary stars like α Her (Deutch 1956) and the ζ Aurigue systems as Reimers described to us during the Workshop. This is important because for symmetric flows $\dot{M}=4\pi RvN_R$, where R is the radial position above which the column density of absorbers (N_R) is measured and v is the mean flow speed.

(10) Emission in important wavelength regions may be nonthermal in origin and thus measure a trace constituent, the number density of relativistic electrons, rather than the amount of outflowing thermal gas. The example I have in mind is the microwave radio emission that Abbott, Bieging and Churchwell (1985) argued is due to synchrotron emission rather than thermal for one quarter of the OB stars, but during the Workshop Pollock argued that the x-ray emission from several OB stars is also nonthermal and thus not a measure of the amount of thermal coronal gas in the winds of these stars.

(11) Several mass loss mechanisms may operate simultaneously in certain stars. Since the mass loss rate for the joint occurrence of several mechanisms can be far larger than if only one occurs, data relevant to one of the acceleration processes may provide quite inadequate information concerning the wind.

(12) Most stars are members of binary systems. In fact, on a statistical basis 4 out of every 3 stars are members of binary systems. Thus the observed radiation may be formed in the atmosphere of an unseen companion. X-ray emission from A stars, for example, in most or all cases likely originates in the coronae of previously unknown K and M dwarf companions (Schmitt et al. 1985).

(13) The geometry of a stellar atmosphere can change with height and this may be important for the mass loss process. For example, solar coronal holes, regions of open magnetic field geometry and the origin of high speed wind streams, diverge with height to cover all but very low latitudes at about $10R_{sun}$.

(14) The correlation of empirical qualities over the many orders of magnitude range in these quantities may not require a common mass loss mechanism. At this Workshop Reimers called attention to two empirical relations: $\dot{M} \approx 5 \times 10^{-13} (L/gR)$, and that the total wind energy (kinetic and potential) per unit area of the stellar surface is roughly constant. These empirical relations appear to be valid both for the ζ Aurigae systems ($\dot{M} = 10^{-7}$ to 10^{-10} M_{sun} yr^{-1}) and the Sun ($\dot{M} = 10^{-14}$ M_{sun} yr^{-1}) for which the mass loss rates are reasonably accurately known. However, it is possible that the mass loss rates for spectroscopic binaries like the ζ Aurigae systems may not be representative of single stars with similar parameters. During the discussion, Kwok pointed out that the Reimers mass loss rate relation predicts $\dot{M}$ rates more than a factor of 10 too low for OH/IR stars for which $\dot{M} = 10^{-4} - 10^{-6}$. Dupree (1986) has compiled $\dot{M}$ estimates for cool giants and supergiants and finds that the first Reimers correlation and the very simple relation $\dot{M} \sim R^2$ appear to fit the data equally well but with large scatter. Nevertheless, the two Reimers correlations appear valid over 7 orders of magnitude in $\dot{M}$ and it is tempting to apply these empirical correlations over the whole H-R diagram and to presume that they require a common mass loss mechanism. I urge that we refrain from such sweeping generalizations until additional evidence emerges to validate them. These correlations may only tell us that there is a common restraining force (gravity) against which the mass loss mechanisms must work. It is interesting that for the cooler stars, which have winds not accelerated by radiation pressure on the gas, the total energy in the wind per unit surface area does not scale as the depth of the gravitational potential well or the stellar luminosity but rather is a constant. This suggests that the flows are self-limiting (i.e. saturated) and thus independent of the specific mass loss mechanism. Thus to understand mass outflow in these stars we must understand the physical processes controlling the output rather that the input (i.e. the acceleration mechanisms) as many different mechanisms could result in the same outflow parameters. Perhaps the

details of the acceleration process become lost in the throat of the nozzle!

(15) As previously mentioned the mass loss process may be far from equilibrium in that $\dot{M}$ depends on the fundamental stellar parameters (T_{eff}, g, R, L, z, V_{rot}, B) and the initial conditions. Thus one must analyze data obtained over a long period of time in order to understand highly variable systems.

(16) The most insidious source of misunderstanding occurs when "theory" and observations agree. By theory I mean here the predictions of a given mass loss mechanism that invariably must rely upon unknown parameters, simplified geometry, and often not include some essential physics. In this case agreement only weakly supports the theory but can stifle studies of competing mechanisms as the problem is presumed to be solved.

3. WHAT IS THE ESSENTIAL PHYSICS OF MASS OUTFLOW?

I would now like to describe what I think are the important physical effects that should be included or considered in future theoretical treatments of mass loss. My list has been stimulated by the papers presented at this Workshop as well as review papers by Holzer and MacGregor (1985), Willson and Bowen (1985), and Kwok (1985). For a more thorough discussion of the acceleration mechanisms themselves and the empirical arguments for deciding which mechanisms are important for each class of star see Castor (1981), Cassinelli (1979), Holzer and MacGregor, and Linsky (1987).

3.1. The Mass Loss Rate and Terminal Velocity are Determined by the Location and Amount of Energy Input Relative to the Critical Point

Parker (1958) first showed that the equations of continuity and momentum conservation in the flow have a solution in which the flow speed is less than sonic below a critical point, is sonic at the critical point, and is supersonic at larger distances from the star. Adding additional terms to the momentum equation such as radiation pressure on dust grains or spectral lines does not alter this general description (cf. Cassinelli 1979), although the inclusion of magnetic wave pressure may result in multiple critical points. For the OB stars and central stars of planetary nebulae the flow speed very far from the star (v_∞) is generally 3-5 times the escape speed from the photosphere ($v_{esc,o}$), so that the kinetic energy imparted to the flow greatly exceeds the potential energy needed to escape the gravitational potential of the star. For the Sun the two terms are roughly equal, while for cool luminous stars $v_\infty \ll v_{esc,o}$ so that most of the energy given to the flow is used to overcome the potential energy. Thus on energetic grounds alone these three classes of winds are quite different and suggest different acceleration mechanisms.

Table 1. Results of Changing the Energy and Momentum Flux Provided to the Flow

Change in Input Parameters	Result on the Flow Parameters
Increase $\dot{E}(R < R_{crit})$	Decrease R_{crit}; Increase T_{crit}, P_{crit}, $\dot{M}$, v_∞
Increase $\dot{\pi}(R < R_{crit})$	Decrease R_{crit}; Increase $\dot{M}$, v_∞
Increase $\dot{E}(R > R_{crit})$	Increase v_∞, T_{wind}; No effect on $\dot{M}$
Increase $\dot{\pi}(R > R_{crit})$	Increase v_∞; No effect on $\dot{M}$

Holzer and MacGregor (1985), among others, have shown that the location where the energy ($\dot{E}$) and momentum ($\dot{\pi}$) per unit time is provided to the gas is all important. Their results are summarized in Table 1. Increasing either $\dot{E}$ or $\dot{\pi}$ below the critical point (R_{crit}) has the effect of moving R_{crit} deeper into the atmosphere to regions of higher density and thereby increasing the mass loss rate ($\dot{M}$). The terminal velocity (v_∞) is also increased either because of the momentum added directly to the flow or the higher temperature (and thus larger v_s) at the critical point when $\dot{E}$ is increased. On the other hand, increasing $\dot{E}$ or $\dot{\pi}$ beyond the critical point in the supersonic region has no effect on $\dot{M}$, because the mass loss rate is determined by the conditions at the critical point and information on the energy or momentum added to the supersonic flow cannot be communicated back to the critical point. Instead, increasing $\dot{E}$ and $\dot{\pi}$ beyond the R_{crit} increases v_∞ while $\dot{M}$ is unaffected.

3.2. What Physical Processes Control the Density at the Critical Point?

The mass loss from a star depends on the density (ρ_{crit}), velocity ($v_{crit} = v_s$), and location (R_{crit}) of the critical point. When the atmosphere is presumed to be spherically symmetric then $\dot{M} = 4\pi R_{crit}^2 \rho_{crit} v_{crit}$. Thus it is vital to understand the physical processes that control ρ_{crit}. The importance of this statement can be illustrated by first assuming that an atmosphere is in hydrostatic equilibrium. In this case the density distribution is

$$\rho(r) = \rho_0 e^{-R/H},$$

where the density scale height is given by

$$H = \frac{kT_{wind}}{\mu g}.$$

For the Sun, empirically $T_{wind} = T_{cor} \approx 2 \times 10^6$ K, so that $H \approx 0.15 R_{sun}$ and $R_{crit} \approx 4 R_{sun}$. Thus for the Sun

$$\dot{M} = 4\pi (4R_{sun})^2 \rho_0 e^{-25} v_{crit},$$

which is a very small number ($\approx 10^{-14} M_{sun}$ yr^{-1}).

This very simple calculation illustrates the crucial role played by the physics that controls the density structure since a small change in H will have an enormous effect on $\dot{M}$. Thus when ρcrit is increased either by dynamical events, turbulent motions, or by the input of momentum by waves, then $\dot{M}$ will increase in proportion to this density. There is ample evidence that the atmospheric densities for cool supergiants far exceeds their hydrostatic equilibrium values. For example, the large photospheric line widths in α Ori (cf. Goldberg 1979) imply that scale heights in M supergiants can be far larger than thermal, and the ζ Aur systems (e.g. Schröder 1985) provide similar evidence. Even the chromosphere of a class III giant like α Boo (Drake 1985) has densities much larger than predicted by hydrostatic equilibrium.

Willson and Bowen (1985) have calculated the density structure of a Mira star atmosphere including a periodic train of acoustic waves generated by a piston located at the base of the atmosphere. The resultant mean densities are orders of magnitude larger than for hydrostatic equilibrium and can be characterized by a dynamic density scale height

$$H_{dyn} \approx \frac{H_{static}}{1 - \gamma^2} \gg H_{static},$$

where γ^2 is the average ratio of kinetic energy provided to a shell of gas by the waves to the gravitational potential energy. Thus the waves lift considerable mass out of the stellar gravitational well – a process that has been called "levitation." Whether or not such waves can provide all of the energy or momentum needed to explain the observed mass loss, they are a major component of the whole picture of mass loss in such stars.

3.3. Several Mass Loss Mechanism Can Work Together

If nature were kind, one mass loss mechanism would be predominant for each type of star and one could obtain detailed information on each mass loss mechanism by studying many stars in each class. Unfortunately, nature is often not kind. As discussed by Noci and Ferrari at this Workshop, the wind expansion velocity of 800 km s^{-1} observed in solar coronal holes cannot be explained only by the thermal pressure gradient of the coronal plasma but also requires momentum deposition beyond the critical point presumably by MHD waves. Also, many properties of winds from OB stars can be explained by radiation pressure on spectral lines when proper account is taken of the finite size of the star and multilevel nonLTE ionization and excitation equilibrium as the Munich group has done. However, even for these stars for which the theory is quite mature, Lamers in this Workshop has called attention to the wide scatter about the mean trends of mass loss rate with stellar luminosity which indicates that some essential physics has not yet been included. Perhaps my analogy with the terrestrial weather is appropriate for these stars. The Mira stars may be a good example for winds accelerated by a combination of acoustic wave pressure, radiation pressure on dust, and thermal pressure gradients.

3.4. What are the Important Time Scales?

Willson and Hill (1979) called attention to the important role that the gravitational return time for ballistic orbits (P_0) plays in the Mira stars for which the observed pulsations indicate that periodic trains of acoustic waves propagate outward through the atmosphere. They argued that when the wave period $P > P_0$, a parcel of gas will return to its initial position before the next wave arrives and, therefore, no potential energy is transmitted to the gas. On the other hand, when $P < P_0$, a parcel of gas cannot return to its initial location before the next wave arrives. Thus the gas acquires potential energy with each successive wave and mass loss is inevitable. Since P_0 increases outwards as the acceleration of gravity decreases, $P < P_0$ at some position in the atmosphere for any realistic P and mass loss must occur.

A second important time scale is the radiative relaxation time

$$\tau_{rad} = \frac{nkT}{n_e n_H P_{rad}(T)} = \frac{c}{\rho}.$$

Note that with increasing distance R from the star, the density decreases and τ_{rad} increases. Willson and Bowen (1985) defined the quantity R_{ad} as that radial position where $\tau_{rad} = P$. When $R \ll R_{ad}$, $\tau_{rad} \ll P$ and the atmosphere can radiate away the thermal energy provided by one wave before the next wave arrives. Thus the atmosphere is effectively isothermal. On the other hand, when $R > R_{ad}$, $\tau_{rad} > P$ and the atmosphere cannot radiate away the thermal energy from one wave before the next one arrives. In this case the atmosphere is adiabatic. The fundamental question for a Mira star with a given P what is thus the location of R_{ad} relative to R_{crit}. The four important cases are described in Table 2. Wood's (1979) calculations show that for cases (1) and (4) there is no continuous mass loss, but rather occasional ejections of matter with a time-averaged mass loss rate of 10^{-12} M_{sun} yr^{-1}. For the adiabatic limit (case 2) he computed unrealistically high mass loss rates of 0.02 M_{sun} yr^{-1} with the wind acceleration due to the thermal pressure gradient in the adiabatic shock. Real flows should be intermediate with nearly isothermal shocks near the base where the densities are high and nearly adiabatic shocks near the top where the densities are low.

Table 2. Four Cases for Mira Star Atmospheres

Location of Energy Deposition	Character of the Flow
(1) At $R<R_{ad}$ $R<R_{crit}$	Input energy is radiated away and $\dot{M}$ is small
(2) At $R>R_{ad}$ $R<R_{crit}$	Input energy is not radiated away and $\dot{M}$ is large
(3) At $R>R_{ad}$ $R>R_{crit}$	Input energy heats the wind but has no effect on $\dot{M}$
(4) At $R<R_{ad}$ $R>R_{crit}$	Input energy is radiated away and has no effect on $\dot{M}$

Willson and Bowen (1985) presented more realistic calculations in which the adiabatic and isothermal approximations are relaxed. Their calculations show that for large mass loss rate Miras ($\dot{M} > 10^{-6}$ M_{sun} yr^{-1}) and all OH/IR stars the mass loss process is due to a combination of pulsations (periodic acoustic waves), which lift matter from the star to where dust can form, and then radiation pressure on the dust (cf. Kwok 1975; Jones et al. 1981; Jura 1986a,b). For small mass loss rate Miras ($\overline{\dot{M} < 10^{-7} M_{sun}}$ yr^{-1}) pulsations alone without radiation pressure on dust appear to be adequate. They also speculate that winds for RR Lyrae and short period Cepheids are thermally driven. However, the phenomenology of Miras is exceedingly complex and such important observations as a stationary layer detected in CO data (Hinkle, Hall and Ridgway 1982) are not yet explained by the theory.

3.5. Instabilities and Phase Changes

When the outflow of matter from a star is unstable for any reason, rapid changes in the physical parameters of the flow which I call "phase changes" may modify the flow in a fundamental way or change its radiative properties. A well known example is the instability in radiatively driven winds, discussed by Lucy and White (1980) and by Lucy (1982), that occurs when small velocity or density perturbations result in local accelerations which doppler shift the gas so that strong absorption lines see brighter emission from the star outside of the line core and thus feel greater radiative acceleration. This produces rapidly moving blobs in the flow with preceding bow shocks and hot coronal gas.

I would like to discuss here a different type of instability first studied in the context of the Sun that should be important for the cooler stars. Ayres and Testerman (1981) and Ayres, Testerman and Brault (1986) showed that the infrared solar CO vibration-rotation bands are inconsistent with a homogeneous atmosphere but suggest instead thermal bifurcation into discrete structures (perhaps magnetic flux tubes) with steep chromospheric temperature rises and cool regions containing CO with little or no chromospheric temperature increases with height. Subsequently Kneer (1983), Muchmore and Ulmschneider (1985), and Muchmore (1986) have explored how cooling in the CO vibration-rotation bands can produce a condensation instability or molecular "catastrophe" in which the initial formation of CO, say by compression, radiatively cools the gas producing more CO (since the association rate is highly temperature-dependent) and thus more radiative cooling. The thermal bifurcation of the solar atmosphere is thus driven by the destabilizing effect of the steep temperature dependence of CO formation and the radiative loss rate. Analogously, the interstellar medium has at least two stable thermal regimes (Field, Goldsmith and Habing 1969).

Stencel, Carpenter and Hagen (1986) and Stencel (1987) have proposed that the CO condensation instability is an example of instabilities that can occur in the atmosphere of M supergiants. Other simple, abundant molecules like SiO, CS, OH and H_2O (Muchmore, Nuth and Stencel 1987) can behave in a manner similar to CO. One plausible scenario is that a chain of molecular "catastrophes" can occur in which cooling by CO and the resultant pressure perturbation produce conditions ripe for SiO condensation (which can occur at temperatures below 3600 K) that triggers formation of other molecules and eventually silicate dust and perhaps also SiO maser emission.

If detailed calculations and observations support this new picture of a cool supergiant atmosphere, then the essential physics of these stars includes thermal instabilities, dynamic phenomena, the presence of very different thermal regimes in close proximity, and several mass loss mechanisms working together. A further problem is that in the atmospheres of M supergiants like α Ori the chemical equilibrium time for such molecules as SiO can be so long (Stencel 1987) that the chemistry is frozen into the flow at some point. These effects could cause atmospheres for these stars to be highly complex and even chaotic. We must even consider the possibility that several metastable modes of atmospheric structure could exist for a given set of stellar parameters.

3.6. Multiphase Atmospheres/Winds

The outer atmosphere of the Sun consists of coronal holes with open magnetic fields and high speed wind, active regions with closed magnetic fields and presumably no wind, and quiet regions with a mixture of open and closed fields and presumably a variable but low speed wind. The Sun thus provides a prototype star with several different phases, in this case controlled by the geometry of the magnetic field. Hybrid stars provide another example with unshifted C IV emission (Brown, Reimers and Linsky 1986) indicating a hot phase with little outflow and the blue-shifted circumstellar absorption feature in the Mg II lines indicating a cool phase with significant mass loss. Zeta Aurigae systems contain several phases including the cool wind from the late-type supergiant, the hot wind and H II region around the B star, shocks where the two winds collide, and perhaps other phases. The presence of highly asymmetric dust and circumstellar gas around α Ori (Beckers 1985; Honeycutt et al. 1980) is consistent with several phases in the cool supergiants. Finally the presence of disks around premain sequence and Be stars clearly indicates the presence of at least two phases.

My purpose in emphasizing the geometric complexity of the outer atmospheres and winds of stars is to alert interpreters of data that the many pieces of information obtained from observing a star should be considered as pieces of several puzzles rather than one.

3.7. What Defines the Geometry?

Since several phases may be present at one time in the atmosphere of the same star, an important question is what physical processes control the geometry. As previously mentioned, the magnetic field alone appears to determine the geometry of solar active regions and coronal holes. Furthermore, the magnetic field inhibits lateral energy transport by conduction and mass flows, so that fields can act as thermal "walls" between the components. Rotation probably has little effect on the solar mass loss rate although it does cause the wind to form an Archimedes spiral structure. For the premain sequence stars the situation is unclear but potential mechanisms separating the disks, bipolar flows and jets from the other components are magnetic fields and the angular momentum of the interstellar medium flowing past the star. Rotation may also play a role in determining the geometry. Finally, Poe and Friend (1987) have proposed a model for Be stars in which magnetic fields and rotation determine the geometry of the wind structures.

3.8. The Role of the Environment

Ordinarily the properties of the interstellar medium near a star play no significant role on the character of the outflow other than defining an outer boundary condition on the pressure and temperature. The premain sequence stars may be an exception. Walter (1986) has identified a class of stars call "naked" T Tauri stars which differ from the classical T Tauri stars in that they display no evidence for a wind and they lie at the outskirts of dark clouds where the extinction and interstellar density are low. The two classes of stars are otherwise identical in age, ultraviolet and x-ray emission, and transient activity. This suggests that the two classes contain very similar stars in the same stages of evolution except that the T Tauri stars lie in a dense interstellar environment which is responsible for the accretion disk and the bipolar outflow. The character of the outflows in other types of stars may also be influenced by their environment; for example, the winds of the central stars of planetary nebulae interact with the slow wind of the pre-existing AGB star (see review by Kwok at this Workshop).

3.9. The Role of Explosive (Transient) Events

Antonucci reminded us several days ago that the kinetic energy in spicules far exceeds that of the solar wind and the net outwardly directed momentum of the spicules is at least a factor of 6 larger than the momentum flux of the solar wind. These facts have encouraged the development of theories in which the spicules somehow supply the energy and momentum of the solar wind. However, the chromospheric downflows outside of spicules essentially balance the spicular upflows, and as Thomas has reminded us there is no clearly understood way in which the spicular motions lead to changes in the density, temperature, or location of the critical point and thus the mass loss rate. If the Sun were viewed as a distant point source, one might infer the presence of dynamical events of high energy from line broadening (macroturbulence) in H_α and other lines and conclude that these dynamic events increase the mass loss rate.

On the other hand, very large scale dynamic events can play a major role in the mass loss process. Solar coronal transients may contribute about 5-10 percent of the time averaged mass loss rate (Howard et al. 1985). Also, ejection of discrete shells at times of high mass loss rate appears to be a major component of the time-averaged mass loss for P Cygni, Be, and premain sequence stars. There is also evidence for discrete mass ejection events called "puffs" for the OB stars. Thus transient dynamic events may be an essential element of the physics of mass loss for certain stars.

3.10. Role of Stochastic Phenomena

Even when large dynamic events are not the predominant contributors to the mass loss, the outflow may be intrinsically variable when the acceleration mechanism is stochastic in character. For example, when periodic acoustic waves provide most of the momentum to the flow as in the Mira stars, waves of larger amplitude can heat the atmosphere more and thus propagate faster and merge with preceding waves, producing a supershock and enhanced mass loss (Willson and Hill 1979).

3.11. Mode-Mode Coupling and Wave Damping Processes

Different types of MHD waves (compressive, transverse, torsional) can couple with each other and with acoustic and gravity modes. For example, transverse and torsional modes compress the adjacent plasma and thus can convert energy into acoustic waves. The interesting physics here is that each mode propagates and is damped differently. Thus the location in the atmosphere where energy and momentum are deposited may depend critically on the efficiency with which the atmosphere can convert wave energy from one mode into another. The physical processes by which different MHD waves are damped requires further investigation since the ratio of the damping length to the stellar radius is critical in determining whether the energy in the wave will produce a hot corona with a small mass loss rate or an extended cool atmosphere with a large mass loss rate (Hartmann and MacGregor 1980; Holzer and MacGregor 1985).

3.12. Departures from Ionization Equilibrium

Far from a star the density can become sufficiently low that the time scales for ionization and recombination are long compared to the time over which the temperature and density change in the expanding plasma. The ionization equilibrium of the outflowing plasma then becomes frozen with properties corresponding to the physical parameters at this position. In the solar wind the ionization balance becomes frozen at a few solar radii (Owocki, Holzer and Hundhausen 1983), although these departures from local ionization equilibria have no known effect on the solar outflow rate. On the other hand, early Copernicus observations of OB stars (e.g. Morton 1976) showed O VI resonance lines with P Cygni-shaped profiles. Since calculations of the ionization equilibrium for these winds including photoionization by the stellar radiation field did not predict appreciable abundance of O^{+5} the "superionization" was attributed to Auger ionization by coronal x-rays. The degree of ionization is of great importance because the acceleration mechanism is radiation pressure on lines predominantly in the ultraviolet. As we heard earlier at this Workshop from Pauldrach, however, there is no superionization in the winds of OB stars and the central stars of planetary nebulae when ionization equilibrium calculations properly take into account nonLTE effects in multilevel atoms. Such calculations (cf. Kudritzki, Pauldrach and Puls 1986) show that the proper amount of O^{+5} is present with no need for Auger ionization by x-rays.

3.13. Role of Nonthermal Phenomena

Relativistic electrons are inferred to be present in the solar corona during flares and continuously in the coronae of dMe stars and RS CVn systems on the basis of microwave synchrotron emission. Synchrotron emission has also been detected from a number of OB stars (Abbott, Bieging and Churchwell 1985), and Pollock at this Workshop argued that the x-ray emission from some of these stars is also nonthermal. As yet the presence of nonthermal particles has not been proposed as important in the mass loss process.

4. CONCLUSIONS

In this somewhat rambling review of the Workshop I have attempted to highlight the problems that can be encountered when deducing mass loss properties for stars and determining which physical processes are likely to be important for understanding the outflows of different types of stars. I could be accused of being unduly pessimistic in outlook as I have chosen to emphasize problems and physical processes that may be difficult to include in future models. If this is pessimism than I plead guilty. However, I believe that even very simple treatments of mass loss that include in turn the different potentially important physical processes are needed to decide which physical processes must be included and which may be excluded for different stars. When such calculations are done, the insight obtained will lead to mass loss calculations that are realistic because they include the essential physics.

I would like to thank the organizers of the Second Torino Workshop for their hospitality and for organizing a comprehensive meeting with a number of excellent invited and contributed presentations. I also wish to thank Dr. Bengt Gustafsson and Dr. Kjell Eriksson of the Astronomiska Observatoriet, Uppsala University for their hospitality and providing the appropriate environment in which this paper was written. This work is supported in part by NASA grants NGL-06-003-057 and NAG5-82 to the University of Colorado.

References

Abbott, D. C., Bieging, J. H. and Churchwell, E. 1985 in The Origin of Nonthermal Heating/Momentum in Hot Stars, NASA CP 2358, p. 47.
Ayres, T. R. and Testerman, L. 1981 Astrophys J., **245**, 1124.
Ayres, T. R., Testerman, L. and Brault, J. W. 1986, Astrophys J., **304**, 542.
Beckers, J. M. 1985, in Mass Loss from Red Giants, ed. M. Morris and B. Zuckerman (Dordrecht, Reidel) p. 57.
Brown, A., Reimers, D. and Linsky, J. L. 1986 in New Insights in Astrophysics, ESA SP-263, p. 169.
Cassinelli, J. P. 1979, Ann. Rev. Astron. Astrophys., **17**, 275.
Castor, J. 1981, Physical Processes in Red Giants, ed. I. Iben, Jr. and A. Renzini (Boston: Reidel) 285.
Deutch, A. J. 1956, Astrophys. J., **125**, 210.
Drake, S. A. 1985 in Progress in Stellar Spectral Line Formation Theory, ed. J. E. Beckman and L. Crivellari (Dordrecht: Reidel) p. 351.
Dupree, A. K. 1986, Ann. Rev. Astron. Astrophys., **24**, 377.
Field, G., Goldsmith, D. and Habing, H. 1969, Astrophys. J. (Letters), **155**, L149.
Goldberg, L. 1979, Quart. J. Royal Astron. Soc., **20** 361.
Hartmann, L. and Avrett, E. H. 1984, Astrophys. J., **284**, 238.
Hartmann, L., Dupree, A. K. and Raymond, J. C. 1980, Astrophys. J. (Letters), **236**, L143.
Hartmann, L., Jordan, C. Brown, A. and Dupree, A. K. 1985, Astrophys. J., **296**, 576.
Hartmann, L. and MacGregor, K. B. 1980, Astrophys. J., **242**, 260.
Hinkle, K. H., Hall, D. N. B. and Ridgway, S. T. 1982, Astrophys. J., **252**, 697.

Holzer, T. E. and MacGregor, K. B. 1985, in Mass Loss from Red Giants, ed. M. Morris and B. Zuckerman (Dordrecht: Reidel), p. 229.
Honeycutt, R. K., Bernat, A. P., Kephart, J. E., Gow, C. E., Sandford, M. R. II and Lambert, D. L. 1980, Astrophys. J., 239, 565.
Howard, R. A., Sheeley, N. R., Koomen, M. J. and Michels, D. J. 1985, J. Geophys. Res., 90, 8173.
Iben, I., Jr., 1985, in Mass Loss from Red Giants, ed. M. Morris and B. Zuckerman (Dordrecht: Reidel), p. 1.
Jones, T. W., Ney, E. P. and Stein, W. A. 1981, Astrophys. J., 250, 324.
Jura, M. 1986a, Irish Astron. J., 17, 322.
Jura, M. 1986b, Astrophys. J., 303, 327.
Kneer, F. 1983, Astron. Astrophys., 128, 311.
Kudritzki, R. P., Pauldrach, A. and Puls J. 1986, in New Insights in Astrophysics, ESA SP-263, p. 247.
Kwok, S. 1975, Astrophys. J., 198, 583.
Kwok, S. 1985, in Relations between Chromospheric-Coronal Heating and Mass Loss in Stars, ed. R. Stalio and J. B. Zirker (Trieste: Osservatorio Astronomico di Trieste), p. 187.
Linsky, J. L. 1987, in Circumstellar Matter (Proceedings of IAU Symp. No. 122), in press.
Lucy, L. B. 1982, Astrophys. J., 255, 286.
Lucy, L. B. and White, R. L. 1980, Astrophys J., 241, 300.
Morton, D. C. 1976, Astrophys. J., 203, 386.
Muchmore, D. 1986, Astron. Astrophys., 155, 172.
Muchmore, D. O., Nuth, J. A. III and Stencel, R. E. 1987, Astrophys. J. (Letters), 315, L141.
Muchmore, D. and Ulmschneider, P. 1985, Astron. Astrophys., 142, 393.
Owocki, S. P., Holzer, T. E. and Hundhausen, A. J. 1983, Astrophys. J., 275, 354.
Parker, E. N. 1958, Astrophys. J., 128, 664.
Poe, C. and Friend, D. 1987, preprint.
Schmitt, J. H. M. M., Golub, L., Harnden, F. R. Jr., Maxson, C. W., Rosner, R. and Vaiana, G. S. 1985, Astrophys. J., 290, 307.
Schröder, K. P. 1985, Astron. Astrophys., 147, 103.
Stencel, R. E. 1987, in Circumstellar Matter (Proceedings of IAU Symp. No. 122), in press.
Stencel, R. E., Carpenter, K. G. and Hagen, W. 1986, Astrophys. J. 308, 859.
Waldron, W. L. 1985, in The Origin of Nonradiative Heating/Momentum in Hot Stars, NASA CP-2358, p. 95.
Walter, F. M. 1986, Astrophys. J., 306, 573.
Willson, L. A. and Bowen, G. W. 1985, in Relations between Chromospheric-Coronal Heating and Mass Loss in Stars, ed. R. Stalio and J. B. Zirker (Trieste: Osservatorio Astronomico di Trieste), p. 127.
Willson, L. A. and Hill, S. J. 1979, Astrophys. J., 228, 854.
Wood, P. R. 1979, Astrophys. J., 227, 220.

PART II: CONTRIBUTED PAPERS

THE LINE SCATTERING RADIATIVE FORCE AS THE REASON OF THE
ACCELERATION AND THE HEATING OF THE HOT STAR WINDS

E.Ya.Vil'koviskii,L.V.Tambovtseva
Astrophysical Institute
480068 Alma-Ata
USSR

ABSTRACT. Moment and energy transport from the star radiation field to the wind's plasma have been examined. It is shown that under certain conditions the wind is heated with the ions accelerated in the wind's plasma.
    From the analysis of the motion and energy equations sistem the following conclusions have been received:
I. A mass loss rate $\dot{M}$ is regulated with the nonstationary process.
2. In some cases the heating of outward wind regions is occured that leads to the appearance of OVI,NV-type ions (by analogy to the empirical warm wind model).

I. INTRODUCTION

The aim of present paper is the reconsideration of the physical processes connected with the interaction of the star radiation field and the expanding envelope and the formulation of the theoretical principles of O-B stars' radial symmetric stationary winds. In our opinion such an approach allows us to construct the "first approximation" theory and later on this basis to consider the contributions of the other processes such as wave propagation, rotation and so on.

2. THE INTERACTION OF THE STAR RADIATION AND THE WIND
    PLASMA

The photosphere radiation flux exchanges energy and moment with the wind plasma mainly by means of Compton effect, photoionization and line resonance scattering. For O-B stars wind it is usually enough to consider at Compton effect only the moment transport (that is reduced to the redetermination of the gravity acceleration $g \rightarrow g_{eff} < g$) and at photoionization only to the plasma heating (although in some cases the photoionization pressure is essential for O-B stars with large mass loss rate and undoubtedly is essential for WR stars).At the resonance scattering only the moment transport to the wind plasma is considered (the radiative line pressure).We shall examine also the heating in such processes as was first pointed by Vil'koviskii (1981).

The main principles of the analysis of the radiative line pressure in an expanding atmospheres is sufficiently known and contained for example in Lucy and Solomon (1970), Castor,Abbott,Klein (1975),Abbott (1982).Here we shall represent only the main relations.We write the line opacity at the resonance line scattering in the form

$$K_j = \frac{1}{\Delta V_D} \frac{\pi e^2}{m_e c} g_j f_j N_j \frac{1}{\rho} \equiv \frac{q_j}{\Delta V_D},$$

where g_j and f_j are Gaunt factor and oscillation force correspondingly, $\Delta V_D = V_j \cdot v_{th}/c$ is the Doppler width of the line, v_{th} is the thermal velocity of ion, $N_j = N_H \alpha n_i n_j$ is the density of the resonancely scattering ions, N_H is the number density of hydrogen atoms, α is a number abundance of A element, n_i is the relative number of scattering ions, n_j is the part of ions in the low state of j-transition, ρ is the density of plasma.The optical depth of the line in the "Sobolev approximation" in an atmosphere expanding with a radial velocity gradient dv/dz is

$$\tau_j = K_j \rho v_{th} \left(\frac{dv}{dz}\right)^{-1} = q_j \lambda_j \rho \left(\frac{dv}{dz}\right)^{-1},$$

where $q_j = K_j \Delta V_D$, $\lambda_j = c/V_j$ is the wavelength of the line.In the "radial streaming" approximation the radiative force per gramm of matter (the radiative acceleration) due to j line scattering is

$$G_j = \frac{B_j}{c} K_j \frac{(1-e^{-\tau_j})}{\tau_j},$$

where B_j is the spectral density of the radiation flux on the frequency V_j.The summary radiative line acceleration over N lines is $G_L = \sum G_j$ .As pointed out by Abbott(1982), G_L may be approximated by power function

$$G_L \simeq A \left(\frac{d\mathcal{E}/dz}{\dot{M}}\right)^{\alpha},$$

where A and α are the koefficients depended on the parameters of a star, $\mathcal{E} = v^2/2$ is the specific kinetic energy of the wind per gramm of matter, $\dot{M}$ is the mass loss rate.

Now let us consider the change in internal (thermal) energy at the resonance scattering.The moment of the scattered photons is directly transported to a small admixture of ions of C,N,O and other elements.These ions transport the moment to the hydrogen plasma of the wind due to the collisions with other ions of the plasma. Therefore there is some "stream" of these ions and the energy dissipation due to this stream (the plasma heating).The power of the heating Q_1 is highly small till the ion velocity relatively the plasma v is small as compared with the thermal proton velocity v_{tp}, but the heating abruptly increases ($Q_2 \gg Q_1$) when $v \geq v_{tp}$. As pointed out by Vil'koviskii (1981), the abrupt increasing of velocity is possible when the radiative force per ion f_{Li} exceeds some value

$$f_{ci} \sim 10^{-15} Z_i^2 N_{10} T_{e4}^{-1} \; dyn$$

where Z_i is the ion charge, $N_{10} = N_e/(10^{10} sm^{-3})$ is the plasma electron density in $10^{10} sm^{-3}$, $T_{e4} = T_e/(10^4 K)$ is the electron temperature in $10^4 K$. In assumption of Plank spectrum of the star the radiative line force per ion for jth resonance line in considered ith stage of ionization can be estimated

$$f_{Li} = \frac{\pi e^2}{m_e c} g_j f_j \frac{2\pi hc}{\lambda_j^3} \left[exp\left(\frac{hc}{kT\lambda_j}\right) - 1 \right]^{-1}.$$

From the equation of continuity $\dot{M} = 4\pi \rho v z^2$ and full hydrogen ionization it follows

$$N_{10} \simeq 0.512 \cdot 10^2 \frac{\dot{M}_6}{(R_{10} \, v_7)},$$

where $v_7 = v/(10^7 sm/c)$ is the velocity of the flow, $R_{10} = R_*/(10 R_\odot)$ is the radius of the star in $10 R_\odot$. The condition $f_{Li} > f_{ci}$ could be rewritten as

$$v_7 \geq 1.2 \; z^2 \dot{M}_6 \lambda_5^3 \frac{1}{R_{10}^2 T_{e4} \, \varphi(T_{eff})},$$

where $\lambda_5 = \lambda_j/(10^{-5} sm)$, $T_4 = T_{eff}/(10^4 K)$. $\varphi(T_{eff})$ involves the dependence on T_{eff}; $\varphi = 1$ when $T_4 = 3$ and $\lambda_5 = 1$. When $f_{Li} > f_{ci}$ the friction force of plasma's protons is small and ions are accelerated up to the value of velocity closed to the thermal electron velocity $v_2 \sim v_{te}$. The power of plasma heating by rapid ions is about $Q_2 \sim \sum N_{i2} v_{i2}^2 f_{Li}$, where N_{i2} is the density of the ions with v_2 velocity. When $N_{i2} \geq 10^{-4} N_p$ (N_p is the number proton density) Q_2 power exceeds the radiative plasma cooling and the plasma is heated up to values of the temperature $T_e \geq 10^5 K$ till the number of resonancely scattering ions is decreasing due to collisional ionization. So the "kinetic" heating mechanism can be the theoretic basis of the earlier offered empirical warm wind model (Lamers and Snow, 1978).

3. FLOW EQUATIONS AND MECHANISM OF REGULATION OF THE MASS LOSS RATE

The equations of motion and energy of stationary spherical symmetric flow can be written in form

$$\frac{d\mathcal{E}}{dz} = \frac{1}{1 - \frac{\mathcal{E}_T}{\mathcal{E}}} \left(\frac{4\mathcal{E}_T}{z} - 2\frac{d\mathcal{E}_T}{dz} - g_{eff} + G_L \right) \quad (1)$$

$$\frac{d\mathcal{E}_T}{dz} = -\frac{2}{3} \mathcal{E}_T \left(\frac{2}{z} + \frac{1}{2\mathcal{E}} \frac{d\mathcal{E}}{dz} \right) + \frac{2}{3\rho\sqrt{2\mathcal{E}}} \left(\chi(N_e, T) + Q + div \, F_\nabla \right)$$

where $\varepsilon = v^2/2$, $\varepsilon_T = a^2/2 = kT/2m_p$, g_{eff} is the effective gravity acceleration, $\chi(N_e,T)$ is the difference between the heating due to photoionization and the cooling due to radiation, Q is the additional heating source (Q is about zero when $f_{Li} < f_{ci}$ and $Q = Q_2$ when $f_{Li} > f_{ci}$), $F_T \simeq 10^{-6} T_e^{5/2} \nabla T_e$ is the heat conduction flux.

The flow equations (I) must be completed with the equations of ionization balance and radiation transfer. The solution of such whole sistem is very complicated problem. Now we are limited ourselves with solution of the simplifying "model" task. Next preliminary conclusions could be made from the analysis of such solutions: for chosen parameters of the star there are solutions of flow equations (I) at the different values of mass loss rate $\dot{M}$ which doesn't exceed some limiting value $\dot{M}_c$ but there are no solutions when $\dot{M} > \dot{M}_c$. Therefore we suppose that the real mechanism of regulation of $\dot{M}$ involves next nonstationary process: when $\dot{M} < \dot{M}_c$ the value of $\dot{M}$ is slowly increased (in fact the flow equations are faintly nonstationary). After $\dot{M}$ reachs the value of $\dot{M}_c$ the break down of flow is occured that leads to the abrupt decreasing of $\dot{M}$, and then the events are repeating. Such a supposition is confirmed in some cases by P Cygni line profiles behaviour.

4. CONCLUSIONS

I. At certain conditions it is possible the heating of outward part of stellar wind flow due to ion's resonance line scattering (the "kinetic" heating).
2. The mass loss rate regulation includes a nonstationary processes.

5. REFERENCES

Abbott,D.C.:1982,Astrophys.J.,259,282.
Castor,J.I.,Abbott,D.C.,Klein,R.I.:1975,Astrophys.J.,195, 157.
Lamers,H.J.G.L.M.,Snow,T.P.:1978,Astrophys.J.,219,504.
Lucy,L.B.,Solomon,P.M.:1970,Astrophys.J.,159,879.
Vil'koviskii,E.Ya.:1981,Astrofisica,17,309.

DUST FORMATION AND MASS LOSS BY WOLF-RAYET STARS

K.A. van der Hucht, SRON Space Research Laboratory, Utrecht
P.M. Williams, Royal Observatory, Edinburgh
P.S. Thé, Anton Pannekoek Astronomical Institute, Amsterdam

ABSTRACT. All late-type galactic Wolf-Rayet stars, i.e. the WC8-10 and WN9-10 types, were observed by infrared photometry over a period of five years. Circumstellar dust emission, at temperatures of the order of 1300 K, was observed from 5 of the 10 WC8 stars, 14 of the 17 WC9 stars, the only WC10 star, the only object classified as WN10, and the related object LSS4005. In all cases, the dust is being continually replenished within the stellar winds. Models of the dust shells, assuming spherical geometry and inverse square law dust density distributions, are presented. Although the problem of dust formation in a 10,000 K plasma is still an enigma, it appears likely that shock-induced density enhancements where $n \geq 6*10^5$ cm^{-3} could trigger the dust formation. The dust formation rate of WC9 stars is of the order of $2*10^{-7}$ M$_\odot$/yr per star. In addition, two WC7+a and one WC8 stars are found to produce dust episodically. Of one of the former it is shown that dust is amorphous carbon.

1. INTRODUCTION

IR photometry of WR stars since the early 1970's, beginning with the work of Allen et al. (1972), reveals two kinds of IR excesses: free-free radiation caused by dense stellar winds ($\dot{M}_{WN} \approx 1-12\times10^{-5}$ M$_\odot$/yr, $\dot{M}_{WC} \approx 2.5-15\times10^{-5}$ M$_\odot$/yr, van der Hucht et al., 1986, henceforth HCW) and thermal emission by hot (T $\approx$ 1300 K) circumstellar dust. This dust emission is observed from the latest subtypes of the WC sequence only. In order to study origin, composition and mass of this hot circumstellar dust around WR stars, we carried out an IR (1.2 - 20 μm) photometric survey of all the 28 known galactic WC8-10 stars and most of the WC7 and late WN stars between 1982 and 1985 with UKIRT and ESO telescopes (Williams et al. 1987a, hereinafter WHT).

2. WOLF-RAYET STELLAR WIND PROPERTIES

Mass loss rates of WR stars can be determined very well from radio continuum measurements, in combination with UV P Cygni profiles, and knowledge of abundances and ionization. In a recent redetermination of mass loss rates of WR stars, HCW used radio and UV data of 12 WN and 10 WC stars published by Abbott et al. (1986), and abundance ratios calculated by Prantzos et al. (1986): He : C : O : Ne : Mg = .31 : .37 : .24 : .03 : .005 by mass. HCW calculated the ionization for an effective temperature range of 19000 to 50000 K for the various subtypes, in line with the results of Schmutz & Hamann (1986). The WC star mass loss rates derived by HCW range from $2.5*10^{-5}$ to $1.5*10^{-4}$ $M_\odot/yr$. Significantly, HCW find that the WC7-9 stars have mass loss rates 2-3 times larger than WC5-6 stars have, and that because of this and their lower terminal wind velocities, the late-type WC stars have 4-7 times larger wind densities. The latter difference may be crucial in explaining why WC8-10 stars display continuous hot dust formation, while WC4-6 stars do not.

3. WC STAR DUST

We find hot dust around 5 of the 10 known WC8 stars, around 14 of the 17 known WC9 stars and around the one and only known WC10 star. In addition, episodic presence of hot dust is found around 2 of the 12 known WC7 stars: the WC7+a star WR137 in 1984 (Williams et al., 1985) and the WC7+O4 system WR140 in 1985, both discussed below. Some typical energy distributions are shown in Figure 1.

Heated dust emission from the two very late WN stars WR122 (WN10) and LSS4005 (WN11?) was reported by van der Hucht et al. (1984). High resolution spectroscopy shows that the latter is better classified as Ofpe/WN9 or B[e] (van der Hucht et al., 1987).

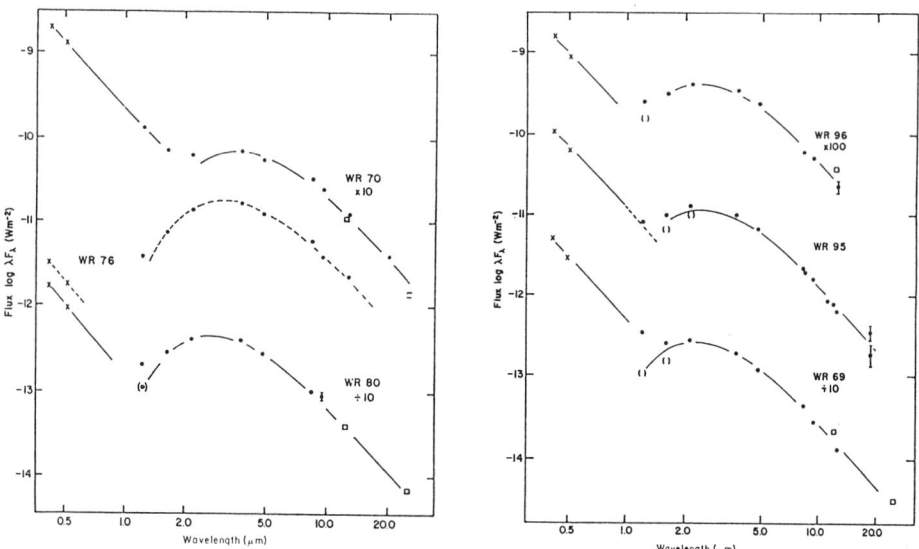

Fig. 1. Energy distributions of one WC8 star (WR70) and five WC9 stars.

The hot dust shells around WC stars can be characterized in the following way: (i) the dust shells are optically thin in most cases; the fraction of stellar luminosity absorbed by the dust and re-radiated in the IR averages about 10%; it exceeds 50% for 4 out of 20 WC8-10 stars; (ii) the IR energy distributions are featureless, except for the interstellar 9.7μ silicate absorption feature, which we find to be related to A_V as $A_V/\tau_{9.7}$ = 19.8±1.7, and consider to be interstellar in origin; (iii) the fact that the IR energy distributions are featureless rules out dielectric grains, and the 1/λ emissivity found for the episodic dust producer WR140 (Williams et al., 1987b) points to amorphous carbon; (iv) from the apparent lack of variations in most cases (at least since the beginning of the IR photometry era) we infer that the dust is being formed continually and at distances of the order of 450 R_* from the stars, well within their ionized stellar winds.

Average circumstellar dust shell parameters are given in Table 1.

Table 1. WC dust shell parameters

average values	WC8 (5)	WC9 (14)	WC10 (1)
L_{IR}/L_*	.12	.14	.71
T (K)	1310	1330	1410
$R_i (R_*)$	800	340	260
ρ (g/cm^3)	3.7×10^{-20}	5.4×10^{-20}	4.9×10^{-19}
ρ/ρ_{gas}	.0001	.0004	
M ($M_\odot$)	2.8×10^{-6}	8.5×10^{-6}	1.1×10^{-5}
$\dot{M}$ ($M_\odot$/yr)	9.3×10^{-9}	1.9×10^{-7}	8.2×10^{-7}

It appears likely that in an environment where the radiation field permits dust formation and where $\rho \geqslant 8 \times 10^{-18}$ g.cm^{-3} (i.e. n $\geqslant 6 \times 10^5$ cm^{-3}), density enhancements trigger the dust formation. The compression could occur in stationary shocks which are certainly present where winds collide in binary systems (see below) and probably present in the decelerating zones where FeIII absorbtion lines reveal an "iron curtain" around WC9 stars (van der Hucht et al., 1982).

The typical WC9 star dust formation rate at its dust shell inner radius is 8×10^{-12} g/cm^3.s. The dust production by all WC stars within 3 kpc from the Sun is 1.4×10^{-6} $M_\odot$/yr, i.e. 3.1×10^{18} g/kpc^2.s. The dust production by all known galactic WR (i.e. WC7-10) stars together amounts to 3.5×10^{-6} $M_\odot$/yr; assuming that we know only one third of the WR content of our Galaxy would imply a galactic WR star amorphous carbon dust production rate of ~10^{-5} $M_\odot$/yr.

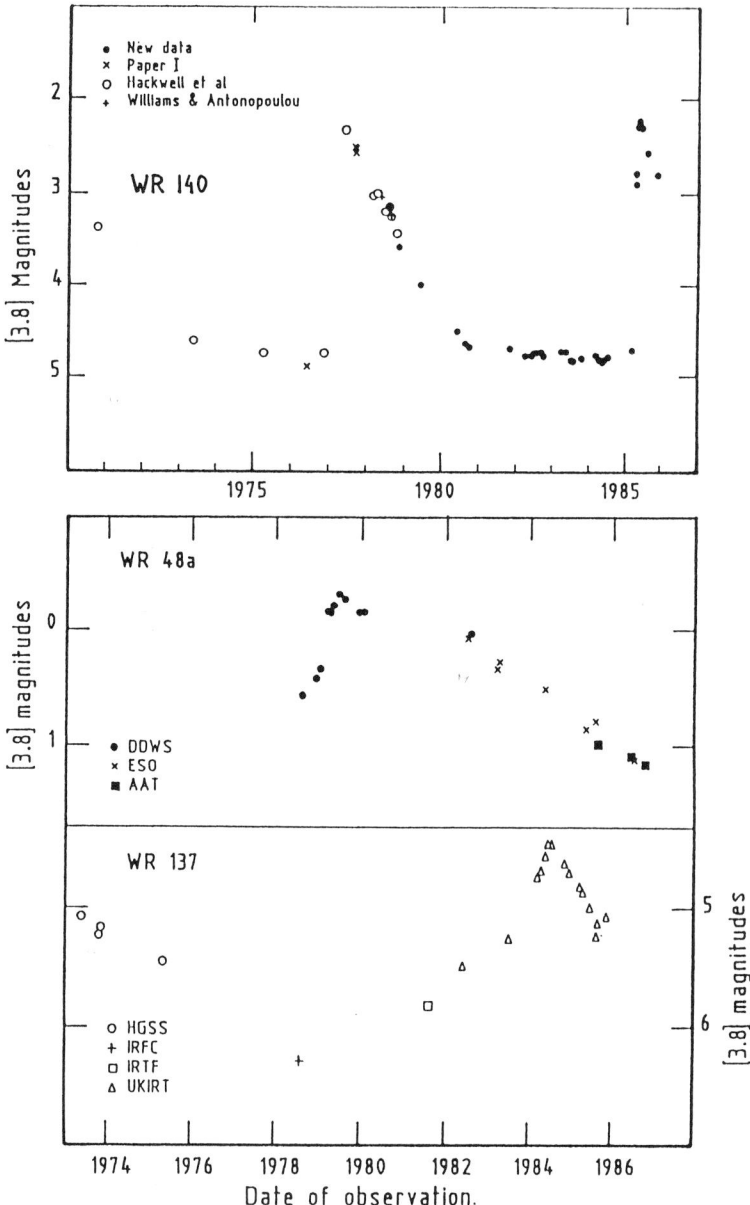

Fig. 2. Light curves at 3.8 μm of WR140 (WC7+O4), WR48a (WC8) and WR137 (WC7+a).

4. VARIABLE DUST EMISSION

Our infrared photometric survey of late-type WR stars shows little evidence for secular variation in mass loss or dust formation rates, apart from the three known variables: WR48a (Danks' WC 9 star, reclassified WC8 by WHT), WR137 (HD 192641) and WR140 (HD 193793), which was the first to be recognized as an episodic dust former (Williams et al. 1978). In 1985, WR140 again brightened sharply in the infrared owing to the formation of circumstellar dust (see Fig. 2a and Williams et al. 1987b). The relation of this event to the previous dust condensation episode strongly suggests that these represent a periodic phenomenon with P = 7.9 y. Phasing published radial velocities of WR140 against this period gives strong evidence for orbital motion. This leads to confirmation by Williams et al. (1987b) that the system is a binary, providing a framework for the interpretation of the observed radio and X-ray variations (see Fig. 3). Moffat & Williams et al. (1987) give as spectral types WC7+O4, with likely masses of respectively 14 $M_\odot$ and 26 $M_\odot$.

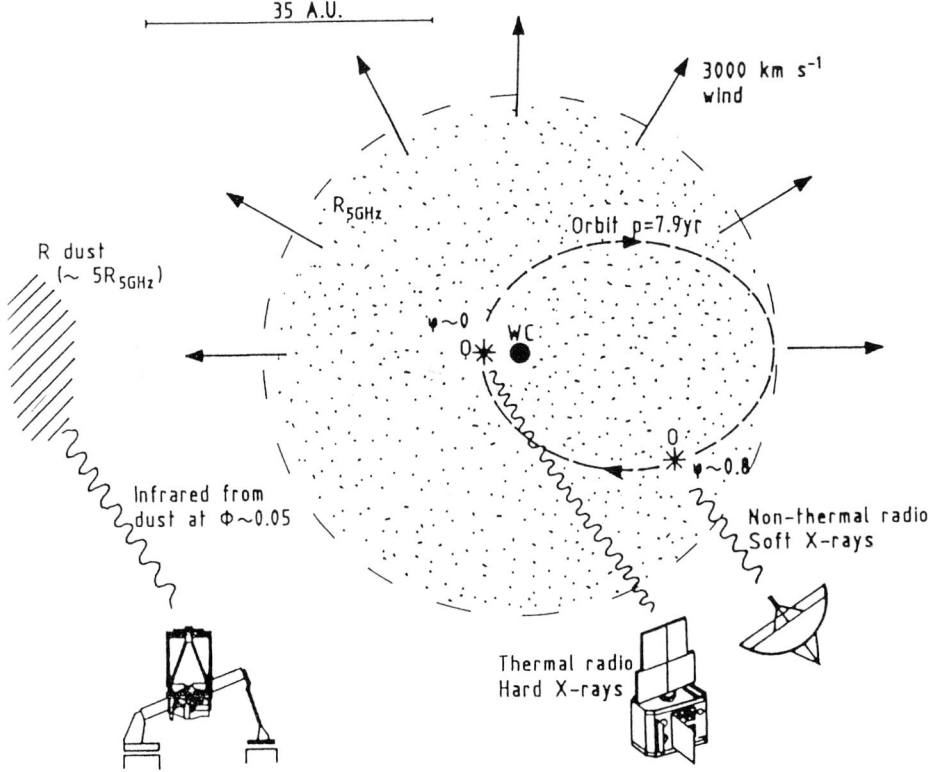

Fig.3. WR140 (WC7+O4): the O star in e = 0.75 orbit in the WC star wind.

Observations of WR48a and WR137 show steady fading since their recent dust condensation events and comparison with earlier photometry suggests a period of ~ 12 y for WR137 (see Figures 2b and 2c).

We conclude that there is a class of long period (~ 10 y) WC+O binaries, presumably of high eccentricity, which appear to shock-compress part of the wind sufficiently near periastron passage to allow dust to condense in it. Movement of a non-thermal source in and out of the WR wind causes X-ray and radio variations observed from WR140. More observations at all wavelengths are needed: we look forward to the return of the non-thermal radio emission from WR140 in about 1990 and its next infrared maximum in 1993.3 (Williams et al. 1987c).

References

Abbott, D.C., Bieging, J.H., Churchwell, E., Torres, A.V.: 1986, Astrophys. J. **303**, 239.
Allen, D., Swings, J.P., Harvey, P.M.: Astron. Astrophys. **20**, 333.
van der Hucht, K.A., Conti, P.S., Willis, A.J.: 1982, in: C. de Loore & A.J. Willis (eds), *Wolf-Rayet Stars: Observations, Physics, Evolution*, IAU Symp. **99** (Dordrecht: Reidel), p. 277.
van der Hucht, K.A., Williams, P.M., Thé, P.S.: 1984: in: A. Maeder & A. Renzini (eds), *Observational Tests of the Stellar Evolution Theory*, Proc. IAU Symp. **105** (Dordrecht: Reidel), p. 273.
van der Hucht, K.A., Cassinelli, J.P., Williams, P.M.: 1986, Astron. Astrophys. **168**, 111 and **175**, 356. (HCW)
van der Hucht, K.A., Williams, P.M., de Loore, C.W.H., Mulder, P.M.: 1987, Astron. Astrophys., in preparation.
Moffat, A.F.J., Lamontagne, R., Williams, P.M., Horn, J., Seggewiss, W.: 1987, Astrophys. J. **312**, 807.
Prantzos, N., Doom, C., Arnould, M., de Loore, C.: 1986, Astrophys. J. **304**, 695.
Schmutz, W. and Hamann, W.-R., Astron. Astrophys. **166**, L11.
Williams, P.M., Beattie, D.H., Lee, T.J., Stewart, J.M., Antonopoulou, E.: 1978, Monthly Notices Roy. Astron. Soc. **185**, 467.
Williams, P.M., Longmore, A.J., van der Hucht, K.A., Wamsteker, W.M., Talavera, A., Abbott, D.C., Telesco, C.M.: 1985, Monthly Notices Roy. Astron. Soc. **215**, 23P.
Williams, P.M., van der Hucht, K.A., Thé, P.S.: 1987a, Astron. Astrophys. **182**, 91. (WHT)
Williams, P.M., van der Hucht, K.A., van der Woerd, H., Wamsteker, W.M., Geballe, T.R., Garmany, C.D., Pollock, A.M.T.: 1987b, in C. de Loore & H. Lamers (eds), *Instabilities in Luminous Early-Type Stars*, Proc. Workshop in honour of Cornelis de Jager (Dordrecht: Reidel), p. 227.
Williams, P.M., van der Hucht, K.A., Thé, P.S.: 1987c, Quarterly J. Roy. Astron. Soc., **28**, 248.

IMPLICATIONS OF VARIABLE MASS-OUTFLOW ON MODELING

V. Doazan[1] and R.N. Thomas[2]
[1] Observatoire de Paris
61, Av. de l'Observatoire
F-75014 Paris, France
[2] Radiophysics Inc.
5475 Western Ave.
Boulder, CO, 80309, U.S.A.

ABSTRACT. It is now recognized that most early-type stars show mass-outflow variability. Using as example the Be stars, we discuss the effects of such variability on the structure of the stellar atmosphere and its implications on stellar modeling.

1. INTRODUCTION

Those theories of stellar winds which have been extensively developed rest on three basic assumptions: (i) the mass-outflow is time-independent, (ii) the origin/acceleration of mass-outflow is due either to thermal gas pressure or to thermal photospheric radiation pressure, (iii) there is no integral coupling between mass-outflow pattern and atmospheric structural pattern, which involves other (nonthermal) fluxes. Departure from any of these three assumptions can produce departure from some or all the others.

In contradiction to the above three assumptions, there is increasing evidence, especially from coordinated visual and far UV observations, that: (a) variable mass-outflows, rather than being exceptional, are quite common at least among early-type stars, (b) some nonthermal contribution to origin and acceleration of the mass-outflow seems required in an increasing variety of stars, (c) there exist remarkable associations between the radial evolution of mass-outflow and an atmospheric structural pattern produced by both the thermal and nonthermal fluxes (including mass) of the star (Doazan and Thomas, 1982, Thomas, 1983, Doazan, 1987).

Long ago, Thomas (1973) concluded that the variety of observations indicating the existence of mass-outflows and chromosphere-coronae could be understood *only* if the basic open system characteristic - mass loss - originated from subatmospheric nonthermal modes (pulsation, rotation, convection), which produced independent fluxes of mass and energy (radiative and nonradiative), cf. Thomas (1982). (Note that the wide-spread discussions of nonradial pulsations as a driving force, or current interest on solar/stellar seismology, are only particular aspects of this conclusion.) The important point is the presence of *some* kind, rather than a specific kind of nonthermal mode, as contrasted to thermal. This suffices to limit the "open system" to "nonthermal open system", and to link the

atmospheric structural pattern to the outward propagation of the associated
nonthermal fluxes. In the same way, any existing time-dependent mass-
outflow must originate in some time-dependent, nonthermal phenomenon - if
there is no corresponding change in the thermal accelerating mechanism -
and its propagation outward will affect the atmospheric structural pattern.
Conversely, an observed time-dependent change in structural pattern
generally implies a change in mass-outflow (we only discuss here mass-
outlfows from single stars.) The example of Be stars is particularly
illustrative of these effects, but they are hardly confined to only such
stars (cf. the above references).

Atmospheric analysis of variable mass-outflow and its atmospheric
effects is a diagnostic tool for inferring subatmospheric nonthermal
variability. Thus, a major focus must lie on those observations which
provide insight on the various time-scales which describe the variable
mass-outflow and the propagation of the flow outward through the several
atmospheric regions. All time-scales of all variable phenomena, both short
and long term variability of the entire atmosphere, must be investigated.
Needless to say, such observations presently exist for only a few stars -
and, among the early-type stars, for only the Be stars.

We have shown that mass-outflow variability has important effects on
the regional atmospheric structure of the Be stars (Doazan and Thomas,
1982, Thomas, 1983, Doazan, 1987). In this paper, we use these stars as a
reference to discuss variability of hot stars. Indeed, the Be stars exhibit
the largest variety and amplitudes of changes among all the observed early-
type stars, including OB stars type of variability.

2. OBSERVATIONAL EVIDENCE FOR VARIABLE MASS-OUTFLOW

Early-type stars generally show variable line-profiles for those lines used
to determine mass loss. But all the OB stars do not show the same variable
behaviour.

(i) In "normal" OB stars the variability has been usually attributed to
changes in the fine structure of the broad asymmetric profiles of highly
ionized resonance lines. These lines often show one, or several, narrow
absorption components, which seem to occur preferentially at high
velocities, although low velocity components are also observed. Far UV
variability in these stars is generally ascribed to the variability of these
narrow components, although the underlying P Cygni profile also varies
(Prinja and Howarth, 1986).

In the Be stars the situation differs profoundly. Generally, the entire
profile of the highly ionized resonance lines strongly varies in strength,
velocity, and shape (see Fig. 1). Variability of the narrow components -
when observed - describes only one aspect of Be stars variability.

(ii) The available far UV observations indicate that the occurrence
/enhancement of narrow absorption components in "normal" OB stars does not
show any systematic time-dependent pattern. They seem to occur in an
erratic way. It is not sure, however, that such an erratic character
reflects a real characteristic of the mass-outflow variability in these
stars. It may well only reflect the inadequacy of the available observations
for delineating organized variability patterns, which may exist.

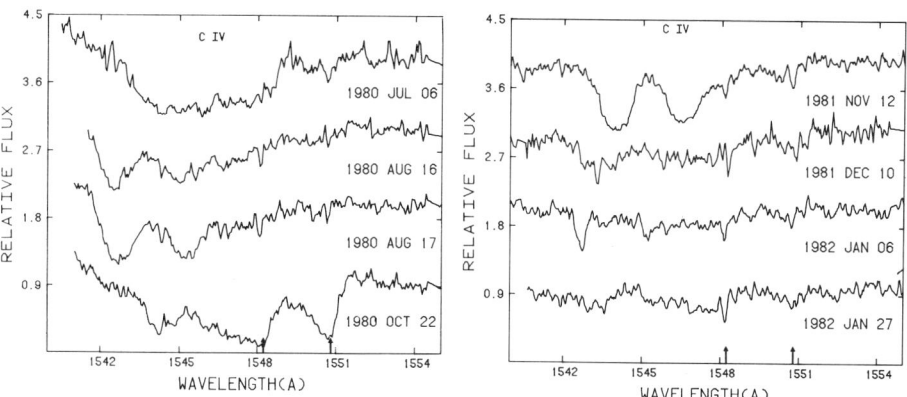

Fig. 1. Variability of the CIV resonance line-profiles of the Be star 59 Cyg (from Doazan et al. 1985). Note the large variations in strength, velocity, and shape. These changes are not simply due to variations of narrow absorption components. Systematic observations made over 8 yrs. show that the strength of the CIV lines delineate a long term variability pattern associated with the long term changes of the Balmer emission-lines.

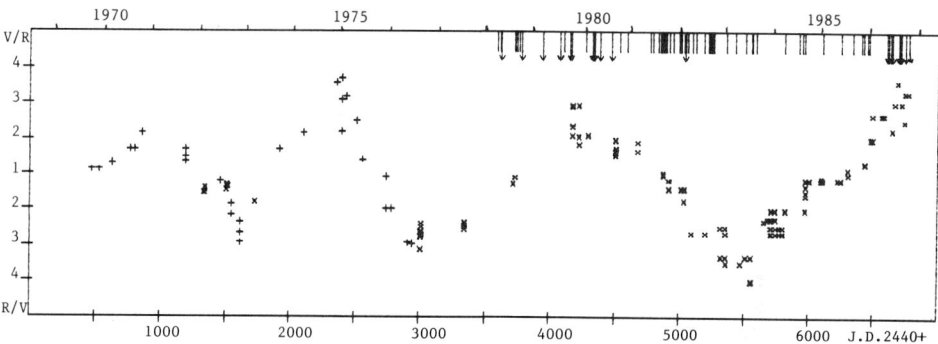

Fig. 2. V/R variations of the Balmer emission-lines in γ Cas. The Hα-emitting envelope undergoes cyclic accelerations and decelerations. The dates of all existing far UV observations made with IUE are indicated by tic marks. Arrows are added when high velocity components are present in the NV, CIV, and SiIV resonance lines. The occurrence of these components is associated with the V/R variations: they are present/more frequent when V/R>1, they are absent/less frequent when V/R<1.

On the contrary, in the Be stars, systematic observations of a few objects, made simultaneously in the far UV and in the visible, have shown that long term, organized, variability patterns of large amplitude are exhibited in the far UV, and that such patterns are associated with those, long-known, exhibited in the visual region by the Balmer emission/ shell lines. For 59 Cyg, the CIV line-strength shows large, systematic long term

changes which are associated with those of the Hα emission intensity
(Doazan et al., 1985, 1987a). For the best-observed Be star, γ Cas, short
term variability observed in the far UV - the occurrence of short-lived
narrow absorption components - is associated with the long term V/R
variations exhibited by the Balmer emission-lines (see Fig. 2) (Doazan,
1986, Doazan et al., 1987b). Finally, for θ Cr B, the CIV resonance lines
exhibit a long term systematic trend toward higher expansion velocities
(Doazan et al., 1986a, b, 1987 c).

Clearly, any Be-star model where the cool Hα-emitting envelope evolves
independently of the hot rapidly expanding region is unable to explain the
observed association between the CIV lines and the Balmer emission-lines.
We have shown that the propagation of a variable mass-outflow, through the
various distinct regions of the atmosphere, produces associated variability
patterns - short and long term, as well as phase-lags - such as those
which are observed (Doazan, 1987).

3. THEORETICAL IMPLICATIONS OF MASS-OUTFLOW VARIABILITY

The mass-outflow variability implied by the observations summarized above
is hardly a small perturbation on a steady outflow. For the Be stars, the
outflow velocities, where most of the absorbing ions are concentrated, can
change by a factor $\simeq$ 2-5, while the number of absorbing ions in the wind
may almost completely vanish, in the far UV as well as in the visible
region. This would imply a change in mass loss from 0 to $\simeq 10^{-8}$ yr.$^{-1}$. For
the OB stars, statistical studies indicate mass loss changes ranging
between 10% and 90% (Prinja and Howarth, 1986). These facts should be put
into perpsective relative to the generally accepted mass loss theory for
hot stars, i.e. the radiatively driven wind theory.

It is useful to comment on the recent improvements brought by
Pauldrach et al. (1986) to the radiatively driven wind theory. The presence
of species of much higher ionization than expected under LTE conditions has
generally been considered as embarrassing for this theory, where the
equations do not include any nonradiative heating terms. Pauldrach et al.
showed that some superionization may be produced by the increasing opacity
in resonance transition, under NLTE and low density gradient due to mass-
outflow. Effectively, this reduces the energy needed for ionizing these ions.
Hence, a greater concentration of highly ionized species will be produced.
But, if this interpretation is correct, we ask: what do the large variations
in CIV line strength - as observed in 59 Cyg and several other Be stars
(Barker and Marlborough, 1985) - and the "disappearance" of CIV lines - as
observed in θ Cr B (Doazan et al., 1986a) - imply? Clearly, they can only
imply a corresponding opacity, or density, variation, i.e. variation of the
mass-outflow, which the theory does not predict. Several changes in the
theory are required in order to agree with the observations.

(i) Contrary to the predictions of the hot star radiative acceleration
theory, the mass-outflow velocity is observed to be variable for a given
star, and the "terminal" velocity may strongly differ for stars of the same
L and g (see Thomas, 1983, Fig. 3-33, Prinja and Howarth, 1986). Using the
Be stars as guide, we suggest that $\dot{M}(t)$, at the lowest atmospheric
boundary, must be specified. Then, the solution for $v(r,t)$ as an initial

value problem must be sought. Because, at the moment, we do not know what values of $\dot{M}(t)$ are imposed by the nonthermal subatmospheric modes – because we know neither the modes nor their time-dependent behaviour – the solution must be sought parametrically. The short and long term variability patterns exhibited by the Be stars, as reflecting the variable mass-outflow in the entire atmosphere, are the best guide we presently have.

(ii) The strong variability in mass-outflow velocity observed in the Be stars – unaccompanied by similar variations of the photospheric radiation field – suggests that such variable acceleration is due to variable radiation pressure from a hot outer-atmosphere – a chromosphere-corona – (Thomas, 1983). This variability arises primarily in variable chromosphere-corona opacity. Therefore, the equations must contain dissipative terms to provide nonradiative heating: variable opacity is a consequence of variable mass-outflow.

(iii) Because the temperature of the Hα-emitting envelope is $\simeq$ 10,000 -20,000 K, and because this temperature lies below the NLTE boundary value for the hotter Be stars, cooling terms must enter the equations. Such a cooling may be of the CO-type, identified in the solar atmosphere.

(iv) Because the Hα-emitting envelope generally undergoes long term cyclic pulsations – i.e. it alternatively accelerates and decelerates – the self-interaction of a variable mass-outflow must be a primary ingredient of any Be star model, it is the cornerstone of our Be star model (Doazan and Thomas, 1982, Doazan, 1987).

REFERENCES

Barker, P., Marlborough, J.M.: 1985, *Astrophys. J.* **288**, 329
Doazan, V.: 1986, *IAU Circ.* No. **4232**
Doazan, V.: 1987, *IAU Colloq. No.92, Physics of Be stars*, ed. A. Slettebak, Dordrecht: Reidel, in press
Doazan, V., Thomas, R.N.: 1982, *The B Stars With and Without Emission-lines*, eds. A.B. Underhill and V. Doazan, NASA SP-**456**
Doazan, V., Grady, C., Snow, T.P., Peters, G.J., Marlborough, J.M., Barker, P.K., Bolton, C.T., Bourdonneau, B., Kuhi, L.V., Lyons, R.W., Polidan, P.S., Thomas, R.N.: 1985, *Astron. Astrophys.* **152**, 182
Doazan, V., Marlborough, J.M., Morossi, C., Peters, G.J., Rusconi, L., Sedmak, G., Stalio, R., Thomas, R.N.: 1986a, *Astron. Astrophys.* **158**, 1
Doazan, V., Morossi, C., Stalio, R., Thomas, R.N.: 1986b, *Astron. Astrophys.* **170**, 77
Doazan, V., Barylak, M., Rusconi, L., Sedmak, G., Thomas, R.N.: 1987a, *Astron. Astrophys.*, in prep.
Doazan, V., Rusconi, L., Sedmak, G., Thomas, R.N., Bourdoneau, B.: 1987b, *Astron. Astrophys.* submitted
Doazan, V., Rusconi, L., Sedmak, G., Thomas, R.N.: 1987c, *Astron. Astrophys.* **173**, L8
Pauldrach, A., Puls, J., Kudritzki, R.P.: 1986, *Astron. Astrophys.* **164**, 86
Prinja, R.K., Howarth, I.D.: 1986, *Astrophys. J. Suppl.* **61**, 357
Thomas, R.N.: 1973, *Astron. Astrophys.* **29**, 297
Thomas, R.N.: 1982, *Astrophys. J.* **263**, 87
Thomas, R.N.: 1983, *Stellar Atmospheric Structural Patterns*, ed. R.N. Thomas, NASA SP-**471**

Highly resolved emission-line profiles of B[e]-supergiants

F.-J. Zickgraf
Landessternwarte Königstuhl
D-6900 Heidelberg 1
FRG

ABSTRACT. First results of high-resolution (R=50000) spectroscopic observations of B[e]-stars of the Magellanic Clouds are presented. The line profiles are compared with theoretical profiles obtained for a disk-like geometry as suggested for the B[e]-supergiants.

1. Introduction

B[e]-stars are characterized by strong Balmer emission-lines (frequ. with P Cygni profiles), low excitation lines of FeII, [FeII], [OI], etc., and a strong IR excess due to thermal radiation of circumstellar dust. The evolutionary stage is not well known. However, some are supergiants like the eight B[e]-stars known in the Magellanic Clouds (MC`s) and several related galactic objects (e. g. Zickgraf et al. 1985, 1986, Wolf and Stahl 1985). For the B[e]-supergiants a two-component wind model was derived by Zickgraf et al. (1985) consisting of an equatorial disk formed by a rotation-enhanced slow, dense and cool wind and a normal high-velocity line-driven wind of an early-type supergiant in the polar region. The narrow low excitation emission lines (FWHM≈30 km/s) and the dust are supposed to originate in the equatorial disk.

2. New observations

Since the emission lines are only sufficiently resolved at a rather high spectral resolution R several B[e]-supergiants were observed in Dec. 1986 with the CES+CCD at the 1.4m CAT at ESO/La Silla with a resolution of 50000, i. e. 6 km/s. Examples for line profiles are shown in Fig. 1. Atmospheric absorption lines of O_2 around 6300 Å are indicated.

3. Discussion

Within the disk model the different spectroscopic appearance of the B[e]-supergiants is due to different inclination angles (Zickgraf et al. 1986). The prototype R 126 of the LMC represents the pole-on, R 50 of the SMC the edge-on, and S 22 of the LMC an intermediate case. Whereas R 126 shows a pure emission line spectrum in the visual in the spectrum of R 50 sharp shell-type absorption lines of TiII and CrII are present (cf. Zickgraf et al. 1985, 1986). The line profiles of R 126 are very narrow. The FWHM is only 8 km/s and hence no structure is visible. R 50 exhibits a split ($\Delta v=12$ km/s) [OI] 6300 line. [FeII] is not split, but slightly asymmetric. FeII shows a central absorption component, which is blueshifted by -5 km/s with respect to the forbidden lines. These facts indicate that line formation around R 50 takes place in different regions. S 22 shows lines of [OI], [FeII], and FeII which all have the same profiles, i. e. steep flanks, strong red and weak blue peaks. Hence in contrast to R 50 the conditions for line formation appear to be similar for the different lines. The related galactic star CD-24/5721 like R 50 shows a split [OI] 6300 line ($\Delta v=15$ km/s) (s. Fig. 2). Theoretical line profiles were calculated using a simple disk model consisting of an optically thin rotating shell with cut off polar callotes (Pöllitsch 1981). The thickness is 2d, inner and outer radius are R and RO, respectively, the rotation velocity is $v(r) \propto (1/r)^{**}m$, the source function is $S \propto (1/r)^{**}n$ with the fit parameters m, n. The stellar radius is set equal zero, hence obscuration can be neglected. The emitted intensity at frequency ν is given by

$$I_\nu \propto \int_V e_\nu dV \qquad (1)$$

with the emitted intensity per unit volume

$$e_\nu \propto (1/r)^{**}n * \delta(\nu - (\nu - \nu_0 \, v(r)/c)) \qquad (2)$$

with ν_0 = line centre frequency, $\delta(x)$ = Dirac δ-function, and V = volume of the disk. The profiles of [OI] of R 50 and CD-24/5721 could be fitted with a rather thick disk (Fig. 2) with d=2*R, and m=1, n=4, representing the case of constant angular momentum. The (projected) rotation velocity at r=R is 35 km/s. This model, however, is not applicable for asymmetric lines. In this case more refined models are necessary including e. g. expansion of the shell and the effects of dust opacity along lines-of-sight.

Acknowledgements. This work was supported by the Deutsche Forschungsgemeinschaft (Wo 269, 2-1).

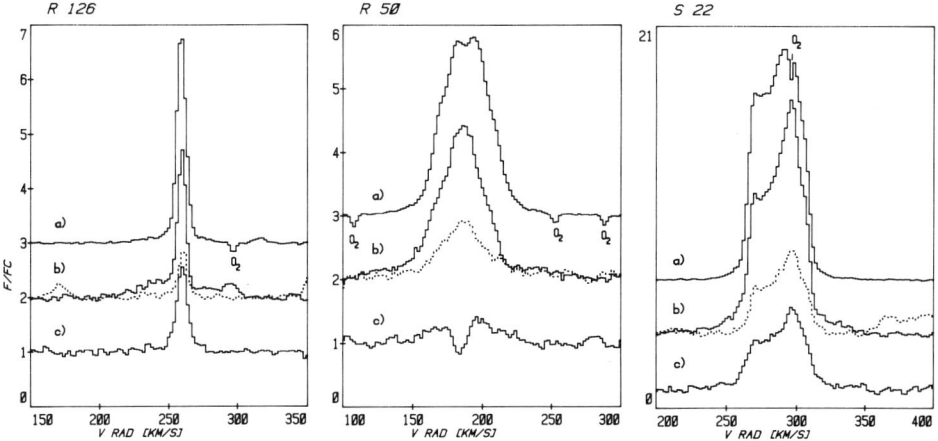

Fig. 1: Line profiles of three B[e]-supergiants of the MC's. a) [OI] 6300, b) [FeII] 4287 and 4277 (dashed), c) FeII 4296. The pole-on object R 126 shows lines with a FWHM of 8 km/s only. The edge-on object R 50 of the SMC exhibits differing line-profiles: [OI] is split by 12 km/s, [FeII] only shows a slight asymmetry, whereas a central absorption component is present in FeII. The intermediate case S 22 of the LMC shows no differences in the profiles of the various lines. All lines show a strong asymmetry with a strong red and a weak blue peak. Note also the steep flanks of the lines.

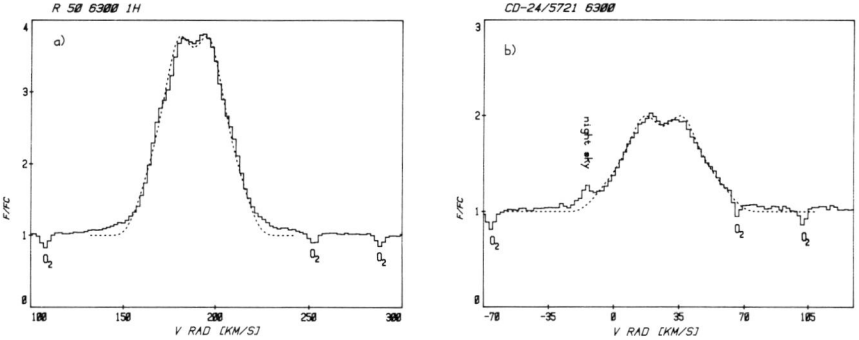

Fig. 2: [OI] 6300 of R 50 and the related galactic object CD-24/5721. The dashed lines represent theoretical profiles calculated with equ. (1) with a rather thick disk. The parameters used are d/R=2, RO/R=7.5, m=1, n=4.

References:

Pöllitsch, G. F.: 1981, Astron. Astrophys. 97, 175
Wolf, B., Stahl, O.: 1985, Astron. Astrophys. 148, 412
Zickgraf, F.-J., Wolf, B., Stahl, O., Leitherer, C., Klare,
    G.: 1985, Astron. Astrophys. 143, 421
Zickgraf, F.-J., Wolf, B., Stahl, O., Leitherer, C.,
    Appenzeller, I.: 1986, Astron. Astrophys. 163, 119

HD 316285: AN EXTREME P CYGNI STAR.

D. J. Hillier
Harvard-Smithsonian Center for Astrophysics.

P. J. McGregor, A. R. Hyland
Mt. Stromlo and Siding Spring Observatories,
Institute of Advanced Studies,
The Australian National University.

Abstract. HD 316285 (CD-27 119444, He3-1482, MWC 272) is an extreme P Cygni star which has been little studied. Extensive spectroscopic observations covering the wavelength range 4000Å to 3.5μm have been obtained to facilitate a detailed investigation. Emission lines due to H, HeI, FeII, FeIII, NI, NII, OI, NaI, MgII, AlII, SiII, [NII], and [FeII] can be identified. The degree of excitation is lower than that of P Cygni. From the strength of [NII]λ5755 relative to [OII]λ7325 we derive a N/O abundance ratio (by number) of greater than unity. Many of the metallic lines (*e.g.*, MgIIλ9218, λ9244) detected in the optical spectrum are due to continuum fluorescence. That is, strong UV lines absorb UV continuum flux and redistribute it to optical (or infrared) wavelengths. Detailed modelling of HD 316285 is currently in progress. Unfortunately, the large reddening ($E_{B-V} \approx 2.0$) and the uncertain distance (roughly 2kpc) add to the ambiguity of the results. Nevertheless, the models indicate that HD 316285 is undergoing a period of rapid mass loss ($\dot{M} \sim 10^{-4} M_\odot$ yr^{-1}). Interestingly, the velocity law may not be as slow as that found for P Cygni.

Introduction

HD 316285 is an extreme P Cygni star discovered by Humason and Merrill (1922). Detailed spectroscopic observations have been made by Merrill(1925), Swings and Struve (1941), and Carlson and Henize (1979). The spectrum is characterized by broad emission lines with blue shifted absorption, Balmer emission to H12, HeI emission, a continuum which is very red, and enhanced iron lines. No change was seen by Merrill over the three years of his observations, nor did he find any variability in its magnitude ($\Delta m_p < 0.1 mag$) on Harvard photographs which covered the years 1890 to 1918. Swings and Struve (1941) concluded that the ejection is more violent than that of P Cygni.

Intrinsic Parameters

The intrinsic parameters of HD 316285 are very uncertain. The values below are typical of those we have been using in order to model the spectral characteristics.

$$V = 9.1, \quad E_{B-V} = 1.9, \quad T_{eff} \leq 20000 K$$

$$R \sim 70 R_\odot, \quad d \sim 2 kpc ?$$

$$\dot{M} \sim 10^{-4} M_\odot \text{ yr}^{-1}, \quad V_\infty = 330 \text{ to } 500 \text{km s}^{-1}$$

New Observations

>Low resolution red spectra (R $\sim$ 1500) - 4800Å to 9800Å.
>Low resolution IR spectra (R $\sim$ 500) - 1.08μm to 3.5μm.
>High resolution spectra of selected emission lines.

Emission lines due to H, He I, N I, N II, O I, Na I, Mg II, Al II, Si II, Fe II, Fe III, [NII] and [FeII] are present in the spectra. In addition, lines due to O II, Mg I and C I could be present. The very rich Fe II spectrum and the non-LTE line formation are the prime reasons for the difficulty in confirming or denying these later identifications. The spectra, together with line identifications and equivalent widths will be published elsewhere.

Continuum Fluorescence

It is commonly believed that many of the optical FeII lines are in emission as a direct consequence of UV continuum fluorescence (*e.g.*, Brandi, Gosset, and Swings 1987). That is, strong UV lines absorb UV continuum flux and redistribute it to optical (or IR) wavelengths. This process also appears to be operating for Al II, Mg II, Si II and possibly other atomic species. A prerequisite is that the optical transitions should be coupled to a level which has an UV transition in the Balmer continuum connecting it to the ground or a low lying excited state.

Consider the MgII($4p\,^2P^o - 4s\,^2S$) doublet at 9226Å. The upper level of this transition can be populated by continuum flux at 1240Å. The Einstein A coefficients indicate that greater than 60% of the continuum photons absorbed by the resonance transition will appear as emission at 9226Å.

As we detect strong emission in the MgII(λ8232) and MgII(λ7890) transitions the fluorescent process must also be operating from the $3p\,^2P^o$ state whose excitation potential is 4.43eV.

Nitrogen Abundance

The stellar evolution models of Maeder(1983) indicate that we should see changes in the surface composition of massive supergiants due to the appearance of CNO processed material at the stellar surface. A comparison of observation with theory is difficult as the interpretation of subordinate emission lines of C, N and O requires detailed non-LTE radiative transfer calculations. However, the detection of [NII]λ5755 in HD 316285 allows us to make some quantitative statements concerning the N/O abundance ratio in this star.

To interpret the forbidden lines we made the following assumptions.

(i) [OII]λ7325 and [NII]λ5755 are formed in the wind where the terminal velocity has been obtained. Reasonable assumption as $N_e \sim 10^7$.

(ii) The wind is isothermal.

(iii) O^+ and N^+ are the dominant ionization states of O and N respectively. There are both observational and theoretical arguments which suggest this is reasonable.

With these assumptions it can be shown that

$$\frac{[\text{OII}]\lambda 7325}{[\text{NII}]\lambda 5755} = 2.6 e^{(\frac{-1.14}{T})} \left(\frac{N_O}{N_N}\right).$$

The flux ratio of [OII]λ7325 to [NII]λ5755 is independent of $\dot{M}$ and the assumed distance, and is only moderately sensitive to the electron temperature. Using the upper limit on the [OII]λ7325 flux of $6.6 \times 10^{-11} erg\,cm^{-2}s^{-1}$, an electron temperature of 10^4K, and the observed [NII]λ5755 flux of $(4.8 \pm 1.4) \times 10^{-10} erg\,cm^{-2}s^{-1}$ ($E_{B-V} = 1.9 \pm 0.1$) we infer

$$\frac{N_O}{N_N} < 0.17$$

which is very different from the solar value of 7.

Whilst the assumptions made will affect the above determination, it is extremely unlikely that HD 316285 could have a solar N to O abundance ratio. We conclude that HD 316285 has $N_N > N_O$, and that we are seeing CNO processed material at the stellar surface. The overabundance of N is qualitatively consistent with the observed optical spectrum which shows a much richer spectrum of nitrogen compared with carbon and oxygen. As theoretical modelling of the permitted emission lines improves, we will be able to refine this result, and shall also be able to examine the C/N and He/H abundance ratios.

Model Calculations

We have been performing detailed radiative transfer calculations using the code previously described by Hillier(1987). This assumes spherical geometry, and can use either the Sobolev approximation or the comoving frame method of Mihalas, Kunasz and Hummer (1975) to do the line transfer.

The calculations indicate that to model the spectrum of HD 316285 (and P Cygni) it is necessary to

(i) include the Lyman lines - they cannot be assumed to be in detailed balance.
(ii) treat the Lyman continuum correctly - there is a strong coupling between the Lyman continuum and Lα.
(iii) use spherical geometry and allow for the effects of the wind on the continuum formation - our models do this, but at this stage we cannot include line blanketing.

At present, the calculations are unable to explain the observed Hydrogen spectrum and the IR continuum shape *simultaneously*. Our models tend to yield a Hβ to Hα ratio which is too small. Further, our models cannot explain the large strength observed for H(15-3) and other Paschen lines. A comparison of model profiles with observation is shown in Fig. 1 for Hα, and Fig. 2 for H(15-3). The parameters of the model are :-

$$R_* = 70 R_\odot \qquad \dot{M} = 1.0 \times 10^{-4} M_\odot \text{ yr}^{-1}$$
$$V_\infty = 350 \text{km s}^{-1}, \qquad L = 7.5 \times 10^5 L_\odot$$
$$T_{wind} \sim 12000 K \qquad N_{He}/N_H = 0.1$$

At depth, an exponential atmosphere with a scale height of $0.02 R_*$ was adopted, whilst the velocity law in the wind is closely represented by the standard velocity law

$$V(r) = 30 + (V_\infty - 30)(1 - R_*/r)$$

Also shown in the accompanying figures is the temperature structure of the model computed under the assumption of radiative equilibrium. For comparison, the grey temperature structure is also illustrated.

References

Brandi, E, Gosset, E., and Swings, J. P. 1987, *Astr. Ap.*, **175**, 151
Carlson, E. D., and Heinize, K. G. 1979, Visitas in Astron., **23**, 213
Hillier., D. J. 1987, *Ap. J. Supp. Ser.*, **63**, 947
Humason, M. L., and Merrill, P. W. 1922, *Publ. A. S. P.*, **34**, 294
Maeder, A. 1983, *Astr. Ap.*, **120**, 113
Merrill, 1925, *Ap. J.*, **61**, 418
Mihalas, D., Kunasz, P. B., and Hummer, D. G. 1975, *Ap. J.*, **219**, 635
Swings, P., and Struve, O. 1941, *Ap. J.*, **93**, 349
van Blerkom, D. 1978, *Ap. J.*, **221**, 186

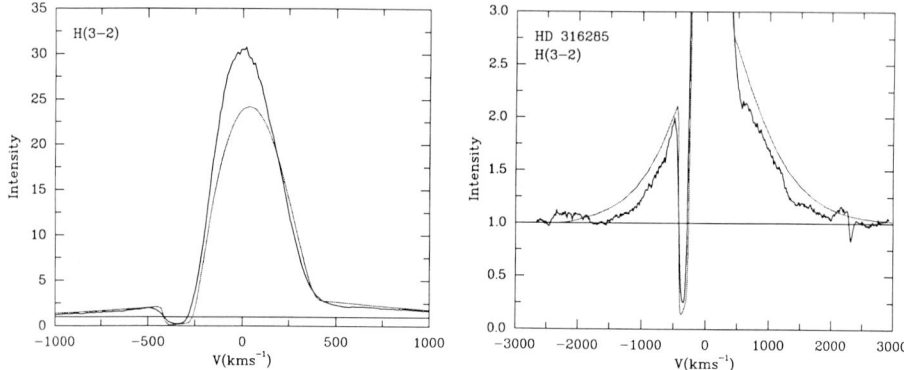

Fig. 1. Comparison of the observed profile, and predicted profile (dotted) for Hα. A turbulent velocity of 30km s^{-1} was used to compute the line profile. The extended high velocity wings are predicted to be much stronger than observed. One possible explanation is that the wind is clumpy - the hydrogen emission scales as the density squared whereas the electron scattering opacity scales as the density.

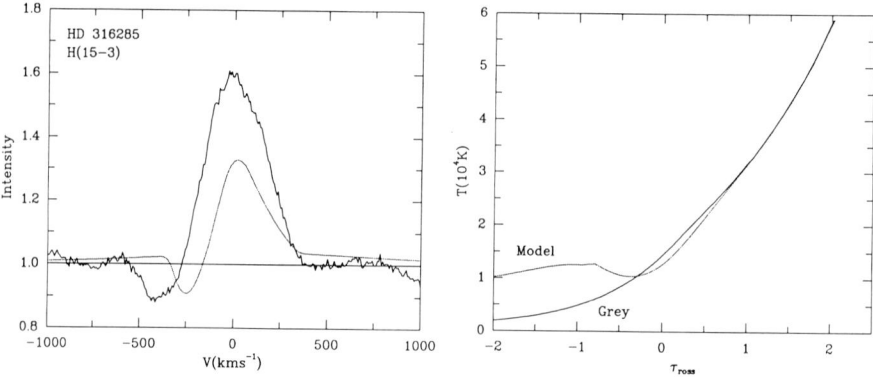

Fig. 2. Comparison of the observed profile, and predicted profile (dotted) for H(15-3)λ8545. The line is blended with a member of the CaII triplet, but the observed Paschen decrement indicates that the contribution by CaII should be small. A 20 level hydrogen atom was used to evaluate the statistical equilibrium equations. The FWHM of this line is similar to that of Hα which would seem to indicate that we cannot have a linear velocity law as determined for P Cygni by van Blerkom(1978)

Fig. 3. The temperature distribution computed under the assumption of radiative equilibrium. For comparison, the grey solution is shown. This was computed using the final non-LTE model opacities.

SOMETHING TO DO WITH THE X-RAY EMISSION OF EARLY-TYPE STARS

A.M.T.Pollock
EXOSAT Observatory, Astrophysics Division
Space Science Department, ESTEC
Postbus 299, 2200AG Noordwijk
The Netherlands

ABSTRACT. The discovery, principally by Abbott and colleagues, of a non-thermal component to the radio emission of a significant fraction of early-type stars suggests an alternative interpretation of the ubiquotous X-ray emission from Wolf-Rayet and OB stellar winds to the usual shock models based on Lucy and White's work. The radio observations show the presence of relativistic electrons and magnetic fields far out in the wind. The electrons probably also account directly for the X-ray emission because first, Compton scattering of photospheric radiation is likely to produce observable amounts of high-energy radiation and second, the brightest non-thermal radio sources, of both the Wolf-Rayet and OB stars, are among the brightest X-ray sources. The acceleration of these electrons may be achieved by current-sheet acceleration (magnetic reconnection) in both binary systems, as oppositely directed fields are pushed together by colliding winds, and single winds, through the dissipation of turbulence.

1. INTRODUCTION

The past few years have seen the unexpected discovery of X-ray and non-thermal radio emission from early-type stars. It was one of the first discoveries of the *Einstein Observatory* (Harnden et al. 1979, Seward et al. 1979) that stars of type B and earlier are without any obvious exceptions X-ray sources whose luminosities scale roughly linearly with their bolometric luminosities (e.g. Pallavicini et al. 1981) although considerable variations are doubtless present (e.g. Long and White 1980). More recently Abbott et al. (1984,1985,1986) discovered that a significant fraction of early-type stars have a non-thermal radio component sufficiently strong to be observable above the thermal emission more normally expected.

Of crucial importance to the interpretation of the measurements is the strength and opacity of the strong stellar winds typical of these objects. Although it is usually thought that both types of emission occur far out in the wind, several authors have argued that mass loss rates lower than those derived with reasonable consistency from a variety of different measurements (whose interpretation is admittedly not entirely straightforward) or unusual ionisation structure would make the wind thin enough to allow both the X-rays (Stewart and Fabian 1981, Waldron 1984) and radio radiation (Underhill 1984) to be interpreted as surface phenomena involving coronae confined near the stellar photosphere. However, the X-ray binaries such as 4U1700 − 37 (e.g. White, Kallman and Swank 1983) that contain a wind-accreting neutron star orbiting near the surface of an early-type star show clearly that the wind is indeed dense and thick, implying in early-type stars generally that the wind itself is the site of both types of emission.

Lucy and White (1980) and Lucy (1982) proposed to explain the X-ray emission through the appearence of multiple shocks in the outer parts of the wind that develop as the limiting state of instabilities to which line-driven winds are subject. White (1985), in a further development, had the idea that the same shocks were also responsible for acceleration of the population of relativistic electrons that plausibly account for the non-thermal radio radiation via the synchrotron mechanism. However, Pollock (1987a) discussed observations of some Wolf-Rayet stars including HD193793, by far the brightest of the early-type stellar non-thermal radio and X-ray sources, and showed that synchrotron-emitting relativistic electrons themselves could be a significant source of X-rays by Compton scattering of photospheric optical and UV radiation independent of the presence or otherwise of shocks. Indeed, for early-type stars in general the observational and "theoretical" evidence is of a more direct connection between the X-ray and radio properties than via the Lucy and White shocks, although the current sparseness of the observations does encourage a cautious view. Nevertheless, it will be argued that relativistic electrons and magnetic fields are general features of massive stellar winds showing, when circumstances allow, as a non-thermal radio component visible against the background of the wind's thermal emission and as the widely observed X-ray emission of these objects, where there is no competing thermal background.

2. OBSERVATIONS

2.1 The Wolf-Rayet Stars

The best coverage so far in both the radio and X-ray regions is of the Wolf-Rayet stars. Abbott et al.'s (1986) distance-limited radio survey of 40 stars with the VLA contained 6 definite or probable non-thermal radio stars, so classified on the basis of one or more of the criteria of excess or unresolved emission, negative spectral index or variability ; the others had the properties expected of thermal wind emission. Pollock (1987b) reported a uniform analysis of all 48 stars, including 5 of the 6 non-thermal radio stars, observed with the X-ray instruments aboard the *Einstein Observatory*. Most of the brightest X-ray stars were either non-thermal radio stars or massive binary systems while single stars and possible low-mass binary systems were generally among the faintest objects.

The 6th and brightest non-thermal radio star is HD193793, an object of exceptional interest that was found with *EXOSAT* to be the brightest "coronal" X-ray source (Pollock 1987a). Williams et al. (1986) discovered that its enormous infrared outbursts define a 7.9-year period that is reproduced in the absorption-line radial velocities, leaving little doubt that the system is a distant, massive binary system.

2.2 The OB Stars

The observations of OB stars are even more patchy. The published non-thermal radio sample numbers 5 stars (Abbott et al. 1985), or about 20% of those observed, and 2 of them have not been observed in X-rays. The 3 which have been are 9 Sgr and the Cyg OB2 stars #s 8A and 9. In the X-ray field several papers have reported *Einstein* observations of a variety of OB stars. Although it is not entirely straightforward to make comparisons because of the different approaches adopted, it is clear that the two Cyg OB2 stars, whose data are given by Harnden et al. (1979), are the brightest OB stellar X-ray sources with luminosities $\gtrsim 10^{34}$ ergs s^{-1}. 9 Sgr is also a strongly detected source.

3. "THEORY" (Pollock 1987a)

If, as seems likely, the non-thermal radio component is from relativistic electrons radiating synchrotron photons in an embedded magnetic field the 5 GHz free-free optical depth is so high that electrons must be present out to great distances from the base of the wind (White 1985, Abbott et al. 1986). In addition to synchrotron losses, the electrons face such formidable radiation densities that observable levels of X-rays may easily be produced by Compton scattering. For a power-law electron slope of -3 the ratio the Compton 1 keV emissivity to the 5 GHz emissivity is given by

$$Q_{1\text{keV}/5\text{GHz}} \approx 4T_{*4}^4 B_*^{-2} \text{X}\mu\text{Jy SmJy}^{-1} \qquad (1)$$

written in terms of the fluxes that would be observed in the optically thin case. $T_{*4}10^4$ K is the star's effective temperature and B_* its surface magnetic field. Most of the detections are at levels not much above the current observational limits of about 0.1 μJy at 1 keV and 1 mJy at 5 GHz. Because free-free absorption at 5 GHz and photoelectric absorption at 1 keV define very roughly equal volumes of a stellar wind in which relativistic electrons are visible equation (1) will give a general idea of the fluxes to be expected from electrons in the wind. It does not require an outrageous combination of the stellar parameters T_{*4} and B_* to reproduce the observations; for surface fields of 10 to 100 G or so and $T_{*4} = 3$ the same electron spectrum is being observed at both 5 GHz and 1 KeV.

4. IMPLICATIONS

The clear implication of the observations and the "theory", however meagre they may be, is of a direct connection between the X-ray and non-thermal radio fluxes : a sufficiently numerous population of relativistic electrons outshines the wind's thermal radio emission and is a luminous X-ray source. What of smaller electron populations ? They would remain hidden from radio gaze by the thermal free-free emission but would shine on in X-rays at the lower levels more normally observed. The only way to escape this conclusion would be for a second X-ray mechanism to take over from the non-thermal component as the same happened quite independently in the radio, an idea of little appeal. The X-ray emission from early-type stars is then a non-thermal phenomenon and testament to the production of relativistic electrons as a normal part of the activity of stellar winds. The possible detection of X-ray emission lines by Cassinelli and Swank (1983) relates to a high temperature thermal component that contributes only a small fraction of the total emission.

5. A SPECULATION ON THE ORIGIN OF RELATIVISTIC ELECTRONS IN STELLAR WINDS

Besides the relativistic electrons the other important aspect of the synchrotron interpretation of stellar wind non-thermal radio emission is its empirical confirmation of the presence of a magnetic field in the outer parts of the wind. Magnetic activity, or more particularly magnetic reconnection or current sheet acceleration, successfully explains some of the high-energy particle phenomena in solar flares and the terrestrial magnetosphere ; Parker (1979) and Syrovatskii (1981) describe the basic thinking. The same type of process is likely to occur in stellar winds. In fact, a pair of colliding highly-conducting magnetic stellar winds would seem to provide the ideal setting for magnetic reconnection. A small dissipative region around the stagnation point on the line joining two early-type stars may be the seat of considerable particle acceleration. This may account for the prevalence of the Wolf-Rayet binaries among the brighter X-ray sources (Pollock 1987b), while the especially spectacular observational properties of HD193793 (WR140) described by Williams *et al.* (1986) may occur as the compression and reconnection of the fields is compounded by the highly eccentric binary motion. A similar but more widespread phenomenon that may occur (Parker 1979), and account for the X-ray emission of single winds, is the ability of magnetic reconnection to provide the main mechanism for the dissipation of turbulence , so that particle acceleration may be occuring locally throughout the wind.

REFERENCES

Abbott,D.C., Bieging,J.H., and Churchwell,E. 1984, *Ap.J.*, **280**, 671

Abbott,D.C., Bieging,J.H. and Churchwell,E. 1985, *Radio Stars*, ed. R.Hjellming and D.Gibson (New York : Plenum), p19

Abbott,D.C., Bieging,J.H., Churchwell,E. and Torres,A.V. 1986, *Ap.J.*, **303**, 239

Cassinelli,J.P. and Swank,J.H. 1983, *Ap.J.*, **271**, 681

Harnden,F.R. *et al.* 1979, *Ap.J.*, **234**, L51

Long,K.S. and White,R.L., 1980, *Ap.J.*, **239**, L65

Lucy,L.B. 1982, *Ap.J.*, **255**, 286

Lucy,L.B. and White,R.L. 1980, *Ap.J.*, **241**, 300

Pallavicini,R. *et al.* 1981, *Ap.J.*, **248**, 279

Parker,E.N. 1979, *Cosmical Magnetic Fields* (Oxford : Clarendon Press)

Pollock,A.M.T. 1987a, *Astr.Ap.*, **171**, 135
Pollock,A.M.T. 1987b, *Ap.J.*, in press
Seward,F.D. *et al.* 1979, *Ap.J.*, **234**, L55
Stewart,G.C. and Fabian,A.C. 1981, *M.N.R.A.S.*, **197**, 713
Syrovatskii,S.I. 1981, *Ann.Rev.Astron.Astrophys*, **19**, 163
Underhill,A.B. 1984, *Ap.J.*, **276**, 583
Waldron,W.L.,1984, *Ap.J.*, **282**, 256
White,N.E., Kallman,T.R. and Swank,J.H. 1983, *Ap.J.*, **269**, 264
White,R.L. 1985, *Ap.J.*, **289**, 698
Williams,P.M. *et al.* 1986, *Instabilities in Luminous Early-Type Stars*, ed. H.J.G.L.M. Lamers and C.W.H. de Loore (Dordrecht : Reidel), in press

THE O 6.5 IIIf STAR BD+60°2522 AND ITS INTERACTION WITH THE SURROUNDING INTERSTELLAR MEDIUM:
CCD-IMAGING AND SPECTROSCOPIC OBSERVATIONS OF THE STELLAR WIND BUBBLE NGC 7635

C. Jaeger, C. Leitherer[1], C. Chavarria K.[2]
Landessternwarte
Koenigstuhl
6900 Heidelberg
Germany

ABSTRACT. We observed the galactic ring nebula NGC 7635, called the bubble nebula, by means of fluxcalibrated narrowband CCD-frames (H_α, H_β, [NII], [OIII]) and highly resolved coude-longslit spectrograms, covering the wavelength range from 6540 to 6750 Å. The results (extinction, exitation, kinematics etc.) yield a picture of NGC 7635 as a wind blown shell, caused by the strong stellar wind from the O 6.5 IIIf star BD+60°2522. An explanation is given on the physical association of the comet like features seen within the bubble nebula with the nebula itself.

1. Introduction

The luminous O-type star BD+60°2522 is embedded in the extended HII-region S 162. Part of S162 is NGC 7635, the striking spherically symmetric bubble nebula surrounding BD+60°2522 (Fig. 1). This star itself is unique in that it is the only known O star apparently associated with warm dust (cf. Fig. 2). The most important parameters are summarized in table I. In an attempt to study the interaction of the central O star with the surrounding HII-region we obtained flux-calibrated narrow-band CCD-frames (H_α, H_β, [OIII], [NII]) and highly resolved coude spectrograms, $\Delta v \approx 7$ km/sec.

2. Observations

The observations were carried out at the Calar Alto Observatory in January, 14./15., 1986. The spectrograms were taken with the 2.2 m telescope while the CCD images were taken with the 1.23 m telescope, using the CCD camera of the Landessternwarte.

3. Results

One of the morphological features is the sharp boundary between the bubble nebula and the nearby HII-region. This boundary represents a drop in density, since the output of stellar Lyman-photons suffices to ionize the whole S 162 complex, i.e. NGC 7635 is density bounded. Notice that BD+60°2522 is displaced relative to the symmetry center of the bubble. Interpreting the bubble nebula as a stellar wind blown shell, this displacement could be understood in terms of a density gradient in S 162. Marked peculiarity within NCG 7635 are the three comet like condensations west of BD+60°2522. They are exactly aligned, having a position angle of (-62+3)° to the north. The extinction, examined by the H_α/H_β-ratio, is found to be very uniform all over the nebula, $E_{B-V} \approx 0.5$ mag, except for the bright westward condensations. There the E_{B-V} reaches values about 1.2 mag, indicating presence of dust.

The ionization structure within NGC 7635 can be studied from the [NII]/H_α and [OIII]/H_β images. [NII]/H_α is largest in the bright condensations (≈ 0.4), where [OIII]/H_β is smallest (≈ 0.5), indicative of different ionization zones. Using the diagnostic diagrams by Sabbadin & D'Odorico (Astron. Astrophys. **49**, 119, 1976), which separate the exitation mechanisms (photoionization in HII-regions, shock-wave heating and exitation in planetary nebulae), one has the following result: The bubble is photoionized, while the bright knots show evidence for shock ionization. A high [NII]/[OIII]-ratio (>10) is indicative for the ionization of dense (neutral) matter. The density structure is derived from the [SII] 6716/6731 ratio obtained from the long-slit spectrograms. The maximum density in the knots is about 10^5 cm^{-3}, whereas the average density outside the knots is $10^2 - 10^3$ cm^{-3}.

The dereddened H_β flux integrated over the bubble nebula without the immersed stars is about 5.2×10^{-10} erg sec^{-1} cm^{-1}. The corresponding total mass of NGC 7635 including the knots - they contribute about 50% - turns out to be 3 $M_\odot$.

An estimate of the nitrogen abundance relative to sulfur can be obtained from the I(6548+6583)/I(6716+6731) ratio. From this ratio we derive the relative ionic abundance N(N$^+$)/N(S$^+$), resulting from the relation given by Benvenuti et al. (Astron. Astrophys.28, 447, 1973). These relative ionic abundances N(N$^+$)/N(S$^+$) can be converted to total abundances using the ionization correcting factors published by Lynds & O'Neill (ApJ **274**, 650, 1983). We find within the limits of uncertainty the [N/S] abundance to be the same on the bubble and on the bright condensations, about 0.4. This value is in good agreement with the [N/S] value for the Perseus arm (Talent and Dufour, ApJ **233**, 888, 1979).

The velocity field can be investigated from our coude-longslit spectrograms. Since we did not find significant radial-velocity variations from line to line eastward of the star, we took the average radial velocity from the [NII], [SII], and H_α lines and derived a mean LSR velocity of (-41 ± 7) km/sec for NGC 7635. On the other hand the bright knots show a significant radial-velocity difference of (12 ± 1) km/sec (Fig.3). Remarkable is the different behavior of the radial velocities of the H and the forbidden lines west of the star, which also appears in the line width. Being on the nebula about 40 km/sec, the FWHM on the knots is not significant higher, 45 km/sec. For the forbidden lines the FWHM has the lowest value on the knots, 20 - 25 km/sec, whereas on the nebula the FWHM reaches 30 km/sec. These kinematic results indicate an origin of H_α and the forbidden lines in different regions in the knots.

4. Discussion

The bubble nebula NGC 7635 can be interpreted as a stellar wind blown shell caused by the strong stellar wind of BD+60°2522 interacting with the ambient interstellar medium. The mechanical power of the stellar wind, $L_w \approx 3.4 \cdot 10^{36}$ erg sec^{-1} is transferred to the ISM leading to an expansion of the swept-up matter by $v_{exp} \approx 30$ km/sec. Then the age of the bubble is of the order 10^4 years. The mass ejected by the star during that time is about 1% of the observed shell mass, i.e. the shell does essentially consist of swept-up ISM. Together with our derived [N/S] abundance this supports the wind-blown shell model, since if the bubble nebula were due to an explosive stellar event, an overabundance of processed material and a higher M_{wind}/M_{shell}-ratio should be expected. Inside the wind-blown shell the typical densities are of the order 10^{-2} cm^{-3}. The high density of the knots (>10^4 cm^{-3}) would encounter severe difficulties for them to exist inside the bubble. If the knots were inside the bubble, they must have formed later than the bubble, i.e. they must have been ejected by the star. This does not seem plausible due to (I) the derived chemical abundances and (II) the orientation of the knots relative to the star. The correspondance of the nebular mass derived from optical and radio measurements and the Balmer-decrement prove that the nebula is optically thin. The dust/gas ratio is the same as in the ISM, 10^{-2} as it is to be expected from extinction. On the other hand the higher dust/gas ratio, > 10^{-1} and the higher A_v for the knots indicate that they are optically thick. So they should be situated on the line of sight behind the bubble, associated with the molecular cloud (globule) and the warm dust detected by Thronson et al. (MNRAS 201, 429, 1982). As already mentioned, the knots show evidence for an shock-ionization front in dense (neutral) matter. The kinematical results provide an explanation for the origin and the existence of this ionization front. On the nebula the FWHM of the H_α and the forbidden lines have typical values, providing the given expansion velocity. On the contrary on the knots the line widths of the forbidden lines show no evidence for expansion but for deccelerated matter, while the H_α-FWHM is the same as on the nebula. This indicates that the emission of the forbidden lines origin mainly from the shockfront at the surface of the globule just lying behind the bubble.

We suggest the following model: The stellar-wind blown shell NGC 7635 expands with a higher velocity than sound speeed, > 20 km/sec, and collides with a globule, which moves with 12 km/sec relative to the bubble. The surface layers of the globule will be compressed, so that a weak shock will form.

Table I.

Parameter	Value	Parameter	Value
T_{EFF}	35000 K	$\log N_L$	49.15
E(B-V)	0.73	U	76 pc/cm²
M_v	-5.6		
R/R_0	13.1	M	$1.6\ 10^6$ M/yr
L/L_0	$3.5\ 10^5$	v	2600 km/sec

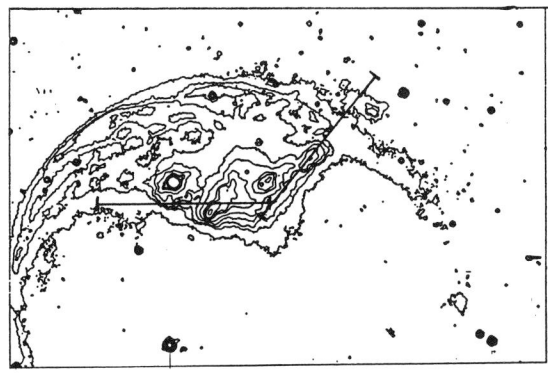

Figure 1. Isocontour plot of the H -frame, showing the northern part of NGC 7635 and a small fraction of S 162. North is up, east to the left. The slit positions are indicated, slit length 75". The lowest contour level is ca. 14.5 mag, steps 0.5 mag.

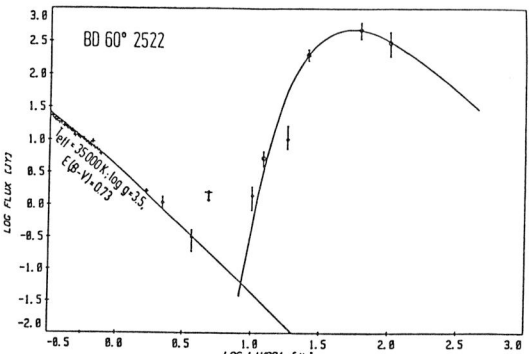

Figure 2. Observed, dereddened energy distribution of BD+60°2522 (symbols) from earth bound and satellite born (IRAS) observations. The solid line represent a model-atmosphere fit for the stellar flux and a black-body fit for the dust contribution ($T^{Dust} = 45$ K)

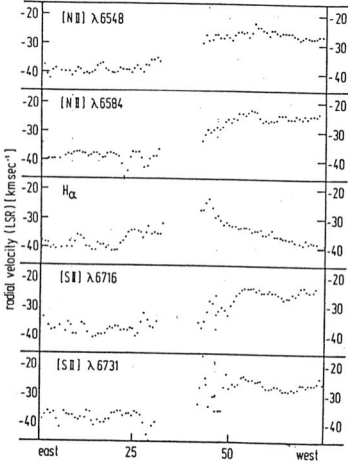

Figure 3. Radial velocities versus slit length (Spectrogram 1) for selected lines.

1) present adress: Joint Institute for Laboratory Astrophysics
    University of Colorado
    Boulder CO 80309

2) present adress: Instituto de Astronomico
    Cindad Universitaria
    Mexico 20

STELLAR WIND FROM Of STARS FROM INFRARED AND RADIO OBSERVATIONS

P.Persi[1], L.F.Rodriguez[2], M.Tapia[3], M.Ferrari-Toniolo[1], M.Roth[3]

1. Istituto Astrofisica Spaziale, CNR, C.P. 67 00044 Frascati, Italy
2. Instituto de Astronomia,UNAM, Apdo. Postal 70-264, Ciudad Universitaria, Mexico,D.F.
3. Instituto de Astronomia,UNAM, Apdo. Postal 877, Ensenada, Baja California, Mexico

ABSTRACT. Infrared and radio observations of the Of stars HD108,HD14947,HD16691,HD47129 are reported.They confirm the presence of thermal stellar wind emission with mass loss rate $1-2 \; 10^{-5} \; M_o y^{-1}$.
  The radio variability on time scale of months of the Of star Cyg OB2 No.5 is shown.This variability is not correlated with the 10 micron excess.

1. INTRODUCTION

Of stars are a class of luminous O-type stars with dense,high velocity stellar winds whose mechanisms of acceleration are not well understood.
  The structure and physical processes in winds can be well studied by combining infrared and radio data.The radio continuum observations represent a direct method of deriving $\dot{M}/v_\infty$ depending solely on observable quantities (Panagia and Felli,1985;Wright and Barlow,1975),while the observed IR excesses combined with the radio measurements can be used to derive the wind velocity law.
  In this paper,we report the results of radio and IR observations of the Of-type stars HD108, HD 14947, HD16691, HD47129,and Cyg OB2 No.5.These stars are known to have a strong IR excess (Persi et al.1983).In addition,a study of radio variability on time scale of months is presented for the radio variable star Cyg OB2 No.5 (Persi et al.1985).

2. RADIO AND INFRARED OBSERVATIONS

The radio continuum observations at 2 and 6 cm were obtained on September 1984 with the Very Large Array (Socorro,New Mexico,USA).During our observing runs the VLA was in D-array configuration.Standard VLA procedures were used.Calibrators were selected from the standard list,and their flux densities were measured relative to the primary calibrator 3C286.

The infrared photometry was obtained on different observing runs between 1983 and 1986 at the 2.1m telescope of the Mexican National Astronomical Observatory at San Pedro Martir(Baja California),and at the 1.5m telescope of the Gornergrat Observatory (TIRGO).InSb photometer systems and the IAS Ge bolometer were used during the observations.

A typical IR-radio energy distribution (HD16691),is shown in Figure 1.

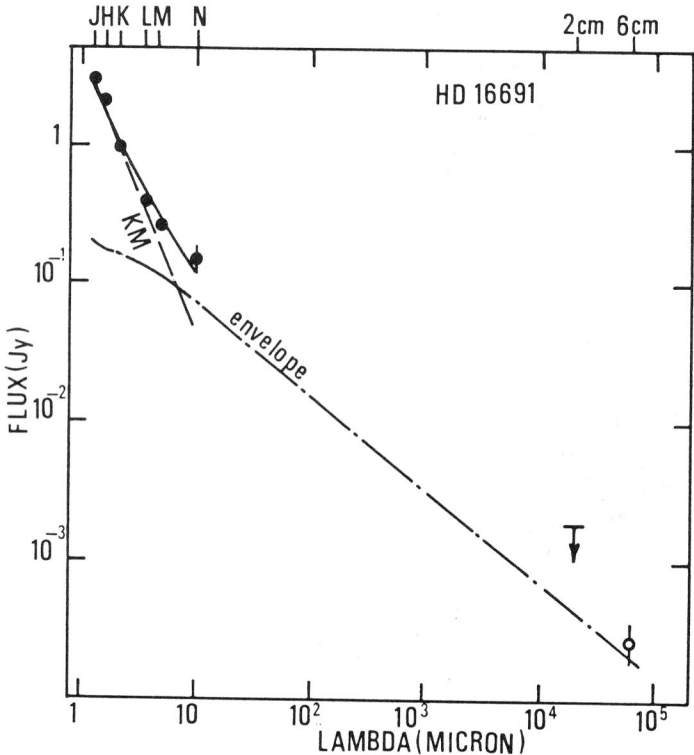

Figure 1.Infrared-Radio energy distribution of HD16691.

3. RESULTS

Mass loss rates were computed for HD108, HD14947, HD16691, and HD47129 (Table 1) from the observed radio flux densities and the corresponding stellar parameters reported by Persi et al.(1983) by using the relationships given by Abbott et al.(1981). The observed spectral indexes computed by comparing the excess emission at 10 μm and the flux at 6 cm (2 cm for HD108) are all in the range 0.7 to 0.85. This confirms that the observed emission originates in stellar winds by thermal Bremmstrahlung processes.

Table 1
Mass loss rates for Of-type stars

Star	Sp.	S(6cm) (mJy)	$\dot{M}$ ($10^{-6} M_\odot y^{-1}$)	S^{exc}(10μm) (mJy)	$\alpha_{10\mu-6cm}$
HD108*	O7If	<0.23	17.5±3.3	195±50	0.81±0.06
HD16691	O5f	0.30±0.07	18.3±3.2	99±30	0.67±0.06
HD14947	O5.5f	<0.5	<7.2	254±90	>0.7
HD47129	O7.5IIIf	0.2(3σ)	17.0±5.0	336±156	0.85±0.10

Note:* The wind parameters are from the 2cm flux of 0.4±0.1mJy.
  + Spectral index calculated from 10 μm to 6cm.

The value of $\dot{M}$ obtained for HD16691 is in agreement with that derived from the 10 μm excess, while the values of $\dot{M}$ for the other Of stars are approximately a factor 2 lower than the corresponding values of $\dot{M}$ derived from IR observations (Persi et al.1983). The IR-radio spectrum of HD16691 has been fitted using the simple model described by Persi et al.(1983), assuming the radio value of $\dot{M}$. The best fit was obtained for a velocity law of the type $v(r)/v_\infty = 0.01 + 0.99(1-R*/r)^\beta$ with $\beta = 1$ (solid line in Fig.1). A higher value of β ($\beta > 2$) is required to fit the IR-radio spectrum of HD108.

4. CYG OB2 No.5

The radio (6cm) flux from this star has apparently increased steadily by a factor of ~3 between 1980 and 1984 (see Persi et al.1985), and later on it has returned to its lower value in timescales of months. At same time, the 10 μm excess, presumably arising from the stellar wind, has remained constant at a average level of 0.84 Jy (Figure 2).

At minimum 6 cm radio flux, the IR-radio spectral index

was found to be 0.7,consistent with thermal radiation from the wind,but the radio spectral index (2cm-6cm) measured when the 6 cm fluxes were at a level of 6-7 mJy,was 0.2±0.1.

At present it seems very difficult to explain the radio variability in terms solely of a stellar wind.

This star, which seems to resemble Cyg OB2 No.9 (Abbott,1985), needs future monitoring at radio and infrared wavelengths.

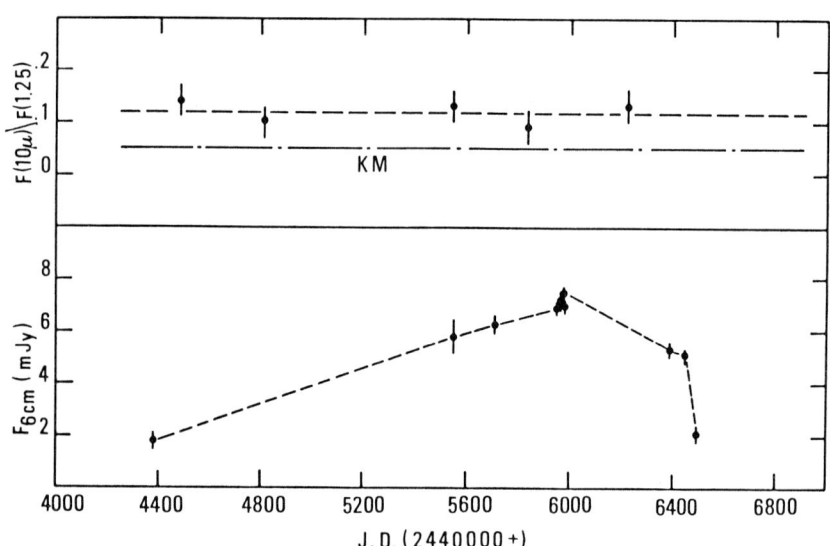

Figure 2. Radio (6cm),and IR (F(10μm)/F(1.25μm)) long term variability of Cyg OB2 No.5. KM (dashed-dotted line) represent the reddened (Av=6.2) Kurucz model atmosphere for Teff=40000 K and log g=4.

REFERENCES

Abbott,D.C.,Bieging,J.H.,Churchwell,E.:1981 Astrophys.J. 250,645
Abbott,D.C.:1985 in" Radio Stars",eds.R.M.Hjellming and D.M.Gibson,Reidel:Dordrecht,p.61
Panagia,N.,Felli,M.:1975 Astron.Astrophys. 39,1
Persi,P.,Ferrari-Toniolo,M.,Grasdalen,G.L.:1983 Astrophys. J. 269,625
Persi,P.Ferrari-Toniolo,M.,Tapia,M.,Roth,M.,Rodriguez,L.F.: 1985 Astron.Astrphys. 142,263
Wright,A.E.,Barlow,M.J.:1975 Mon.Not.Roy.astr.Soc. 170,41

STELLAR WINDS OF MASSIVE STARS IN M31

L.Bianchi(1), J.Hutchings(2), P.Massey(3)
(1)Osservatorio Astronomico di Torino, I-10025 Pino Torinese (TO)
(2)Dominion Astrophysical Observatory, Victoria, Canada
(3)Kitt Peak National Observatory, NOAO, U.S.A.

ABSTRACT. We observed with IUE and in the optical some OB supergiants and one WR star in M31. The UV resonance lines have lower outflow velocities than similar galactic stars and no P Cygni profiles, suggesting that mass loss mechanisms could be different among local group galaxies, due to different metal abundances. Model atmosphere fits to the UV continuum give values of Teff and Lbol consistent with the optical photometry and spectral classification.

1. INTRODUCTION

Optical and UV spectroscopy of OB supergiants and giants in the LMC and SMC have shown that global differences exist between the winds of these objects and those of galactic stars of similar spectral type (e.g. Hutchings 1980, 1982; Garmany and Conti 1985). In particular MCs stars have lower outflow velocities and weaker resonance lines than galactic analogous objects, proving a dependence of the mass loss characteristics on the metallicity.
We are extending this study to the most luminous hot stars of the nearby galaxies M31 and M33. Following an extensive optical survey work by one of the authors (Massey 1985, Massey et al. 1986), that allowed the identification, and spectral classification, of the hottest objects, we were able to observe also with IUE few of the most luminous stars in these galaxies. Results on the first stars observed have been published by Massey et al. (1985). We report here new observations of two more stars in M31.

2. THE OBJECTS

We observed an O-type star in the OB78 association of M31, OB78-277 of Massey et al. (1986), classified by these

authors as O6-7 on the basis of MMT spectra. We obtained both Short Wavelength (SWP) and Long Wavelength (LWP) spectra with exposure times of the order of 16 hours.
We also obtained a very long SWP observation of a WR star in M31, OB68-WR2 of Massey et al. (1986), whose optical spectrum suggests a classification of WN7.

3. THE CONTINUUM UV ENERGY DISTRIBUTION

3.1. OB78-277

We first corrected the UV spectrum for the galactic foreground extinction towards M31, with $E(B-V)=.11$. Comparison of the continuum emission with model atmospheres would however give a higher temperature in the LW range than in the SW, indicating that an additional extinction within M31 is present. Assuming that the extinction law in M31 is similar to the galactic one, we would need to adopt a large intrinsic reddening to match the Teff derived from the slope of the UV spectrum with the spectral type of this star, but the results from the model fit would be inconsistent with the optical photometry ($V=17.35$, $U-B=-1.15$, $B-V=0.06$, Massey et al. 1986).
The most consistent result is found by assuming a small amount ($E(B-V)=0.05$) of LMC type reddening in M31. Fig.1 shows the dereddened spectrum and superimposed the best fit model atmosphere with $Teff=35000$ K, though a temperature of $Teff=40000K$ would also reproduce fairly well the observed slope.
Though the absolute flux level of the SW and LW spectra do not exactly match (probably due to the difficulty of centering such faint objects with the Blind-offset technique in the IUE slit), extrapolation of the SW model fit to the visual range give $V=17.25$, in very good agreement with the optical photometry. The colors of the model, when "reddened" for the extinction derived above, also agree fairly well with the observations by Massey et al. (1986).
The value of Teff derived from the continuum slope, and the observed magnitude which corresponds to $Mbol=-10.24$ in M31 (applying a $B.C.=-3.389$ of the model atmosphere) match the spectral classification quoted above.

3.2. OB69-WR2.

The SW spectrum (Fig.2) of this very faint star ($V=18.2$, Massey et al. 1986) is rather noisy and therefore the value of the effective temperature can be determined with larger uncertainty than in the previous case. Again, the best fit consistent with both the IUE and the ground-based data is obtained by dereddening the spectrum with an LMC-type

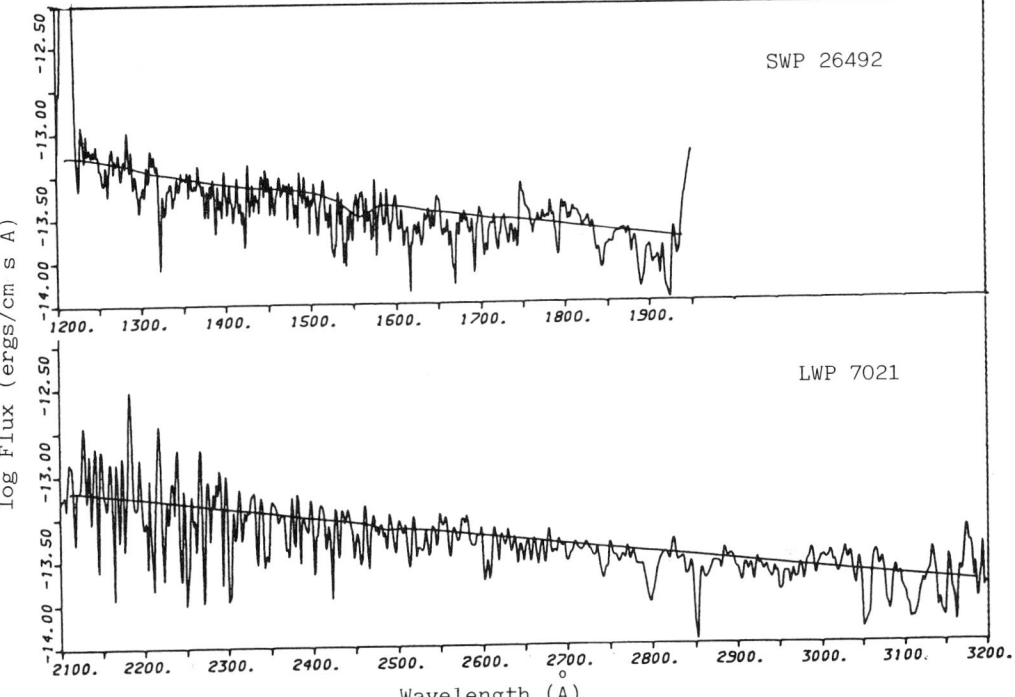

Figure 1. The SW (top) and LW (bottom) IUE spectra of OB78-277 in M31 corrected for interstellar extinction. Superimposed is a Kurucz (1979) model atmospere of Teff=35000K.

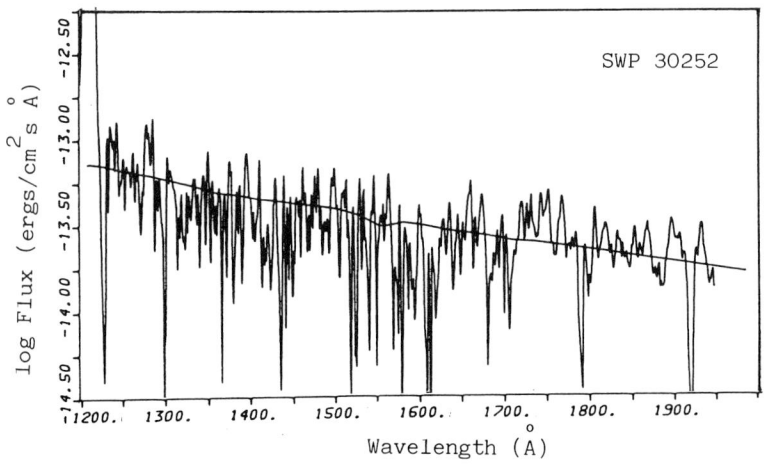

Figure 2. The dereddened spectrum of OB69-WR2 in M31. A Kurucz(1979) model of Teff=50000 K is also shown.

extinction law for E(B-V)=0.15 in the M31, after correcting for the foreground extinction, and a model atmosphere with Teff=50000K. Lower temperature models also reproduce reasonably well the continuum slope, but the fit would give inconsistent V-values when extrapolated to the optical band. Such high temperature is also consistent with the spectral type derived from the visual spectrum.

4. THE LINE SPECTRA

The line spectra of these stars are discussed in some detail in a forthcoming paper (Hutchings, Massey and Bianchi, 1987). In this aspect the UV spectra are rather unusual. In both stars the typical P Cygni profiles of hot star winds (e.g. NV, HeII) are absent. In OB78-277 the strongest lines are very likely of interstellar origin (SiII 1260, OI 1300, CII 1335, MgII 2800) while in the UV spectrum of the WR star there are no strong absorption features, nor there are emission lines, which are typical of similar galactic objects, and which are seen in the optical spectrum of this star (Hutchings et al. 1987).

The absence of strong emission in the resonance transitions of ions which are the most abundant in the envelopes of hot galactic stars was found also in our earlier observations of stars in M31 and M33 (Massey et al. 1985), and it is likely related to different metal content in the stellar winds.

Since the metallicity in M31 is supposed to be higher than in the Galaxy, it could be that these stars are more similar to the luminous galactic stars with extremely high mass loss rates (but low outflow velocity) like P Cygni, rather than to the MC stars.

REFERENCES

Garmany, C., Conti, P., 1985: Astrophys.J., 293, 407
Hutchings, J., 1980: Astrophys.J., 235, 413
Hutchings, J., 1982: Astrophys.J., 255, 70
Hutchings,J., Massey, P., Bianchi,L., 1987: preprint
Kurucz, R., 1979: Astrophys. J. Suppl., 40, 1
Massey, P., 1985: P.A.S.P., 97, 5
Massey,P., Hutchings,J., Bianchi,L.:1985, Astron.J., 90, 2239
Massey,P., Armandroff,T., Conti,P., 1986: Astron.J., 92, 1303

HYDROMAGNETIC WINDS FROM PARTIALLY OPEN MAGNETOSPHERES

Kanaris Tsinganos
Department of Physics, University of Crete
Heraklion, Crete, Greece

ABSTRACT. We present a class of global analytic solutions of the steady MHD equations for an axisymmetric stellar wind. The magnetic field is in a partially open configuration, with an equatorial region of closed fieldlines trapping the higher density plasma in magnetostatic equilibrium, a polar region of open fieldlines through which the wind escapes, and a purely radial field beyond a certain radial distance. For such given form of the magnetosphere we analytically solve the MHD equations for the flow speed, plasma density and pressure, as well as for the heating/cooling sources neccessary to support such an outflow. The solutions illustrate some interesting relationships between the physical parameters of hydromagnetic winds and may guide more sophisticated numerical models of wind-type astrophysical outflows.

1. THE MODEL

It is quite possible for the heating responsible for the existence of a hot stellar corona to be such that the resulting pressure distribution exerts lateral forces on the magnetic flux tubes keeping some of them open, at the same time that drives a wind outflow along them. This is the case, for example, in the solar corona and wind, wherein we can found regions with the magnetic field closed and in static equilibrium, as well as regions where the magnetic field is open and the wind escapes. Consider then a magnetosphere with such properties,

$$\mathbf{B} = (2F_1/r^3 + 2F_2 + 2F_4 r^2)\cos\theta \hat{r} + (F_1/r^3 - 2F_2 - 4F_4 r^2)\sin\theta \hat{\theta} \quad (1)$$

where F_1, F_2, F_4 are constants. This magnetic field is the superposition of a dipole magnetic field (F_1), a uniform magnetic field (F_2), and a non-potential magnetic field (F_4) that diverges at large r. To avoid this diverging behavior at large r we may relate the constants F such that at some distance $r*$ the field is radial and remains so for all $r>r*$. Two such cases are the following,

$$2F_2 = 1, \quad 2F_1 = 4/5, \quad 2F_4 = -3/10 \tag{2a}$$

and

$$2F_2 = 1, \quad 2F_1 = 3, \quad 2F_4 = 0 \tag{3a}$$

The magnetic field is then,

$$\frac{\mathbf{B}}{\gamma_1 B_0} = \left[1 + \frac{4(r*)^3}{5(r)^3} - \frac{3}{10}\frac{(r)^2}{(r*)^2}\right]\cos\theta\hat{r} - \left[1 - \frac{2(r*)^3}{5(r)^3} - \frac{3}{5}\frac{(r)^2}{(r*)^2}\right]\sin\theta\hat{\theta} \tag{2b}$$

and

$$\frac{\mathbf{B}}{\gamma_2 B_0} = \left[1 + \frac{2(r*)^3}{r^3}\right]\cos\theta\,\hat{r} - \left[1 - \frac{(r*)^3}{r^3}\right]\sin\theta\,\hat{\theta} \tag{3b}$$

where

$$\frac{1}{\gamma_1} = 1 + \frac{4(r*)^3}{5(ro)^3} - \frac{3(ro)^2}{10(r*)^2}, \quad \frac{1}{\gamma_2} = 1 + \frac{2(r*)^3}{(ro)^3} \tag{4}$$

and where B_0 is the radial magnetic field at the stellar base ro.

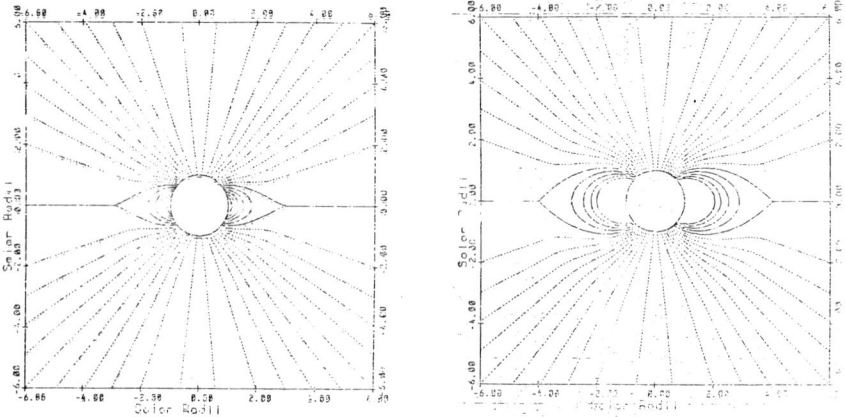

Figures (1a-1b): Plots of the fieldlines of two magnetospheres with partially open magnetic lines, the nonpotential magnetic field Eq. (2), and the potential field Eq. (3).

Thus the space around the star consists of three regions: (1) the static region of closed fieldlines around the equator and up to r*, (2) the flow region of open fieldlines around the pole and up to r*, and (3) the radial magnetic field region in all space outside $r*$. In the nonradial field region (2) the density ρ is assumed spherically symmetric, $\rho=\rho(r)$, the pressure P and the rate of heating σ are assumed nonspherically symmetric, $P = P_o + P_1(r)\sin^2\theta$ and $\sigma = [\sigma_o(r) + \sigma_1(r)\sin^2\theta]\cos\theta$, respectively, while the flow is along the fieldlines, $\mathbf{v} = W(r)\mathbf{B}$.

2. RESULTS

(a). Outflow with Finite Terminal Speed. The case of a non-potential magnetic field, Eq. (2b) and Fig. (1a), gives a steep acceleration of the sub-Alfvenic flow speed at the base $r=ro$ to about $v=3v_A$ at $r=2.7ro$, followed by a gradual deceleration down to $v=2v_A$ at $r=4ro=r*$. Beyond $r*$ the flow continues to decelerate to a terminal speed $v_\infty =v_A$, the Alfven speed. There is strong heating near the base ro, such that the pressure gradient is so steep as to accelerate the flow rapidly. However, at about $r=2.7r*$ the heating becomes negative (cooling) resulting to the plasma weight taking over the pressure gradient to decelerate the flow.

In the separatrix dividing the static from the dynamic regions, the continuity of the total pressure requires that the plasma pressure is continuous as well. Since in the accelerating part of the flow the pressure gradient exceeds the plasma weight, it follows that for hydrostatic support inside the static region we must have higher density there than outside (at the same radial distance). Exactly the opposite happens at the deccelerating part of the flow where the density inside is lower than the density outside. Finally, we note that at $r=r*$ we have a weak discontinuity in the plasma parameters resulting in a sonic front at $r*$.

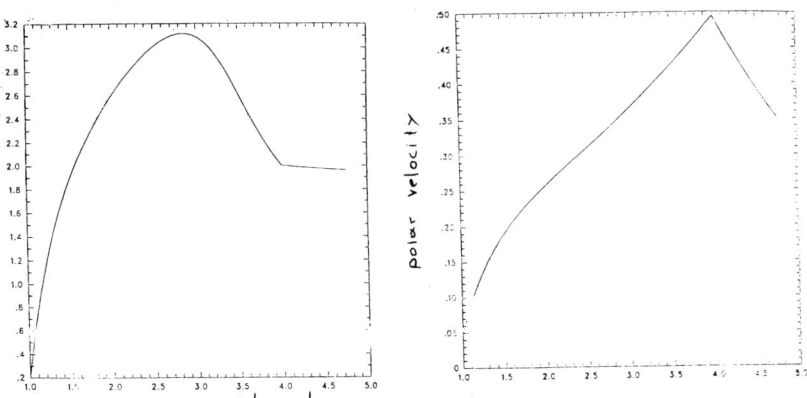

Figure 2: Velocity radial profiles for the two magnetospheres of Eqs. (2) and (3).

(b). Outflow with Zero Terminal Speed - Formation of a Current Sheet.
The case of a potential magnetic field, Eq. (3b) and Fig. (1b),
gives a steep acceleration of the flow speed from $r=r_0$ all the way
up to the tip of the streamer at $r*$, followed by a gradual
deceleration to a zero asymptotic speed at the radial fieldlines
region. The heating in $r<r*$ results in a pressure gradient larger
than the plasma weight and this outward force is responsible for
the acceleration.

In the previous case we had a continuous magnetic field across the
separatrix dividing the static from the dynamic regions. The
result was to have lower density inside than outside, close to the
tip of the static region. However, we can obtain a solution with
higher density inside the static region, all the way up to $r*$, if
we allow lower magnetic field intensity just inside the separatrix
than outside, i.e., an electric current sheet at the separatrix.
The formation of this current sheet is inevitable and can be
understood as follows: the temperature T in the closed fieldlines
zone should be higher than outside where the fieldlines are open
and the plasma is cooled by the expanding flow. In other words,
the pressure height inside is higher than outside because of the
higher T, while outside the pressure drops faster according to a
generalized Bernoulli law. The two pressures across the
separatrix do not match and this jump in pressure must be
compensated by a jump in the magnetic pressure, i.e. we have the
formation of a current sheet.

3. CONCLUSIONS

The basic conclusions from the two classes of solutions that we
briefly presented here are: (i) suitable heating/cooling
distributions at the magnetosphere of an object may provide the
necessary pressure gradients first to open some fieldlines to
infinity and second to drive a wind-type flow along them, (ii)
that at the separatrix of closed and open fieldlines, higher
density inside the static zone is accompanied by a current sheet
formation at this separatrix, and (iii) that acceleration or
deceleration of the wind as well as its terminal speed, depend on
the particular geometry of the magnetosphere. A detailed
presentation of these solutions may be found in Low and Tsinganos
(1986) and Tsinganos and Low (1987).

Acknowledgements. I would like to acknowledge Dr B.C Low's of the
High Altitude Observatory, USA, invaluable contribution in
obtaining the results presented in this paper.

REFERENCES

Low, B.C. and Tsinganos, K.: 1986, Astrophys. J., **302**, 163
Tsinganos, K. and Low, B. C.: 1987, Astrophys. J. (to be submitted).

HELICOIDAL ASTROPHYSICAL OUTFLOWS

Kanaris Tsinganos and Georgia Vlastou
Department of Physics , University of Crete
Heraklion, Crete, Greece

ABSTRACT. We present three classes of analytic solutions of the
basic hydrodynamic equations that we assume to govern the flow of
plasma in the atmosphere of a rotating astrophysical object. In
the first class, we have accelerating helical wind-type outflows
wherein a finite speed at the stellar surface accelerates
monotonically outwards to an asymptotic speed at large radial
distances, with a sinusoidal dependence in the latitude
corresponding to a maximum flow at the equator and no flow at the
poles. In the second, we have decelerating helical outflows
wherein a finite initial speed decelerates monotonically to a lower
speed at infinity, with a latitudinal dependence such that the flow
is concentrated at the poles. As a special subcase of this class
we obtain an accretion-type helical inflow. Finally, the third
class of solutions with a spherically symmetric radial flow speed
in a non-spherically symmetric plasma density may be regarded as a
superposition of the previous two classes, i.e., it accelerates
close to the base and decelerates further away to a constant
asymptotic speed. The solutions are written in terms of
dimensionless physical parameters in order to facilitate their
direct application to specific astrophysical flows, such as in
winds from massive pre-main sequence objects and T Tauri stars, or,
bipolar flows in young stellar objects.

1. BASIC EQUATIONS

Consider the two familiar laws expressing conservation of mass and
momentum in the steady dynamical equilibrium of inviscid
compressible fluids in the spherically symmetric gravitational
field of a central object,

$$\nabla \cdot (\rho \mathbf{V}) = 0 \quad , \quad \rho (\mathbf{V} \cdot \nabla) \mathbf{V} = -\nabla P - \frac{\rho G M}{r^2} \hat{r} \quad (1)$$

together with an equation of state $P = (k/m)\rho T$, where ($\mathbf{V}$) denotes
the velocity, (ρ, P, T, m) the density, pressure, temperature and
mean molecular weight of the fluid, respectively, (k,G) Boltzmann's
and Newton's constants, respectively, and M the mass of the central
gravitating object.

Before we proceed with solutions of Eqs. (1), we define some dimensionless physical parameters in terms of which our solutions will be expressed. It is convenient to introduce first some fixed radius r_o as the radius of the base of the atmosphere and normalize all physical quantities in terms of their values at this base r_o. Thus, express the radial distance r in terms of the dimensionless distance $R = r/r_o$, the ratio of the azimuthal to the radial speeds at the base by λ, the ratio of the escape speed to the radial base speed by v, and the dimensionless density by D,

$$R = \frac{r}{r_o} \quad , \quad V_{esc} = \left[\frac{2GM}{r_o} \right]^{1/2} \quad , \quad D = \frac{\rho}{\rho_o} \quad , \quad \lambda = \frac{V_\varphi(R=1)}{V_r(R=1)} \quad , \quad v = \frac{V_{esc}}{V_o} \, . \quad (2)$$

2. ACCELERATING HELICAL WIND-TYPE OUTFLOWS.

Consider first the case of a helical flow with a sinusoidal dependence on the colatitude θ, in a spherically symmetric density $D(R)$,

$$V_r(R,\theta) = V_o Y(R) \sin\theta \quad , \quad V_\varphi(R,\theta) = \frac{V_1}{R} \sin\theta \quad , \quad D(R) = \frac{1}{YR^2} \, . \quad (3)$$

With this spherically symmetric density and a dimensionless pressure $Q(R,\theta) = Q_o(R) + Q_1(R)\sin^2\theta$, the continuity equation is automatically satisfied, while the r- and θ-components of the force balance equation become,

$$Q_1 = \frac{\lambda^2}{YR^4} \quad , \quad \frac{dQ_1}{dR} + \frac{2}{R^2}\frac{dY}{dR} - \frac{2\lambda^2}{YR^5} = 0 \quad , \quad \frac{dQ_o}{dR} + \frac{v^2}{YR^4} = 0 \, . \quad (4)$$

It is evident from Eqs. (4) that in this model the spherically symmetric pressure term Q_o supports entirely the plasma weight, leaving exclusively to the nonspherically symmetric pressure component Q_1 the role of acceleration and lateral force balance. The fact that we do have accelerating flow in this model becomes evident from Eq. (4) wherein the balance of forces in the θ-direction gives $Q_1 > 0$, i.e., higher pressure at the equator than at the poles. Then, since at the poles $\theta=0$ we have static equilibrium conditions and the gradient of Q_o supports the plasma weight, as we move to the equator at the same radial distance R, both the centrifugal force and the gradient of Q_1 that act outwards accelerate the plasma. Formally, this accelerating motion can be seen by combining Eqs.(4) and solving the resulting single differential equation for $Y(R)$,

$$\frac{dY}{dR} = \frac{Y}{R} \frac{6\lambda^2}{2Y^2R^2 - \lambda^2} \quad , \quad Y = \left[\frac{Y^2 + 5\lambda^2/2R^2}{1 + 5\lambda^2/2} \right]^3 \, . \quad (5)$$

It follows that if at the base $\lambda^2 < \lambda^2_{max} = 2$, $dY/dR > 0$ and $Y_{asy} \longrightarrow (1+5\lambda^2/2)^{3/5}$. The maximum asymptotic speed is therefore $V_{asy} \approx 6^{3/5} V_o \approx 2.93 V_o$. For values of λ greater than λ_{max} the density increases outwards and the solutions are unphysical.

In Fig. (1a) we display the radial dependence of the radial flow speed $Y(R)$ for several values of λ up to λ_{max}. It can be seen from these plots that the higher the value of λ, the larger the asymmetry in the pressure, Q_1 and the higher the acceleration and asymptotic speed, as expected. This behavior becomes evident if we recall that the asymmetric pressure term Q_1 is solely responsible for the acceleration which increases with λ. Note in particular the steep acceleration of the velocity profile close to the base. At large distances the density D drops like the square of the radial distance R while the pressure Q_o like the cube of R, i.e., we have a polytropic relation $P \approx \rho^{\Gamma}$ between pressure and density with $\Gamma = 3/2$. We find thus the heated $\Gamma = 3/2$ spherically symmetric Parker solar wind solution.

A similar situation is encountered if instead of the azimuthal rotation V_φ, we have a radial magnetic field that depends on R and θ,

$$B(R,\theta) = (B_o / R^2) \cos\theta \; \hat{r} , \qquad (6a)$$

and with V_r, $D(R)$ and $Q(R,\theta)$ given by Eqs.(3) as before. Now the role of the θ-component of the centrifugal force to balance the θ-component of the pressure gradient is played by the θ-component of the Lorentz force,

$$\frac{4\pi J_\varphi}{c} = (\nabla \times B)_\varphi = \frac{B_o \sin\theta}{r_o R^3} \; , \; \left[\frac{J \times B}{c}\right]_\theta = \frac{J_\varphi B_r}{c} = \frac{B_o^2}{8\pi r_o R^5} \sin 2\theta , \qquad (6b)$$

where J is the electric current density associated with B. We obtain similarly an accelerating helical flow with the asympotic value of the velocity depending on the ratio of the Alfven speed and radial flow speeds at the base. A detailed presentation of the model expressed by Eqs. (6) has been given recently by Low and Tsinganos (1986).

3. DECELERATING HELICAL OUTFLOWS

Consider next the case of a flow concentrated at the polar caps of a rotating magnetosphere,

$$V_r(R,\theta) = V_o Y(R) \cos\theta , \qquad (7)$$

and with the azimuthal speed $V_\varphi(R,\theta)$, density $D(R)$ and pressure $Q(R,\theta)$ given by expressions (3), as in the previous model. As we shall see, this simple change $\sin\theta \rightarrow \cos\theta$ in the latitudinal dependence of V_r will drastically change the character of the solution, i.e., from an accelerating outflow that we obtained in the previous model, to an decelerating outflow now. With this separation of the r- and θ-variables the equations denoting mass and momentum flux conservation reduce to:

$$Q_1 = \frac{\lambda^2}{YR^4} \; , \; \frac{dQ_1}{dR} - \frac{2}{R^2}\frac{dY}{dR} - \frac{2\lambda^2}{YR^5} = 0 \; , \; \frac{dQ_o}{dR} + \frac{2}{R^2}\frac{dY}{dR} + \frac{v^2}{YR^4} = 0 . \qquad (8)$$

Eliminating Q_1 we can find $Y(R)$ as the solution of the following ordinary differential equation,

$$\frac{dY}{dR} = -\frac{Y}{R}\frac{6\lambda^2}{2Y^2R^2+\lambda^2} \quad , \quad Y = \left[\frac{Y^2 - 5\lambda^2/2R^2}{1-5\lambda^2/2}\right]^3 , \quad \lambda^2 = \frac{2}{5} \quad , \quad (9)$$

while if $\lambda^2 = 2/5$, $Y = 1/R$. Note that the radial flow speed depends on λ alone and for $\lambda^2 < 2/5$ the asymptotic speed $Y_{asy} > 0$, while for $\lambda^2 = 2/5$, $Y_{asy} = 0$. For larger values of λ such that $\lambda^2 > 2/5$, the density increases outwards, $dD/dR > 0$, and the solutions are unphysical. In Fig. (1b) we display the radial dependence $Y(R)$ of the radial speed $V_r(R,\theta)$, for various values of λ between zero and $\lambda_{max} = (2/5)^{1/2}$.

The pressure in this model is larger at the equator than at the pole ($Q_1 > 0$) and this feature is related to the following aspect of the dynamical equilibrium. The θ-component of the centrifugal force is in the θ-direction and is balanced by the θ-component of the pressure gradient which ought to act then in the -θ-direction. For consistency we need a higher pressure at the equator than at the poles, and for this reason we obtained $Q_1 > 0$ in Eq. (8). Then, since at the equator the flow speed is zero and the weight of the plasma is supported by the sum of the pressure gradient and the centrifugal force, as we move towards the pole (at the same radial distance) the weight remains the same but the centrifugal force is zero there and the pressure is lower. A decelerating force appears as a result. Note that at large distances we obtain again the polytropic $\Gamma = 3/2$ spherically symmetric Parker solution. As before, a similar situation is encountered if instead of the azimuthal rotation we have a radial magnetic field that depends on θ, as the one given by Eqs. (6), (Low and Tsinganos (1986).

The case $\lambda^2 = 2/5$ deserves special attention as an accretion-type flow due to its similarity in the θ-dependence and radial asymptotic values, with the physical situation assumed to occur at the polar caps of the rotating magnetosphere of an accreting object. The asymptotic radial speed, rotational speed, density and pressure become zero in this case,

$$V_r = \frac{\cos\theta}{R} \quad , \quad D = \frac{1}{R} \quad Q_1 = \frac{1}{R^3} \quad , \quad Q_0 = \frac{v^2}{2R^2} - \frac{2}{3R^3} \quad , \quad (10)$$

while the temperature T varies inversely proportional to the radial distance R, as well.

4. ACCELERATING/DECELERATING SPHERICALLY SYMMETRIC HELICAL OUTFLOWS.

In the previous two classes of solutions the density has been assumed spherically symmetric while the radial flow speed θ-dependent. In order to explore the properties of a complementary case consider the following case of a spherically symmetric radial flow speed in a nonspherically symmetric density,

$$V_r(R,\theta)=V_c Y(R) \;,\; V_\varphi(R,\theta)=\frac{V_1}{R}\frac{\sin\theta}{[1+\omega\sin^2\theta]} \;,\; D(R,\theta)=\frac{[1+\omega\sin^2\theta]}{YR^2} \;. \quad (11)$$

We obtain then the following solutions of the hydrodynamic Eqs.(1)

$$Q_1 = \frac{\lambda^2}{YR^4} \;,\; \frac{dQ_1}{dR} + \frac{2\omega dY}{R^2 dR} - \frac{2\lambda^2}{YR^5} + \frac{v^2\omega}{YR^4} = 0 \;,\; \frac{dQ_2}{dR} + \frac{2dY}{R^2 dR} + \frac{v^2}{YR^4} = 0. \quad (12)$$

Eliminating $Q_1(R)$ from Eqs.(12) we obtain the following differential equation for $Y(R)$,

$$\frac{dY}{dR} = \frac{Y}{R}\frac{6\lambda^2/\omega - \omega v^2 R}{2Y^2R^2 - \lambda^2/\omega} \;. \quad (13)$$

This expression for $Y(R)$ is similar to expression (5) of the previous accelerating wind-type outflow if we substitute $\lambda^2 \longrightarrow \lambda^2/\omega$, except for the additional term $-\omega v^2 R$ in the numerator. We expect therefore that for $6\lambda^2/\omega \gg \omega v^2$, $Y(R)$ will be increasing to a finite value and further away when the term $-\omega v^2 R$ takes over, $Y(R)$ should decrease to a lower asymptotic value. The maximum asymptotic value of $Y(R)$ is obtained for $\lambda^2/\omega \approx 2$. On the other hand, if $6\lambda^2/\omega < \omega v^2 R$ we have a decelerating flow speed right from the base $R = 1$. The relative strength therefore of the constants λ^2/ω and ωv^2 will determine whether and where the radial flow speed accelerates or decelerates, giving the corresponding character to the solution.

In Figs. (1c-1d) we display the radial flow speed for $v=1$, and several values of λ for each of two values of ω: 1 and 10, respectively. Thus, in Figs. (1c) with $\omega=1$ and $\lambda = 0.1, 0.6, 1.0$ and 1.4 the deviations from a spherically symmetric density distribution are rather small and the behavior of the solutions is similar to our previous accelerating wind-type flows, as shown in Sec. 3 and 4. For example, the maximum value of λ that gives physically accepted solutions, is 1.4 as in Fis. (1a). In Fig. (1d) with larger deviations from the spherically symmetric density distribution, $\omega=10$, the term $-\omega v^2 R$ dominates even at small R resulting in a decelerating behavior of the flow speed as in the previous second case, Figs. (2). The maximum asymptotic speed is obtained for $\lambda^2_{max}/\omega \approx 2$, i.e., $\lambda_{max} \approx 4.5$, while for $\lambda > \lambda_{max}$, $dD/dR > 0$ and the solutions are unphysical. In all such cases we observe that the flow speed accelerates to a constant value unless the value of $6\lambda^2/\omega$ is smaller than ωv^2 wherein it decelerates right from the base; and the maximum asymptotic speed is obtained for $\lambda^2/\omega \approx 2$. We conclude that this class of solutions may be regarded as a superposition of the two previous classes of solutions where we had a purely accelerating or decelerating profile for the flow speed.

Acknowledgements: The research reported in this article was supported, in part, by the University of Crete, Department of Physics and by NATO grant 221/87.

REFERENCES

Low, B.C. and Tsinganos, K.: 1986, Astrophys. J., 302, 163.
Tsinganos, K. and Low, B. C.: 1987, Astrophys. J. (to be submitted).

Figure 1: Velocity radial profiles for different values of the parameter λ in helical outflows: (a) accelerating, Sec. 2, (b) decelerating, Sec. 3, and (c)-(d) decelerating, Sec 4.

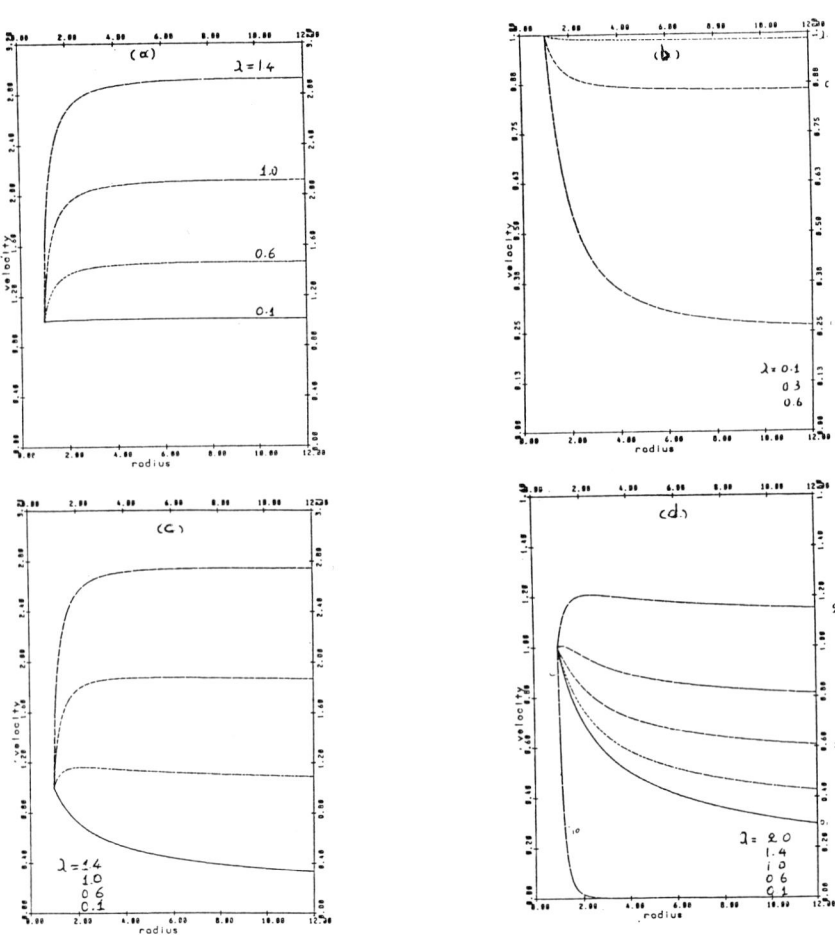

LOCAL THERMONUCLEAR RUNAWAYS ON WHITE DWARFS

Marina Orio
Dept. of Physics, Israel Institute of Technology,
Haifa 32000, Israel

Giora Shaviv
Dept. of Physics, Arizona State University, Tempe AZ 85287
(on leave from Dept. of Physics, Israel Institute of
Technology)

ABSTRACT. The question of local thermonuclear runaways (LTNR) on the surface of white dwarfs is investigated. 1-D equations of energy and of motion are applied, including the accretion process in the continuity equation. The accretion is assumed to be spherically symmetric and to create non spherically symmetric finite gradients in temperature density or composition. We found that the TNR tends to occur locally when the mass and the luminosity of the white dwarf and the accretion rate are high. For certain sets of parameters the size of the LTNR is a significant part of the WD total surface area. The propagation of the TNR may take from few hours to months, producing a slow rise to maximum in novae (unexplained by models so far). Hydrodynamical flows caused by temperature in homogeneities or by rotation can explain the asymmetries observed in nova shells.

This work is directed to understand how the thermal and hydrodynamical flows propagate in the envelope accreted from a companion in a binary system in order to explain several features of the outbursts of novae. The geometry of the ejected shells does not appear to be spheri-symmetrical from photographic plates and from the splitting of several spectral bands and lines (see for instance Mustel & Boyarchuck 1968, Weaver 1974, Rosino and Chincarini 1963).
    A few authors attempted an interpretation in terms of a mechanism that occured after the outburst itself: magnetic fields (Mustel & Boyarchuk 1969), interaction with the secondary (Hutchings 1972), with the disk (Sparks & Starrfield 1973), Raleigh Taylor instabilities in the expanding shell (Chevalier & Klein 1978). Only Warner, 1972, justified this with the mechanism that Rose, 1969, considered responsible for the outburst: oscillations of the white dwarf.
    Gallagher and Anderson, 1978, proved that the inhomogeneities of HR Del must have been present since the beginning of the outburst. The matter can actually be accreted without spherical symmetry and not homogeneously from the companion.

Durrisen, 1977, studied how accretion can occur through the disk: more material should accumulate at the equator. Kippenham & Thomas, 1978, calculated the structure of accretion belts that should be formed and found that molecular friction could not be sufficient for relaxation of the material to a spherical structure. Significant shear mixing should occur and, due to the interest roused by the problem of abundances, the studies of accretion went on in the last few years with a number of works. It has been found that the mechanism of distribution of angular momentum should be more efficient than it was previously thought. For Livio and Truran, 1987, formation of an Ekman layer makes material distribute to the poles. It is thus reasonable to assume spherical envelopes.

Shara, 1984, points out, however, at the fact that even slightly non symmetric accretion can cause transverse temperature gradients, leading to a local thermonuclear runaway (LTNR). He calculated that the time scale for the TNR is shorter than the time scale of the thermal conductivity whenever the density in the burning layer reaches the value of 10^4 g/cm^3, commonly obtained at the onset of the runaway in calculation of nova models. According to an analytical estimate the LTNR is favoured by high accretion rates, high masses and luminosities of the WD. Propagation of the flash was not foreseen, but a "vulcanic outpouring", capable of explaining perhaps DNe. Fryxell & Woosley (1982) indicated that Shara underestimated the velocity of the burning front, not taking convection into account.

In this work we investigated if the TNR occurs locally before thermal equilibrium can be established and, if it does, analysed how it propagates. The scale of the inhomogeneities we considered is comparable with the radius of the WD, so we took curvature into account. Our model follows the propagation of the thermal and hydrodynamical flows along the meridian. An element of meridional length is $D(1) = R*\sin(\Delta(\theta)) = R*\Delta(\theta)$, where θ is the polar angle, varying from $0°$ to $90°$, and the flows are supposed to be symmetrical at the angles $\alpha = -\theta$ and $\alpha = 90° + \theta$. We solved four equations integrating all the relevant quantities over the volume of a ring of material of the burning layer. We considered the equation of continuity:

$$\frac{\partial \rho}{\partial t} = -\nabla \cdot (\rho \vec{v}) + \rho' a \tag{1}$$

(where ρ' = increase of density due to accretion);
equation of momentum considering $\vec{v} = (v_r, v_\theta, v_z) = (0, v_\theta, 0)$:

$$\rho \frac{\partial v_\theta}{\partial t} = -\rho(v_\theta \nabla_\theta) v_\theta - \nabla_\theta P - \rho \omega^2 (R)_\theta \tag{2}$$

the equation expressing the variation of the chemical composition:

$$\frac{\partial x}{\partial t} = -\frac{\varepsilon}{Q} \tag{3}$$

(where ε = energy generation rate, Q = energy liberated per gram, and a term for advection is here neglected, since it was found to be very small in all the cases we examined); and finally the equation of energy:

$$\rho T \frac{DS}{Dt} = \eta \rho \varepsilon - \nabla \cdot \vec{F} \tag{4}$$

where $(1 - \eta)$ is the fraction of generated energy that flows radially. To calculate η analytical expressions for the p-p and CNO burning were used instead of a nuclear network, but when the CNO time scale exceeded the value $\tau = 1000$ sec, typical for the $\beta +$ decaying nuclei that regulate the process, we assumed:

$$\varepsilon_{CNO} = \frac{1}{1000} \text{ erg g/sec} \tag{5}$$

The flux was calculated considering both the radiative and conductive opacities and all the thermodynamical quantities were calculated according to Kovetz & Shaviv (1976).

The free parameters of the system were η, the mass accretion rate $\dot{m}$, the mass of the white dwarf m(wd) and ω, the angular velocity around the polar axis. Only in the last set of models $\omega \neq 0$. The initial conditions and the perturbation were chosen in each case.

Substituting the continuity equation in the equation of energy and writing as a function of the temperature we obtained:

$$\rho c_v (\frac{\partial T}{\partial t} + \frac{v_\theta}{R} \frac{\partial T}{\partial \theta}) = -P(\nabla \cdot \vec{v})_\theta + \varepsilon \rho \eta - \frac{1}{R} \frac{\partial F}{\partial \theta} + P \frac{\rho' a}{\rho} \tag{6}$$

It can be easily checked that the time scale for heating of the material caused by the hydrodynamical flows is at least two orders of magnitude greater than all the other time scales of the problem. Preliminary calculations with initially only a thin envelope of density 100 g/cm^3 showed, however, that the material can be heated to high temperature in a limited zone, still at low density and far from degeneracy. So we expect that radial expansion occurs, cooling it and making our one dimensional calculation impossible. To avoid this difficulty we divided the problem in two parts. In a first set of models the initial density was $\rho_i = 100$ g/cm^3. Supposing that the importance of the hydrodynamical flows is minimized by small radial expansions, we calculated the evolution of the thermal flows only, to obtain a clear idea of this part of the physics of the problem. In a second series of models the initial density was $\rho_i = 10^4$ g/cm^3, so the inhomogeneities were in already degenerate material, that does not cool expanding radially. For such cases we followed the hydrodynamical flows with an implicit scheme.

The chemical composition was taken to be initially X(H) = 0.65, X(He) = 0.24, X(CNO) = 0.10. For the first part of the work we simulated a very hot (luminous) WD dwarf considering an initial temperature T(in) = 15 10^6 K, a less hot one with T(in) = 5 10^6 K. Accretion was considered assuming a linear increase in density due to a given $\dot{m}$, the density in the burning layer being 10 times the average density of the envelope. This was found to be in good agreement with the calculations of Prialnik, 1986.

An initial temperature gradient was assumed in zone extended 3° or 45° from the pole or from the equator. With an accretion rate 10^{-9} - 10^{-10} $\dot{m}$/yr an initial gradient of a few hundreds K extended to 3° is sufficient for development of LTNR in the initially perturbed zone only for the hot WD. With the same parameters the initial gradient must be of 2 orders of magnitude higher for the case with T(in) = 5 10^6 K. We also found that, once ignited, the flash propagates to the adjacent zones on a time scale due to the conduction. This time scale is regulated by the η parameter and varies from 10 hrs with η = 1 to 14 weeks with η = 0.01 (see Fig. 1). With η = 0.10 propagation of the flash takes a time of the order of a week and we consider that a LTNR can be a reasonable explanation for the slow rise time of many novae. In full scale calculations the time for the rise to maximum can be of a few hours hours only, because it depends on the convection turnover time scale. Before the runaway, however, the temperature profile developes more smoothly with smaller value of η (see Fig. 2). As a consequence, for each set of parameters there is a minimal value of at which thermal equilibrium is reached before local development of the TNR.

Table 1 shows the minimal D(T) needed in a polar cap extended to 3° for development of a LTNR extended only to this zone. It increases of 3 orders of magnitude as the accretion rate increases of 5. Table 2 shows how, with a fixed D(T), less zones are involved in the LTNR as $\dot{m}$ decreases. This means that a period of hibernation (Shara et al 1986) can avoid the possibility of a LTNR. Also in agreement with Shara's forecast we found that the minimal Δ(T) required for a LTNR extended a certain portion of the star increases varying m(WD) from m = 0.85 m_o 1.3 m_o. The competition between conductivity and energy generation rate is definitely favoured when high densities are reached sooner, as it is possible accreting on a smaller surface. The difference is, however, of less than 1 order of magnitude, because the surface in which thermal equilibrium must be established to avoid the LTNR is smaller for a more massive WD.

If the initial Δ(T) is extended to θ = 45°, the temperature profile becomes smoothly decreasing from zone to zone during the evolution and the LTNR turns out to be extended to a smaller area than in the case of initial Δ(T) in 3° only (this tends to thermal equilibrium with the adjacent zones.) We also found that for each set of parameters there are minimal density or composition gradients leading to a LTNR. They generate temperature gradients, in fact, since the first phases of evolution.

Starting calculations in a gas of ρ = 10000 g/cm^3 and temperature T(in) = 15 10^6 K - already quite degenerate - we followed the hydrodynamical flows. An initial perturbation at the pole or at the equator is smoothed out in a time of the order of 0.1 sec, but motion towards the opposite region starts, eventually heating the limited portion of the star and leading to a LTNR. With Δ(T) = 500 K at the equator we obtained a LTNR in a zone extended from θ = 6° to θ = 30°, while with the same Δ(T) at the pole from θ = 66° to θ = 78°.

Models with this initial density distribution

$$\rho(\theta) = \rho(0°)\exp(\frac{\text{kinetic energy}}{\text{thermal energy}} \cdot a) \qquad (7)$$

(1 < a < 10, a is a free parameter), had only for an angular velocity ω << ω(keplerian), to avoid significant effects due to non conservation of angular momentum implied calculating only the meridional flows on the surface.

Table 3 summarizes some results. We found that when the density becomes high enough at the equator, significant pressure builds up to push the material back towards the pole, causing in the end a LTNR in a zone that is extended to about 25° for ω = 10^{-3}·ω(keplerian) and tends to be more displaced towards the pole decreasing "a". The LTNR zone is wider for higher ω. Our results indicate that rotation can be the explanation of the peculiar spatial structure of novae, whose ejected material seems to concentrate around the equator and/or in polar caps.

accretion rate = 10^{16} G/sec
ΔT=100000 K in first zone
R(WD)=$7 \cdot 10^8$ cm

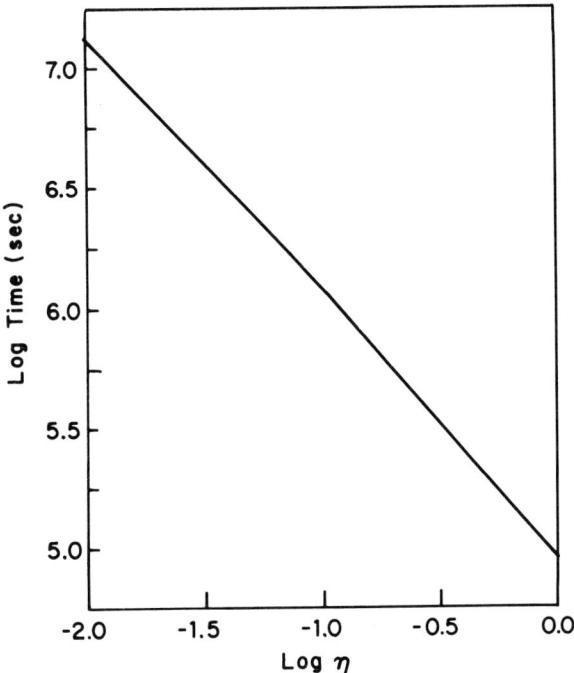

Fig. 1 Time for propagation to all the surface vs. η for a WD of m ≃ 0.85 m_o. ρ_{in} = 100 g/cm^3, $\dot{m}$ = 10^{10} g/sec. Initially T = 10^5 K in a polar cap extended to 3°.

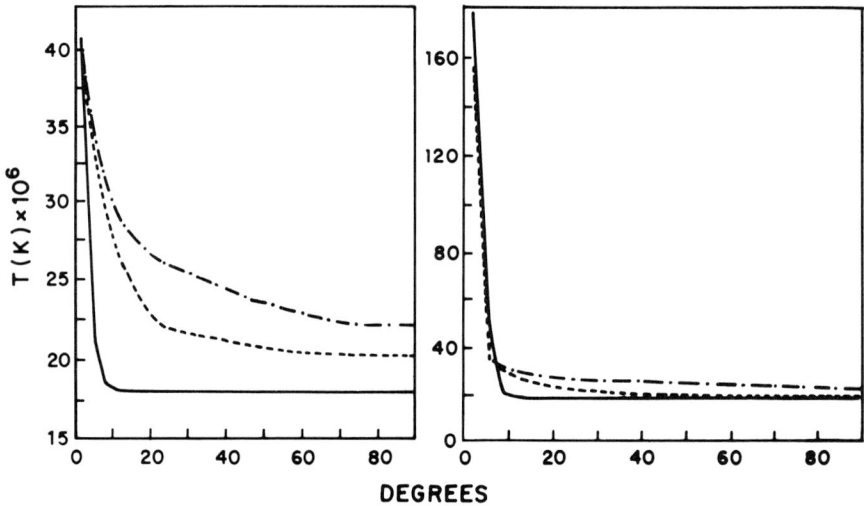

Accretion rate = 10^{16} g/sec
ΔT = 100000 K
R(WD) = $8 \cdot 10^8$ cm
solid line for η = 1.00
dashed line for η = 0.10
dashed-dotted line for η = 0.01

Figure 1. Temperature profiles on the surface for three models with different values of η. The time of evolution and the density are different in each case, but we show the profile at the onset of the runaway (2a) and during the LTNR (2b).

TABLE I Minimum ΔT needed in a polar cap extended to $\theta = 3°$ for a LTNR only in this zone, vs. accretion rate.
$m(WD) \simeq 0.85\ m_o$

$\dot{m}$ (g/sec)	ΔT
10^{16}	500K
10^{15}	800K
10^{14}	2000K
10^{13}	250000K
10^{12}	350000K

TABLE II Extension of surface ignited vs. $\dot{m}$
for $\Delta T(in) = 10^5 K$ in a polar cap
extended to 3° for $m(WD) = 1.3\, m_o$.

$\dot{m}$ (g/sec)	Extension of TNR
10^{12}	from 0° to 90°
10^{13}	from 0° to 6°
10^{14}	from 0° to 3°
10^{15}	from 0° to 3°
10^{16}	from 0° to 3°
10^{17}	from 0° to 3°

TABLE III

Rotation

"a" parameter	ω (sec^{-1})	Result
10	$\sqrt{2} \cdot 10^{-3}$	LTNR from $\theta = 75°$ to $\theta = 90°$
7	$\sqrt{2} \cdot 10^{-3}$	LTNR from $\theta = 72°$ to $\theta = 90°$
5	$\sqrt{2} \cdot 10^{-3}$	LTNR from $\theta = 3°$ to $\theta = 48°$
1	$\sqrt{2} \cdot 10^{-3}$	LTNR from $\theta = 3°$ to $\theta = 41°$
1	$\sqrt{2} \cdot 10^{-2}$	LTNR from $\theta = 3°$ to $\theta = 48°$

References

Chevalier, R. Klein, R.I., 1978, Ap. J. 219 994.
Durisen, R.H., 1977, Ap. J. 213, 145.
Gallagher, J.S., Anderson, C.M., 1976, Ap. J., 203, 625.
Fryxell, B.A., Woosley, S.E., 1982, Ap. J., 261, 332.
Hutchings, J.B., 1972, MNRAS, 158, 177.
Kippenhanhn, R. Thomas, H.C., 1978, Astron. Astrophys., 63, 265.
Livio, M., Truran, J.W., 1987, preprint.
Mustel, E.R., Boyarchuck, A.A., 1970, Astrophys. Space Sci., 6, 183.
Rosino, L., Chincarini, G., 1963, Contributi dell'Osservatorio Astrofisico di Padova in Asiago, no. 139.
Shara, M., 1984, Ap. J., 261, 649.
Shara, M., Livio, M., Moffat, F.J., A., Orio, M., 1986, Ap. J., 311, 163.
Weaver, H.F., 1974, Highlights of Astronomy, ed. G. Contopolous, 3, 509, Dordrecht, Reidel.

IMPROVED FIRST ORDER MOMENT METHOD FOR DETERMINATION OF MASS-LOSS RATES

D. Hutsemékers[1], J. Surdej[2]

Institut d'Astrophysique, Liège, Belgium

Also, [1] aspirant and [2] chercheur qualifié au Fonds National de la Recherche Scientifique (FNRS)

ABSTRACT. The first order moment of P Cygni line profiles has been computed for the case of doublets and subordinate line transitions. This improved method can be used for deriving more accurate mass-loss rates of planetary nebula nuclei, broad absorption line quasars, etc.

1. THE FIRST ORDER MOMENT OF P CYGNI LINE PROFILES: A POWERFUL METHOD

The first order moment of P Cygni line profiles,

$$W_1 \propto \int (\frac{E(\lambda)}{E_c} - 1)(\lambda - \lambda_0) d\lambda$$

allows very simple and direct determination of mass-loss rates, even for unresolved profiles.

For optically thin lines, the first order moment is directly proportional to the mass-loss rate $\dot{M}$, i.e. $W_1 = W_1^o$ where $W_1^o \propto \dot{M}\bar{n}$ ($\bar{n}$ refers to the average fractional abundance of the relevant ion). This relation holds irrespective of the velocity and opacity laws, of the presence of rotation or collisions as well as of any Sobolev-type approximation.

If the line is not unsaturated enough, mass-loss rates may be derived from "log W_1 - log W_1^o" diagrams, computed for realistic opacity and velocity laws. These diagrams also give an estimate of the uncertainty on $\dot{M}$.

Contamination by a superimposed symmetric line (as a nebular one) as well as imprecisions of the terminal velocity v_∞, have no influence on the mass-loss rate estimates obtained with this method.

2. IMPROVED METHOD FOR REAL LINES

Because the P Cygni-type lines observed in the ultraviolet have either a doublet structure or originate from excited levels, the first order moment method has been improved for a 3-level atom model, taking into account the mechanisms of radiative coupling and photoexcitation, respectively.

Examples of "$\log W_1 - \log W_1^o$" diagrams are shown in Figs. 1,2. In both cases, W_1 remain proportional to $\dot{M}\bar{n}$ for optically thin lines. The preliminary results of Fig. 2 show that the curves computed for resonance doublets are not too much dependent on $\Delta v / v_\infty$, if they are normalized to the asymptotic value W_1^t. This value only slightly depends on the velocity laws. Note that W_1 must be measured using the effective wavelength of the doublet:

$$\lambda_d = (\lambda_{12} f_{12} + \lambda_{13} f_{13}) / (f_{12} + f_{13})$$

3. APPLICATION TO PLANETARY NEBULA NUCLEI (PNN)

W_1 values for the OIV and OV subordinate lines observed in PNN have been reported in Fig. 3 (values from Cerruti-Sola, Perinotto, 1985, Ap. J., 291, 237, hereafter CP). An average curve of the 18 models has been adopted (see Fig. 1). The error bar corresponds to the dispersion of the curves due to unknown velocity and opacity laws.

This figure clearly shows that no line is actually unsaturated and that the departure from the linear relation may be greater than one order of magnitude. We conclude that all $\dot{M}$ derived by CP for PNN have been systematically underestimated by about one order of magnitude.

4. CONCLUSIONS

We have presented "$\log W_1 - \log W_1^o$" diagrams which allow direct $\dot{M}$ determinations from the analysis of resonance doublets and subordinate line profiles.

Because the P Cygni-type lines observed in the ultraviolet are generally not unsaturated, these improvements cannot be neglected. For such profiles, departures from the linear relation between W_1 and $\dot{M}$ may be as high as one order of magnitude.

We also conclude that high signal to noise spectroscopic data of faint, i.e. unsaturated, P Cygni profiles are very necessary in order to provide accurate but still inexistent mass-loss rate estimates.

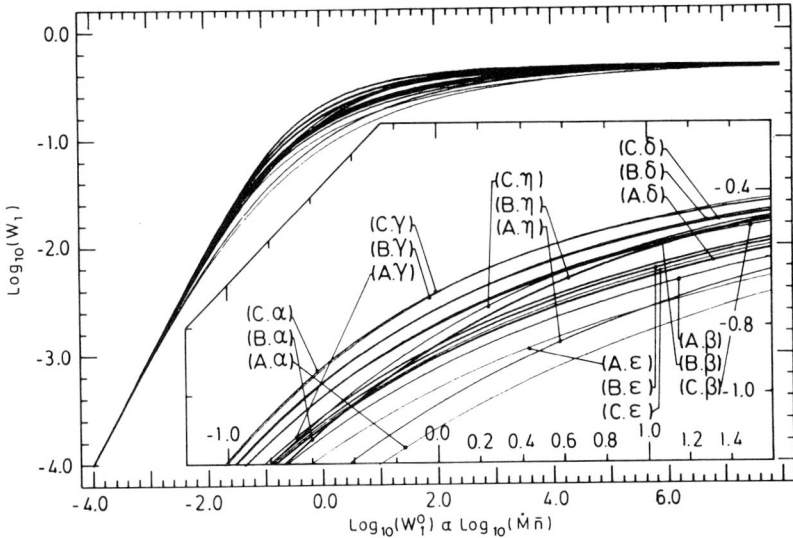

Figure 1.: "log W_1-log W_1^o" diagrams obtained for subordinate lines. These have been computed for 18 realistic combinations of opacity (α-η) and velocity laws (A-C)

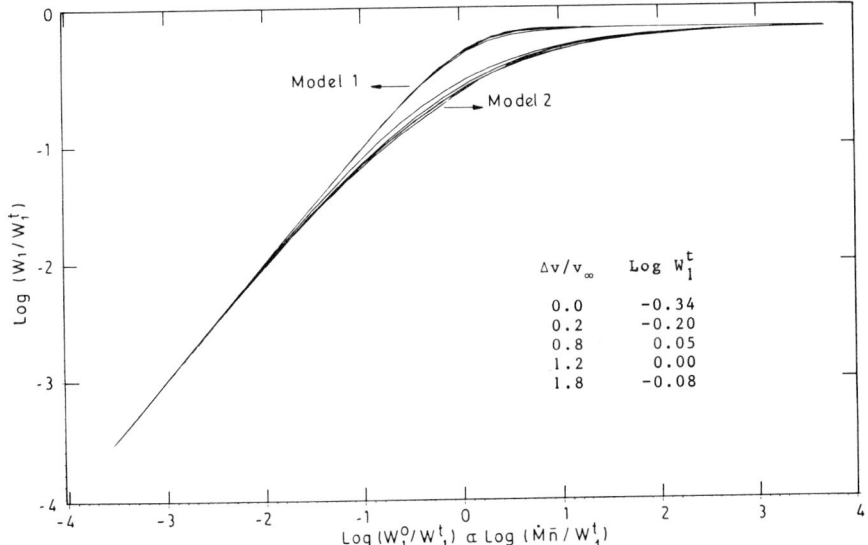

Figure 2.: "log W_1-log W_1^o" diagrams obtained for resonance doublets. Two extreme models are presented, each computed for different values of $\Delta v/v_\infty$.

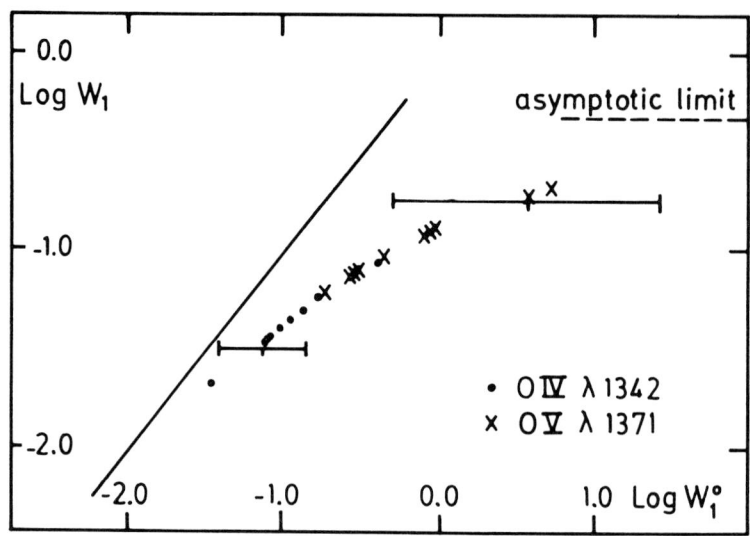

Figure 3: The W_1 values measured for the OIV and OV line profiles observed in PNN clearly show a definite departure from the linear relation between W_1 and $\dot{M}$.

REFERENCES

CASTOR, LUTZ, SEATON, 1981, MNRAS, 194, 547.
SURDEJ, 1982, Astrophys. Space Sc., 88, 31.
SURDEJ, 1983a, Astron. Astrophys., 127, 304.
SURDEJ, 1983b, Astrophys. Space Sc., 90, 299.
SURDEJ, 1985, Astron. Astrophys., 152, 361.
HUTSEMEKERS, SURDEJ, 1987, Astron. Astrophys., 173, 101.
HUTSEMEKERS, SURDEJ, 1987, in preparation.

SHELL FORMATION AND MASS LOSS IN THE PLANETARY NEBULAE A78

A. Manchado, A. Mampaso,
Instituto de Astrofisica de Canarias
38200 - La Laguna,
Tenerife,
Spain.

S.R. Pottasch,
The Kapteyn Laboratorium,
Postbus 800, 9700 AV Groningen, The Netherlands.

High resolution bidimensional spectroscopy of A78 in the He II 4685 Å, Hβ λ 4861 Å, [OIII] λ 4959 Å and [OIII] λ 5007 Å lines was obtained at the Observatorio del Roque de los Muchachos (La Palma), using the 2.5m Isaac Newton Telescope with the bidimensional detector IPCS (Image Photon Counting System). Using the 2400 gr mm^{-1} grating in first order the dispersion of 7.5 Å mm^{-1} gave an effective wavelength resolution of 0.43 Å which corresponds to 27 Km s^{-1}. These spectra confirmed the different structure of the nebula in these lines (see, Jacoby, 1979; Jacoby and Ford, 1983).

The resulting velocity maps suggest different episodes in the history of the nebula, with an external hydrogen-rich layer expanding at low velocity (35 ± 10 km s^{-1}) showing little structure and extending from approximately 35 arcsec to 55 arcsec from the center. The [OIII] and also the He II maps show, however, two velocity-structured ellipsoidal-shaped inner shells with expansion velocities of 73 ± 10 km s^{-1} and 41 ± 10 km s^{-1} for the intermediate and inner one respectively.

These data, together with recent low resolution optical data for the nebula (Manchado et al., 1987) allow us to calculate, assuming a distance of 1.7 Kp (Cohen et al., 1977), the mass of each layer, using the measured Hβ flux the n_e and the T_e (e.g. Pottasch, 1980).

The resulting masses for each shell are, 0.1 $M_\odot$ for the more external hydrogen-rich one, 0.1 $M_\odot$ for the intermediate oxygen rich and 0.024 $M_\odot$ for the inner oxygen rich. Using the expansion velocity of each of the three shells we calculated their eyection time scales, finding 1800, 3500 and 11900 years for the inner, intermediate and external one respectively. Further details will be published in a subsequent paper.

Acknowledgements

The Isaac Newton Telescope is operated on the island of La Palma (Spain) by the Royal Greenwich Observatory in the Observatorio del Roque de los Muchachos of the Instituto de Astrofisica de Canarias where we received the customary excellent support from all the staff.

References

Cohen, M., Hudson, H., O'Dell, S., Stein, W.A.: 1977, Mon. Not. Roy. Astron. Soc. **181**, 233.
Jacoby, G.H.: 1979, Pub. Astron. Soc. Pac. **91**, 754.
Jacoby, G.H., Ford, H.C.: 1983, Astrophys. j. **266**, 298.
Manchado, A., Pottasch, S.R., Mampaso, A.: 1987, Astron. Astrophys. (submitted).
Pottasch, S.R.: 1980, Astron. Astrophys. **89**, 336.

NARROW-BAND IMAGING OF PLANETARY NEBULAE WITH THE
CANADA-FRANCE-HAWAII TELESCOPE PHOTON-COUNTING CAMERA

C.T. HUA
Laboratoire d'Astronomie Spatiale du CNRS
Allée Peiresc
13012 Marseille - France

ABSTRACT.

Faint secondary structures are detected for some planetary nebulae (PN's) thanks to the Canada-France-Hawaii Telexcope (CFHT) photon-counting camera (PCC) used in the direct imaging mode with narrow-band interference filters centred at Hα, Hβ, [NII]6583, [OIII]4363, 5007 and HeII 4686. Small-diameter objects such as IC 351, 2165 ; NGC 6790, 6881, 6884, 6886, M1-1, M1-4, M1-65, Vy 2-3 were imaged at f/36 focus, whereas more extended nebulae ($\emptyset > 10$ ") were observed at f/8: A2, M1-75, IC 289. Generally, forbidden emission-line images exhibit inhomogeneous structure while the hydrogen distribution is rather smooth, with faint outer emission evidenced. These outer envelopes could result from multiple-shell formation process. Absolute fluxes were estimated for various wavelengths as well as rough estimates of the ionized mass which would correspond to the total mass since, according to Seaton, planetaries with R ($\dot{n}$) > 0.1 pc could be optically thin.

1. INTRODUCTION

Empirical classifications of PN's are commonly based upon available observations of the nebular shapes through more or less adequate angular resolution and, unfortunately not always selective enough, monochromatic imagery. Most earlier morphological studies were performed with 50-70 Å wide filters, thus including for example in the red, the Hα and two [NII]6548,6583 emission lines, which depend differently on the ionization structure.

It is generally accepted that the Hα and radio maps of PN's do have similar appearance originating from ionized hydrogen, corresponding to the general distribution of gas throughout the nebular volume, whilst ionic species such as [OIII] or [NII], when observable, rather outline the distribution in ingomogeneities. Our objective aims to study the ionization structure in small, and presumably young, planetaries, as well as the matter distribution.

across the nebular volume in order to better comprehend the physical link between PN formation and the evolution of the progenitor, mass of which could play an important role in the "monochromatic" morphologies displayed by the nebulae.

2. OBSERVATION

Our purpose was to establish a relevant set of morphological data using a sensitive device, and related to large and small angular diameter planetaries. The selective interference filters are chosen in such a manner to avoid as much as possible the continuum contribution, as well as to escape the wavelength blending with neighbouring emission lines. The PCC description was already given in previous reports (see Hua et al., 1987).

Extended nebulae with angular diameter larger than 10" are observed with an image-scale of 0.245 arcsecond/pixel (at the f/8 of the CFH 3.60m telescope), whereas more compact PN's are studied using the f/36 aperture ratio of the same telescope operated at Mauna-Kea, Hawaii. The quality of the site usually permits a subarcsecond "seeing".

3. RESULTS AND DISCUSSION

Hereafter we give separate comments for some individual objects.

IC 351

A closed shell structure is exhibited in the H alpha emission by this N+ weak planetary nebula. The [OIII]5007,4363 pictures show two conspicuous north-south condensations. A mean electron temperature of 13000 + 1000 K was derived from the I (4363)/I (5007) intensity ratio (Hua and Grundseth, 1986).

IC 2165

The H alpha image also exhibits a closed shell whereas the [NII] picture displays four condensations of about 1" in diameter well separated and located within two symmetrical lobes, with respect to the centre suggested by the H alpha image (see Hua and Grundseth, 1985).
SKIP 1

NGC6884

The roundish H alpha image is definitely smaller than the [NII]6583 picture. In addition, the hydrogen inner two-lobe structure obviously differs from the nitrogen distribution.

NGC 6886

This nebula was assigned successively 6" and 9" by Bauschinger (1911, Strassb. Ann. 4,79) and Curtis (1918, Publ. Lick Obs., 13,55). In fact,

the H alpha image shows an elliptical shape with a 8.5 arcsecond
major axis around a brighter circular (4"5 in diameter) disk.

More detailed information will be made available in a forthcoming
paper in Astronomy and Astrophysics.

REFERENCES

Hua, C.T., Grundseth, B., 1985. Astronom. J., 90, 2055
Hua, C.T., Grundseth, B., 1986. Astronom. J., 92, 853
Hua, C.T., Grundseth, B., Nguyen-Trong, T., 1987. Astrophys. Lett.,
    26, 167.

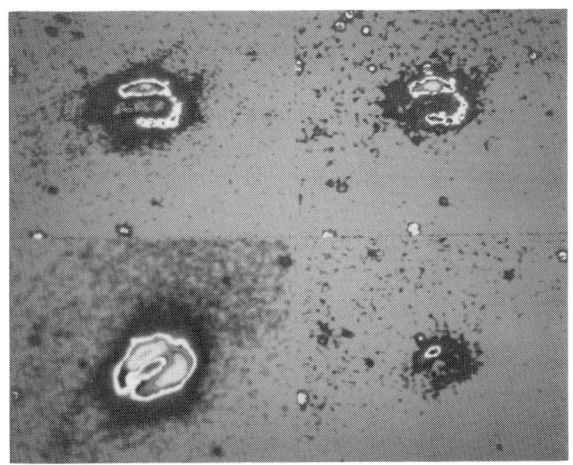

M1-75 (left right)
[NII], Hα
[OIII], Hβ

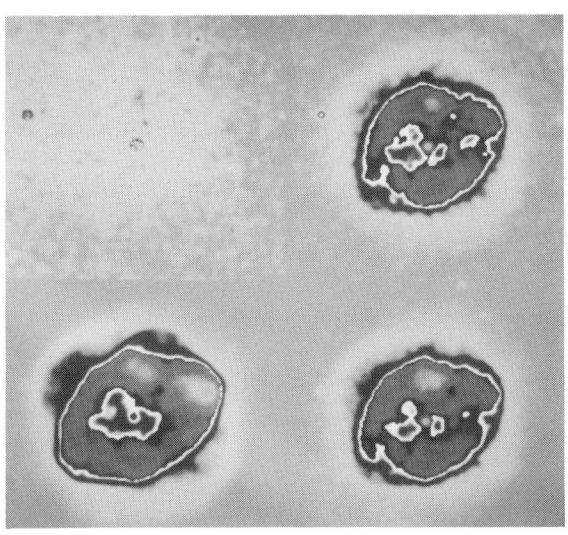

IC 289 (left to right)
[NII], Hα
[OIII], He II 4686

V645 CYGNI AND THE DUCK NEBULA

R. W. Goodrich
Lick Observatory
University of California
Santa Cruz, CA 95064
U.S.A.

ABSTRACT. The region around the infrared source V645 Cygni has been studied with a variety of optical techniques. Several indications of mass outflow have been discovered in this young star. Imaging has revealed a long, linear feature which resembles morphologically the emission jets found near other young stars. In V645 Cygni, however, the feature is shown to be predominantly reflection. A high-velocity [S II] system has been found between the two brightest parts of the nebula, and a Herbig-Haro object has been found to the south of the nebula.

1. INTRODUCTION

The nebulosity around V645 Cygni was first described in detail by Cohen (1977) who presented spectra of three regions within the nebula and including the stellar-like knot N0 (his notation). Deeper CCD imaging (Goodrich 1986) has revealed the existence of even more extensive nebulosity including a long, narrow feature. In deep images the nebula somewhat resembles the outline of a duck, with the narrow feature as the duck's "beak"; hence the name "Duck Nebula."

Morphologically the "beak" resembles the emission jets found by Mundt and other workers near other young stars (see Mundt's review article in this volume). However images taken with a broad-band red filter and a narrow-band Hα filter indicate that the beak is predominantly reflection. This is confirmed with low-resolution (14 Å) spectra which show that the spectra of the knot N0 and the beak are very similar, with the beak being redder and actually having less Hα, [S II], and [O I] emission.

The long-slit CCD spectra show variations of the equivalent widths of some of the emission lines across the face of the nebula and it is of interest to map this "excess" emission. Broad-band red and narrow-band [S II] filters were used to create a map of the [S II] equivalent width across the nebula. One prominent feature of this map is a relatively compact, high equivalent width knot to the south of the nebula. This knot lies along the axis of the "beak," and low resolution spectra show the emission lines characteristic of a Herbig-Haro object.

Some more interesting data come from high-resolution (1 Å) spectra in the Hα-[S II] region. The slit was placed across the knots N0 and N1 of Cohen. There appear in the spectrum a pair of lines at the stellar velocity of [S II] $\lambda\lambda$6717, 6731 in both N0 and N1, but in addition there appear a pair of lines offset by 0.5 arcseconds from N1 in the direction of N0. They are blueshifted by 244 ±12 km/sec, a velocity typical of HH ejection velocities. However, without a better idea of the overall spectrum and the morphology of this high-velocity gas it is premature to conclude that this emission does in fact arise in an HH object close to the star.

The absorption spectrum of the brightest knot N0 has been classified as that of an A5e star by Humphreys, Merill, and Black (1980; hereafter HMB80). While Cohen (1977) and Harvey and Lada (1980) calculate a kinematic distance to V645 Cygni of 6 kpc using a Schmidt rotation law Rickard (1968) demonstrates that there is a peculiar velocity in part of the Cas-Per spiral arm in this direction and hence the proper kinematic distance to V645 Cygni should be 3.5±0.5 kpc. The A5 spectral type may be compared with the observed colors of N0 and N1 to derive a reddening. Confirmed by measuring the equivalent width of the diffuse interstellar band at 6614 Å this value is E(B-V) = 0.83, and as Goodrich (1986) pointed out this is mostly interstellar. Taking this reddening into account and combining the infrared fluxes from HMB80 and IRAS the total luminosity of V645 Cygni is 5×10^4 $L_\odot$, consistent with an A0 I star.

2. DISCUSSION

The "beak" of the Duck Nebula morphologically resembles the emission jets found in other young stars. However it is predominantly reflection and hence its relationship to the emission jets, if any, is unclear. One possibility is that it is the cavity which a now-extinct jet has bored into the parent molecular cloud. This seems unlikely, however, since the near face of the cavity must be thick enough to contain the jet and yet thin enough for us to see through it to the cavity. Observations of emission jets do not indicate large amounts of dust which might have been left behind by an extinct jet, and of course the time scale for dust formation within the low-density jet material is prohibitively long.

On the other hand the "beak"-star-HH object line (along p.a. 5°) defines a natural axis for the bipolar outflow, and this would indicate that the beak is in fact where an emission jet might be expected. Early CO maps indicated that the N0-N1 line at p.a. 135° better defined the molecular outflow axis. More recent high-resolution CO observations by Rodriguez et al. (preprint), however, show that while the outer CO contours lie more or less along the N0-N1 line the inner contours do in fact follow a more nearly N-S axis. These latter authors independently arrived at the same conclusion as Goodrich (1986), namely that there appears to be a twist in the outflow axis, perhaps indicating the presence of some sort of a warped disk.

Currently V645 Cygni is the only known example of a long, highly collimated reflection nebulosity. It is clearly of interest to watch for more of these objects with the hope of defining their possible relationship to the emission jets and other manifestations of the outflows surrounding young stars.

3. ACKNOWLEDGEMENTS

The author would like to thank the many people who contributed observing time, discussions, and other types of help to this project, in particular Drs. G. Herbig, J. S. Miller, B. F. Jones, and M. Cohen. He would also like to express his gratitude to the organizers and participants of the Second Torino Workshop for the opportunity to share data, theories, and points of view.

4. REFERENCES

Cohen, M. 1977, Ap. J., 215, 533.

Goodrich, R. W. 1986, Ap. J., 311, 882.

Harvey, P. M., and Lada, C. J. 1980, Ap. J., 237, 61.

Humphreys, R. M., Merrill, K. M., and Black, J. H. 1980, Ap. J. Letters, 237, L17 (HMB80).

Rickard, J. J. 1968, Ap. J., 152, 1019.

MASS LOSS FROM PLANETARY NEBULAE : IC 418

S.Cerrato**, M.Baessgen**, L.Bianchi*, M.Grewing**

* Astron. Obs. Torino, 10025 Pino Torinese (To), Italy
** Astronomical Institute Tübingen, D-7400 Tübingen, FRG

ABSTRACT. The ESO 2.2m telescope equipped with the B&C CCD spectrograph has been used to search for emission from regions well outside the bright shells of Planetary Nebulae. From the intensity distribution of these halos, the current mass of the envelope as well as the mass loss rate of the Red Giant progenitor of the nucleus can be inferred.
In addition, the present mass loss rate of the central stars can be derived from the P-Cygni profiles of the UV-resonance lines in the IUE spectra of the nuclei. Results from both methods are presented and compared to theoretical predictions.

1. THE METHOD

High-quantum yield low noise optical detectors allow to study the surface brightness of planetary nebulae to much fainter limits than has previously been possible. It is of particular interest to use this technique to trace out the emission from faint halos around the more prominent bright nebular shells. Such halos should be remnants of the extended wind spheres created during the AGB-phase of the progenitor stars (see e.g. Kwok et al. 1978), and from their current structure one can re-derive the mass loss rate at this earlier evolutionary stage, provided one is willing to make some simplifying assumptions. These concern the geometrical shape of the halos (spherical symmetry), the constancy of the mass loss rate during the AGB phase, and the constancy of the terminal wind velocity. With these assumptions one can calculate the density distribution in the wind sphere for various assumed mass loss rates. Assuming furthermore that hydrogen is fully ionized in this wind and that the temperature of the plasma is in the order of 10^4 K, we can predict the expected surface brightness distribution, compare it to observations, and thereby determine dM/dt for the progenitor star.

Making use also of the capability of the IUE satellite to observe the ultraviolet spectra of the central stars of planetary nebulae which contain a number of strong resonance lines, we can furthermore determine the mass loss rate from the current nuclei. This technique relies on the theory of radiation pressure driven winds which is used to compute theoretical line profiles to match the observed P Cygni lines. In addition, from the UV spectra we obtain an estimate of the radiation temperatures of the nuclei which -if we also know their distances- allows us to place the objects in the HR-diagram, thereby enabling us to compare the stellar parameters with evolutionary models.

2. APPLICATION TO IC 418

In the following we describe the application of these methods to IC 418. This nebula is shown in Fig.1 to illustrate its high degree of spherical symmetry. It is a low excitation nebula (excitation class 3) surrounding a central star with T=30000-35000 K. In Fig.2 we show the de-reddened low resolution ultraviolet spectrum, and in Fig.3 an optical spectrum of the central region in order to illustrate the fact that the radiation temperature of the nucleus and the excitation state of the nebula are indeed fairly low. From fits of model atmosphere spectra (Kurucz 1979) we obtained T=32500±2500 K in good agreement with values quoted earlier in the literature (e.g. Pottasch et al. 1978). Given the fact that IC 418 obviously is a fairly young object (expansion age less than 3000 yrs., see below), we expect the temperature of the nucleus to still increase in the future, with the luminosity remaining practically unchanged for some time, i.e. we believe that the object is still on the horizontal part of its evolutionary track. If this is correct, the current luminosity of 5500 Lo is also indicative of the luminosity of the central star when it left the AGB. We shall return to this later.

First, we shall address the mass loss from the current nucleus. In Fig. 4 we show portions of the high-resolution IUE spectrum in the neighbourhood of the C IV and Si IV resonance doublets. These profiles illustrate the fact that the nucleus of IC 418 has a wind with a terminal velocity of about 1200 km/s. Superimposed on the observations are theoretical line profiles calculated from the radiation pressure driven wind theory. These yield an estimate of the current mass loss rate of about $3 \cdot 10^{-7}$ Mo/yr. Adopting an age of IC 418 of about 2600 yrs. which follows from its current size ($1.3 \cdot 10^{17}$ cm) and an expansion velocity of 17 km/s (see e.g. the Lyα profile in Fig. 5), the derived mass loss rate is clearly far too low to have significantly changed the nebular mass which is estimated to be M(neb)≈0.1 Mo from the total Hβ emission.

Let us now turn to the mass loss from the progenitor. As stated before, the current luminosity of the nucleus should also be indicative of the progenitor's luminosity before it turned away from the AGB. Using L=5500 Lo and $T_{eff}=10^{3.8}$ $(L/Lo)^{0.08}$ on the AGB, we can plug these numbers into the red giant mass loss relation (Kudritzki and Reimers 1978) to calculate the expected mass loss rate. We find M > $1.2 \cdot 10^{-6}$ Mo/yr where we have used the current mass of the nucleus of 0.6 Mo as a lower limit to the mass of the progenitor at the end of its evolution on the AGB.

How can we verify this mass loss estimate ? To this end we have studied the Hα emission from the halo of IC 418. On the assumption of a constant expansion velocity (10 km/s), and a constant mass loss rate during the AGB phase, we can derive from the current intensity distribution in the halo an estimate of M during the AGB phase. We find for IC 418 M= 7 10^{-6} Mo/yr which agrees rather well with the above estimate for the AGB wind.

Expanding at a velocity of 10 km/s, the AGB wind would have lasted for $>9 \cdot 10^3$ yrs. (the exact value depending on where we assume the outer boundary of the halo to be). This is not an unreasonable timescale. During this time 0.06 Mo would be expected to accumulate in the entire wind sphere about a factor of two less than the mass estimate of the bright nebular shell.

In conclusion we can therefore state that in the case of IC 418 there may still be a need for a superwind phase, but the parameters derived here seem to be not too far off from the basic features of the two-wind model by Kwok et al. (1978).

REFERENCES

Kudritzki,R.P., Reimers,D. 1978 Astron. Astroph. 70,227
Kurucz,R.L. 1979 Astroph. J. Suppl. Ser. 40,1
Kwok,S. ,Purton,C.R., Fitzgerald,P.M. 1978 Astroph. J. 219,L125
Pottasch,S.R., Wesselius,P.R., Wu,C.-C., Fieten,H., van Duinen,R.J. 1978 Astron. Astroph. 63,297

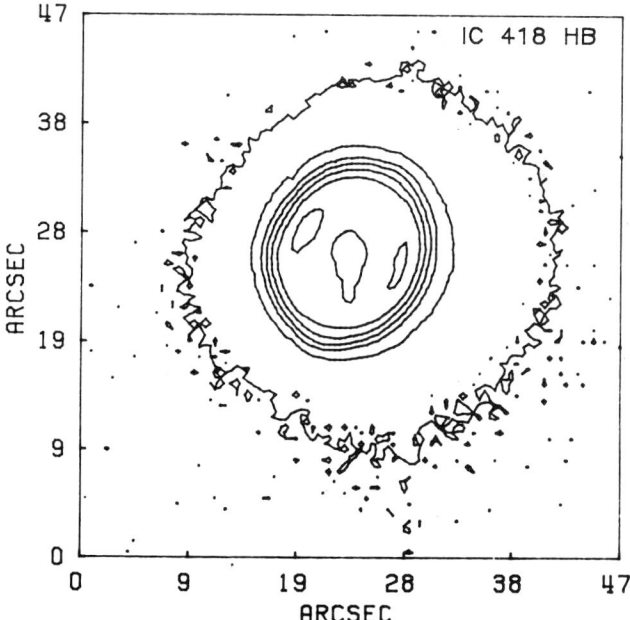

Figure 1. The CCD image in Hβ of IC 418 illustrates the high degree of spherical symmetry.

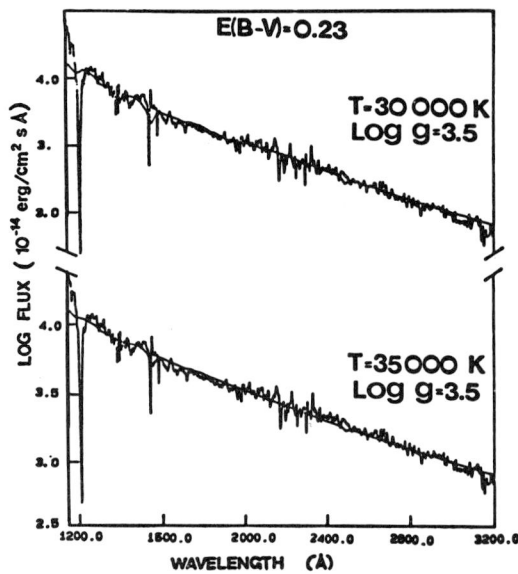

Figure 2. Dereddened low-resolution UV-spectrum; superimposed are two Kurucz model atmospheres.

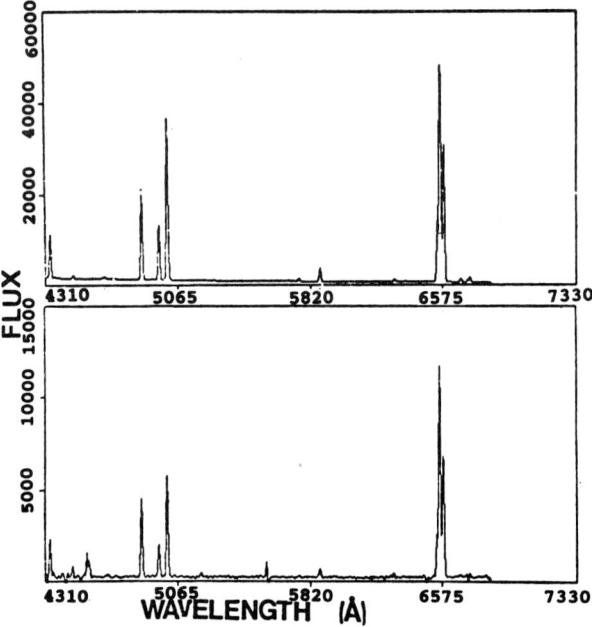

Figure 3. Raw optical spectra of the bright part (a) and of the halo (b) of IC 418.

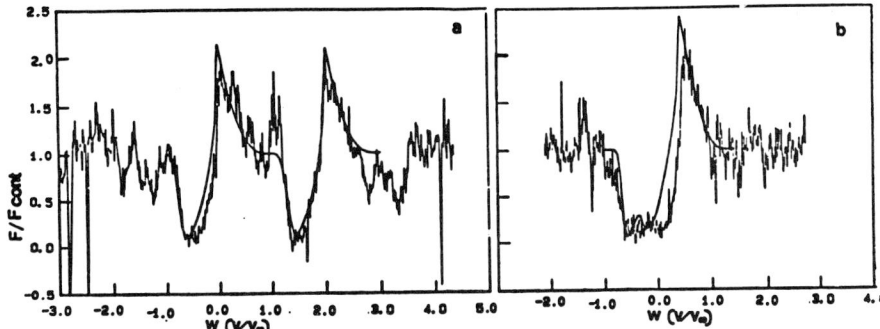

Figure 4. Sections of the IUE high resolution spectrum with superimposed theoretical P-Cygni profiles.

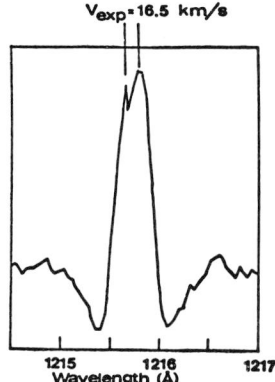

Figure 5. IUE high-resolution spectrum at Lyα.

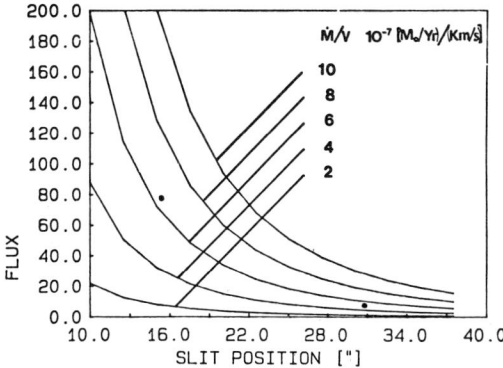

Figure 6. The dots represent density determinations in the halo of IC 418, derived from Hα-line observations. Superimposed are theoretical predictions for various M/V values.

THE KINEMATICAL PROPERTIES OF PLANETARY NEBULAE

L.Bianchi, C.Falcetta
Astronomical Observatory of Torino, 10025 Pino Torinese, Italy

ABSTRACT. The ESO 1.4m CAT in combination with the Coudé Echelle Spectrograph (CES) has been used to obtain high resolution spectra of Planetary Nebulae. With these data we studied the velocity fields of some extended nebulae and we compiled a catalogue of expansion velocities, which is presented here.

1. INTRODUCTION.

Among the most fundamental data to test theories of the hydrodynamical evolution of the nebular envelopes are the measurements of the expansion velocities, and of the detailed kinematical structure of the nebular shells. Since the nebular velocities are rather low (few tens of Km/s), a very high spectral resolution is needed.
We present here a catalogue of expansion velocities of Planetary Nebulae (PNe) based on very high resolution optical line profiles, and discuss some statistical properties.
For some extended objects we also obtained detailed line profile maps. Combining these with monochromatic images from which the density profiles can be derived, we can infer the velocity patterns within these objects. Some preliminary results have been shown by Bianchi et al. (1986) and a more detailed analysis of the observed line profiles will be presented in a forthcoming paper.

2. THE DATA.

We used the ESO 1.4m CAT with the Coudé Echelle Spectrograph to obtain very high resolution (r.p.=10^5) profiles of H-alpha and [NII] 6548, 6584 of several PNe. The actual

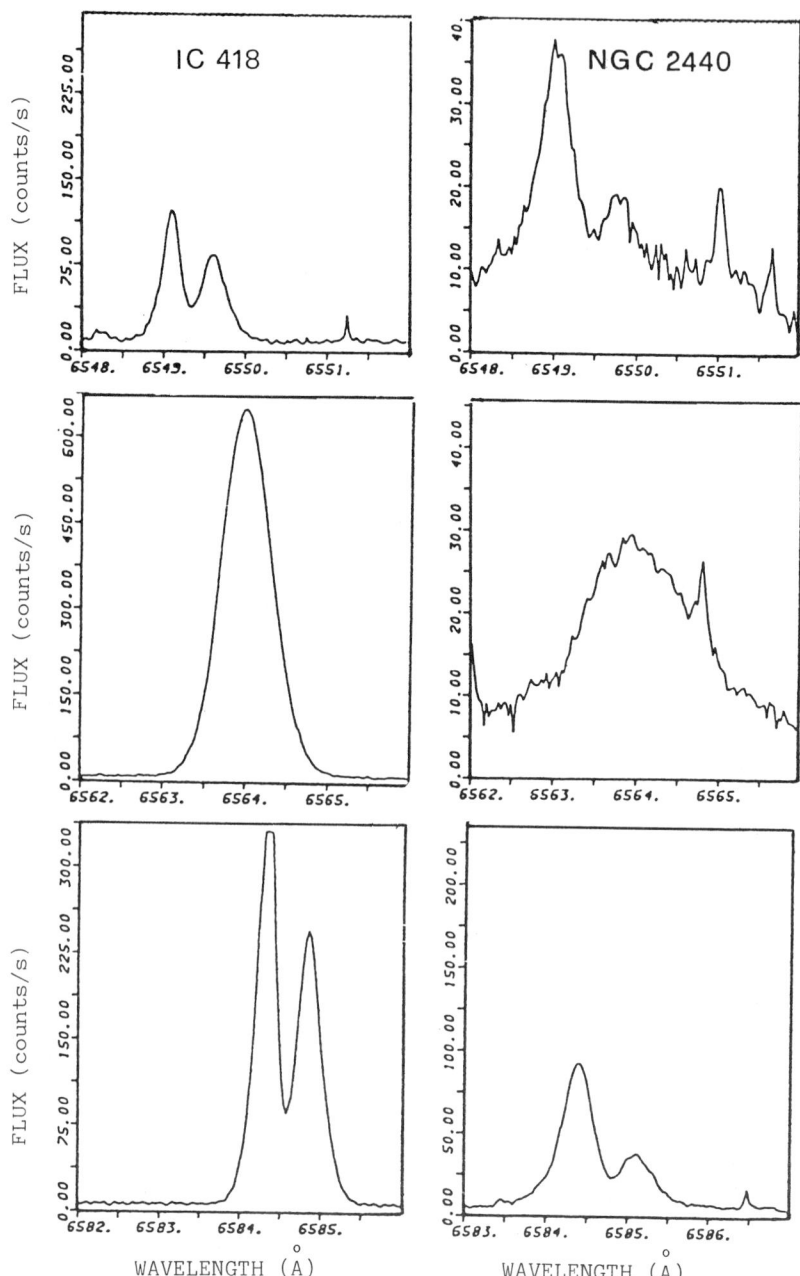

Figure 1. Examples of the observed lines: The H-alpha and [NII]6548,6584 profiles are shown for the central region of IC 418 and of NGC 2440.

resolution, measured from the width of the narrowest lines of the Thorium calibration spectrum, corresponds to about 3 km/s at H-alpha. The spectra have been cleaned for the read-out noise, calibrated in relative intensity by means of flat fields and in wavelength with the ESO IHAP and MIDAS systems. The final accuracy of the wavelength calibration is better than the resolution. The velocity scale has then been transformed into the Heliocentric reference frame for each object. Some profiles are shown in Figure 1.
The scientific analysis has been performed at the Observatory of Torino with codes developed for this purpose. The splitting of the lines, the systemic velocity of the objects and the width of each component were measured by fitting the observed profiles with two gaussian components (examples are shown in Figure 2). A code has also been developed to reproduce the observed velocity profiles with assumed radial velocity laws within the nebula, and using the radial density law derived from the projected density profiles of the monochromatic images. Results from this analysis will be reported in a forthcoming paper.

3. THE CATALOGUE OF EXPANSION VELOCITIES.

To compile the catalogue of expansion velocities we used the spectra taken in the centre of each object, to avoid geometrical projection effects. Table 1 gives the results for the PNe observed in our first observing run. Another group of objects has then been observed and more data are being measured.
For some objects for which previous measurements existed in the literature, we find good agreement with our results, especially when the measurements refer to the same ions. In fact, if a velocity gradient is present in the nebular envelope, ions having different spatial distribution of their relative abundance within the object will obviously show different expansion velocities (see e.g. Bianchi et al.1986).

4. DISCUSSION AND CONCLUSIONS.

For all the objects listed in Table 1 we have taken distance estimates from the literature, by choosing a mean value from a comparative study of all the existing measurements, to convert the observed angular size into linear radius. In Figure 3 we plotted the measured expansion velocities versus diameter of the nebulae, since a relation between these quantities was claimed by different authors (e.g. Sabbadin, 1984; Sabbadin et al. 1984; Robinson et al. 1982). All these authors find an increase of the expansion velocity

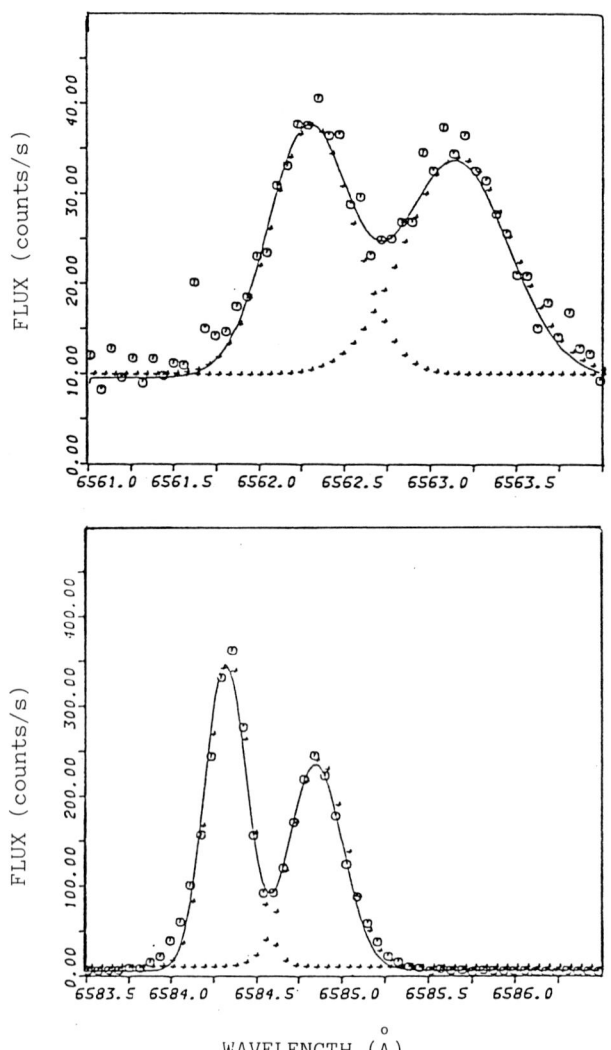

Figure 2. Examples of measurements of the two components of each profile are shown for one line of IC 418 and one of NGC 3242. The circles are the observed data points, the small crosses each gaussian component of the best fit, and the continuous line the sum of the two gaussians, i.e. the best fit of the overall profile.

Table 1. The measured expansion velocities.

N.	P.K.	Name	Exp.Velocity(km/s) H-alpha	[NII]
1	206 -40 1	NGC 1535	18.9	--
2	215 -24 1	IC 418	7.4	11.9
3	234 +2 1	NGC 2440	18.7	17.3
4	285 -14 1	IC 2448	11.5	--
5	261 +32 1	NGC 3242	19.3	--
6	294 +4 1	NGC 3918	19.	26.
7	327 +10 1	NGC 5882	17.3	23.5
8	315 -13 1	He 2-131	8.9	9.9
9	10 +18 2	M 2-9	32.1	--
10	334 -9 1	IC 4642	13.3	16.2
11	345 -8 1	He 2-274	8.3	14.9
12	1 -6 2	SwSt1	9.1	9.4
13	9 -51 1	NGC 6629	7.7	--
14	2 -13 1	IC 4776	10.2	17.3
15	27 -9 1	IC 4846	13.1	--
16	54 -12 1	NGC 6891	8.1	--
17	37 -34 1	NGC 7009	18.4	--

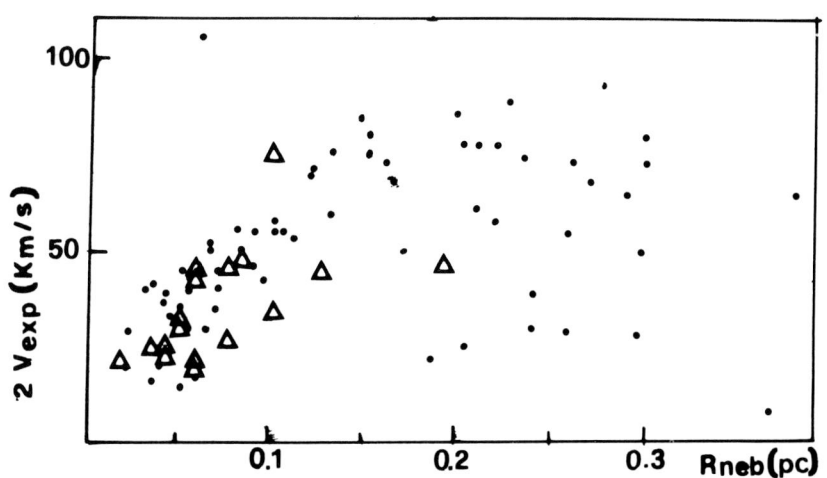

Figure 3. The double of the measured expansion velocities is plotted against the nebular radius (triangles). The dots are the data from Sabbadin et al. (1984) reported for comparison.

with radius (up to a certain radius, beyond which the nebula is probably getting dispersed into the interstellar medium). The steepness of the relation seems to depend on the mass of the nebular shell, as can be expected theoretically (e.g. Fearch and Salpeter 1975), and probably also on the spectral type of the central star.
Our data are shown in Fig.3 together with the results of Sabbadin et al.(1984) for comparison, and it seems that we find the same trend.
If the size of the nebular shell can be interpreted as an indication of the age, the fact that larger nebulae (that is, older nebulae) expand faster could mean that the expansion velocity does not remain constant after an initial ejection but it is affected (i.e. the shell is accelerated) by the pressure of the supersonic wind of the nucleus in its post-AGB evolutionary phase. This is likely especially for the most luminous PN nuclei.

REFERENCES

Bianchi,L.,Grewing,M., Falcetta,C., Baessgen,M., 1986: in "Planetary and Proto-Planetary Nebulae", ed. A. Preite-Martinez, in press
Fearch, B.L., Salpeter, E., 1975: Astrophys.J., 202, 195
Sabbadin,F., 1984: M.N.R.A.S., 209, 889
Sabbadin, F., Bianchini,A., Hamazaoglu,E., 1984: Astron. Astrophys., 136, 193
Robinson, G., Reay,N., Atherton,P., 1982: M.N.R.A.S., 199, 649

THE 'S' PROCESS NUCLEOSYNTHESIS IN LOW MASS STARS AND THE NEUTRON SOURCE $^{13}C(\alpha,n)^{16}O$.

G.Picchio[1], M.Busso[1], R.Gallino[2]

1) Osservatorio Astronomico di Torino, Italy
2) Istituto di Fisica Generale dell'Universita', Torino, Italy

ABSTRACT. The heavy element synthesis induced by the large neutron fluxes released by the reaction $^{13}C(\alpha,n)^{16}O$ in low mass planetary nebulae progenitors experiencing thermal pulses is examined. The mechanism of semiconvection proposed by Iben and Renzini (1982) for the formation of the initial ^{13}C is assumed to be effective. A new reaction network has been created, incorporating updated values for the cross-sections of $\alpha-$, $p-$ and n-captures on 350 nuclei and including 80 branching points along the neutron capture path. As a preliminary investigation we followed 14 thermal pulses. The results of the computations and their relevance in explaining the formation of the heavy n-rich elements are briefly discussed.

1. Introduction.
In planetary nebula progenitors of low and intermediate mass (LMS and IMS, respectively) the last stages of evolution are strongly affected by the combined effect of mass loss and of various processes of remixing (see e.g. Iben, 1977; Chiosi, 1986).

In particular, during the AGB evolution, those stars are thought to loose a consistent fraction of their envelope by stellar winds (Reimers, 1975; Renzini, 1981). The ejected mass is likely to be enriched in various isotopes synthesized by nuclear processes occurring in intermediate burning regions (Renzini and Voli, 1981; Becker and Iben, 1979), so that nucleosynthesis and mass loss combine in making these stars important contributors to the chemical evolution of galaxies (Truran and Iben, 1977; Iben and Truran, 1978; Iben, 1983).

The pulsed nature of the He-shell during the double shell phase was recognized to offer a suitable mechanism for the formation and the ejection of the s-elements in a solar system distribution, through the occurrence of repeated neutron exposures (Ulrich, 1973). The major neutron source responsible for that process was originally identified in the reaction $^{22}Ne(\alpha,n)^{25}Mg$, which efficiently takes place during the high temperature thermal pulses of IMS (Truran and Iben, 1977; Cosner et al., 1981).

However, observations do not show evidence of IMS as luminous as those predicted by theoretical models (see e.g. Habing, 1987); moreover, many revisions of important thermonuclear reaction rates (Caughlan et al., 1985) and of n-capture cross sections (Bao and Käppeler, 1987), imply a strong reduction in the efficiency of the reaction $^{22}Ne(\alpha,n)^{25}Mg$ for the formation of s-elements with $A > 120-130$ (Busso et al.,1987).

It has long been recognized that the reaction $^{13}C(\alpha,n)^{16}O$ may offer an alternative neutron source (Cameron, 1955; Ulrich, 1973) if some remixing mechanism brings proton fluxes from the envelope into the C-rich region. The occurrence of such a mechanism has indeed been found in LMS during AGB phases (Iben and Renzini, 1982a,b), due to the development of semiconvective regions above the He-shell. A first analysis of the synthesis of n-rich elements in LMS, induced by α-captures on ^{13}C, has been performed by Malaney (1986 a,b), using a restricted reaction network extending up to $A \simeq 80-90$.

In this note we report the preliminary results of a reanalysis of the whole problem, adopting a complete rection network, up to the heaviest nuclei.

2. Computations and results.

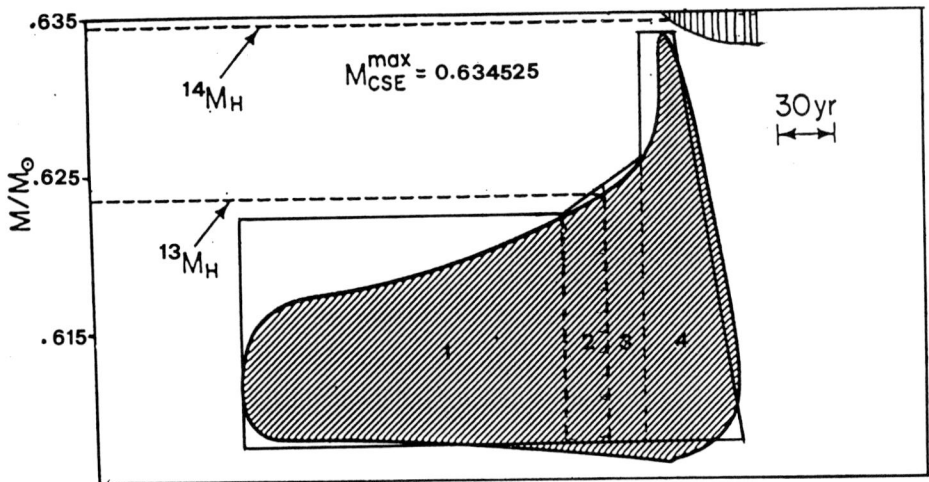

Fig. 1. The adopted profile of the pulse, superimposed on the original model by Iben and Renzini (1982b).

Our numerical code takes in account α, p and n captures on a nuclear network including 350 isotopes, extending from He to Po, and incorporates 80 branching points along the n-capture path, due to α-decays, e-captures and (n,p) or (n,α) reactions. The pulse model is taken from Iben (1982) and Iben and Renzini (1982a,b). After each pulse, their mechanism for the ingestion of material enriched with ^{13}C and ^{14}N is assumed.

The thermal pulse has been schematically divided into 4 zones, as shown by Fig. 1. In the first zone, the convection remixes matter already processed by the previous cycle, but whose initial composition does not include ^{13}C. In the second zone, the ingestion of new material containing primary ^{13}C and ^{14}N in various proportions, as described by Iben and Renzini (1982b), occurs. The rate of ingestion has been assumed to be 10^{-3} $M_\odot$/yr (see Malaney, 1986a,b). After ingestion, ^{13}C efficiently burns, thus releasing strong neutron fluxes. In the third zone, the pulse spreads up to regions enriched with ^{14}N, previously synthesized by the H-burning shell, but where ^{13}C is not present. Here, the residual abundance of ^{13}C is consumed down to very low concentrations. Finally, in the fourth zone convective mixing grows in mass, while the temperature at the base of the He shell increases from $1.5 \cdot 10^8$ up to $3 \cdot 10^8$. Additional neutrons may then be released by the reaction ^{22}Ne(α,n)^{25}Mg, fresh ^{22}Ne being supplied through α-captures on ^{14}N.

We have followed 14 pulses, and the main results are briefly summarized in Table 1 and in Figs. 2 and 3.

Table 1 shows the evolution with pulse number of the maximum neutron density (N_n) at He-shell bottom and of the corresponding number of neutrons captured per heavy seed (n_c).

Table 1

Pulse no.	2	4	6	8	10	12	14
$N_n/10^8$	2.7	1.5	1.1	0.8	0.7	0.6	0.5
n_c	8.6	7.2	6.8	6.5	5.7	5.0	4.4

Fig. 2 shows the evolution of the concentrations of various intermediate and heavy elements, in terms of overproduction factors with respect to the initial composition. The figure underlines that the lower is the mass of the nucleus, the faster it reaches its asymptotic abundance; after 14 cycles, the heaviest species, like Hg and Pb, have not yet reached equilibrium.

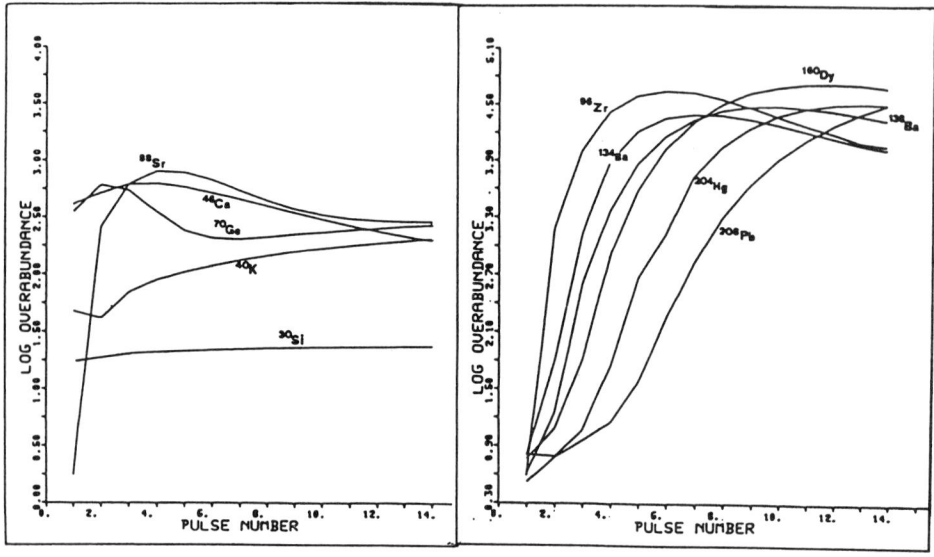

Fig. 2. Evolution of overabundances with pulse number.

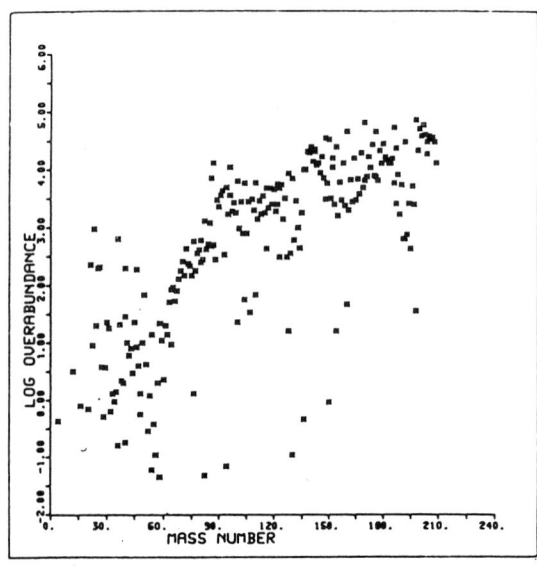

Fig. 3. Overabundances after the last computed pulse.

Fig. 3 shows the resulting composition at the end of the 14th cycle. As is evident in the figure, large overproductions may be obtained, expecially for the heaviest nuclei (A>120) that were not efficiently produced neither by massive stars (Busso and Gallino, 1985), nor by intermediate mass stars (Busso et al., 1987).

We notice that the neutron path corresponding to the high n-densities of Table 1 does not match a typical "s" irradiation, so that several isotopes traditionally ascribed to the "r" process are produced.

The results of our computations show that low mass AGB stars do efficiently contribute to the production of the heavy n-rich elements through a complex capture mechanism (n-process) that is intermediate between the classical "slow" and "rapid" exposures.

REFERENCES

Bao,Z.Y. and Käppeler,F.:1987, Atomic Data and Nuclear Data Tables (in press).
Becker,S.A. and Iben,I.jr.:1979, Ap. J. 232, 831.
Busso,M. and Gallino,R.:1985, Astr. Ap. 151, 205.
Busso,M., Picchio,G., Gallino,R. and Chieffi,A.:1987, Ap. J. (submitted).
Cameron,A.G.W.:1955, Ap. J., 121, 144
Caughlan,G.R., Fowler,W.A., Harris,M.J. and Zimmermann,B.A.:1985, Atomic Data and Nuclear Data Tables, 32, 197.
Chiosi,C.: 1986, in 'Nucleosynthesis and Chemical Evolution', 18th Course of the Swiss Society of Astrophysics and Astronomy, Obs. Geneve press, p. 199.
Cosner,K., Iben,I.jr and Truran,J.W.:1980, Ap. J. 238, L91
Habing,H.J.:1987, Proceedings of the workshop 'The late stages of stellar evolution', Calgary (in press).
Iben,I.jr:1977, in 'Advanced Stages of Stellar Evolution' 7th Course of the Swiss Society of Astrophysics and Astronomy, Obs. Geneve press, p. 3.
Iben,I.jr.:1982, Ap. J. 260, 821.
Iben,I.jr.:1983, Ap. J. 275, L65.
Iben,I.jr. and Renzini,A.:1982 a, Ap.J. 259, L79
Iben,I.jr. and Renzini,A.:1982 b, Ap.J. 263, L23
Iben,I.jr. and Truran,J.W.:1978, Ap. J. 220, 980
Malaney,R.A.:1986 a, Mon. Not. R. Astr. Soc. 223, 683
Malaney,R.A.:1986 b, Mon. Not. R. Astr. Soc. 223, 709
Reimers,D.:1975, in 'Problems in Stellar Atmospheres and Envelopes', p.229, eds. Baschek, B., Kagel, W.H. and Traving, G., Springer
Renzini,A.:1981, in 'Physical Processes in Red Giants', p.165, eds. Iben, I.jr. and Renzini, A., Reidel
Renzini,A. and Voli,M.:1981, Astr. Ap. 94, 175
Truran,J.W. and Iben,I.jr.:1977, Ap. J. 216, 797
Ulrich,K.:1973,in 'Explosive Nucleosynthesis', p.139, eds. Schramm, D.N. and Arnett, W.D., Univ. of Texas Press

SPECTROSCOPIC STUDY OF GAS OUTFLOWS IN THE BIPOLAR NEBULA S106

A. Riera[1], A. Mampaso[1], P.J. Phillips[2], J.M. Vilchez[1],
A. Manchado[1]
1. Instituto de Astrofisica de Canarias, 38200-La Laguna, Tenerife, Spain.
2. Queen Mary College, London University, Mile End Road, London E1 4NS, UK.

ABSTRACT. High resolution bidimensional optical spectroscopy in $H\alpha$, [NII] $\lambda\lambda$ 6548, 6583 Å and [SII] $\lambda\lambda$ 6717, 6713 Å lines of the bipolar nebula S 106 are reported. The resulting $H\alpha$ velocity maps show strong variations in the emission line profiles along the nebula, a situation which is duplicated in other lines, and suggest a sub-arcsec scale structure for some zones. The clumpiness of the nebula is furthermore evidenced by the presence of line splitting with typical relative velocities of about 90 km s^{-1}. We discuss a possible model for this nebula taking into account its morphological, physical and kinematic characteristics.

1. OBSERVATIONS AND RESULTS

A kinematic study of the bipolar nebula S106 has been made from high resolution bi-dimensional spectroscopy, obtained at the 2.5m Isaac Newton Telescope at the Observatorio del Roque de los Muchachos (La Palma, Canary Islands) using the 235mm camera and the Image Photon Counting System (IPCS) bi-dimensional detector, with a grating of 1800 grooves mm^{-1} optimized in the V band. The 1 arcsec wide slit was orientated at a position angle of 28° north-east covering a total extension of approximately 3'. With this configuration the effective spectral resolution achieved was equivalent to 25 km s^{-1} FWHM and the spatial resolution attained was 1.5 arcsec.

The spatial profiles centered at $H\alpha$ [NII] and [SII] show the strong difference in brightness between the two lobes and the previously reported complex structure of the Southern lobe. In this, we distinguished two regions from their intensity emission and kinematic characteristics: the brightest part of the lobe, hereinafter called SA, and the tail-shaped region, referred to here as SB, which extend from 9 to 39 arcsec and from 66 to 78 arcsec with respect to the central obscured star. Region SA presents three local maxima and strong line emission variations with position, pointing to changes over angular distances of less than 1.5 arcsec.

From our data, we derived a positive velocity excess of 33 km s^{-1}

for the Nothern lobe with respect to region SA which, in its turn, has a negative velocity excess of about -12 km s^{-1} with respect to the emission tail-shaped area SB, in good agreement with the results from Solf and Carsenty (1982).

In the Nothern lobe, the Hα lines present widths at half intensity ranging from 45 to 60 km s^{-1}. In the Southern lobe we observed a different behaviour in each selected area: region SA shows corresponding values between 45-90 km s^{-1} and reaches 120 km s^{-1} at 20 arcsec from the central exciting star, while area SB shows a nearly constant width of about 45 km s^{-1}. Furthermore in area SA the Hα and [NII] lines are split into two components with a separation of about 90 km s^{-1}, as shown in Figure 1.

The existent small scale structure, the line splitting detected and the strong density variations derived from the [SII] 6731 Å/6717 Å ratio (from a few hundred cm^{-3} to values higher than 10^4 cm^{-3}) for the Southern lobe close to the central optical gap, point to the presence of ionised clumps with typical sizes of about .01 pc.

These results, together with low resolution spectroscopic data and NIR mapping of the nebula (Riera et al., 1987) point to a model in which the emitting material is flowing from the central source in the direction of the lobes, as suggested before by Solf (1980), Hippelein and Münch (1981) and Solf and Carsenty (1982). The velocity field observed in the Nothern lobe could be explained by emitting gas out-flowing with conical symmetry, while, in the Southern lobe we have found condensations out-flowing with conical symmetry in the region close to the centre of the nebula and with cylindrical symmetry in the southernmost zone.

These clumps could be due to previously existant condensations in the ambient molecular cloud where the bipolar nebula is embedded or, alternatively, may be the result of Rayleigh-Tayler type instabilities in the shell of gas swept-up by the stellar wind.

References

Hippelein, H., Münch, G.: 1981, Astron. Astrophys. **99**, 248.
Riera, A., Mampaso, A., Phillips, P.J., Vilchez, J.M., Manchado, A.:
    1987, in preparation.
Solf, J.: 1980, Astron. Astrophys. **92**, 5.
Solf, J., Carsenty, U.: 1982, Astron. Astrophys. **113**, 142.

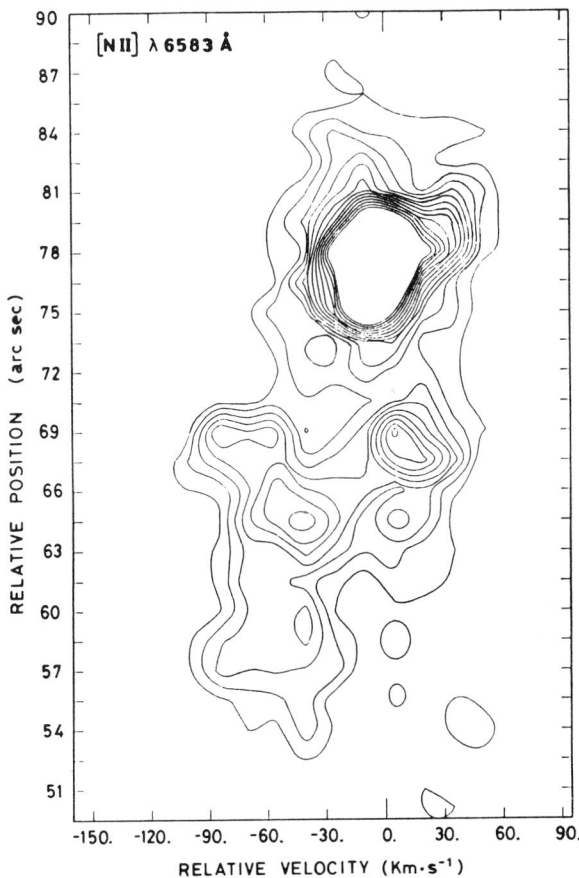

Figure 1: Intensity contour map in a position-velocity representation of the Southern lobe at NII line. The peak flux is 1.92 10^{-14} erg/cm^2/s. The contours range from 3% to 30% of the peak with steps of approximately 7%.

CCD IMAGES OF THE SERPENS BIPOLAR NEBULA[*]

A.I. Gomez de Castro[1] and C. Eiroa
Observatorio Astronomico Nacional
Alfonso XII, 3-5
28014-Madrid
SPAIN

ABSTRACT. We present the preliminary results at $\lambda = .9\mu m$ (I band) of a more extended near infrared and optical study of the Serpens dark cloud. An amount of 5 stars associated with a cometary-like nebulosity and two bipolar nebulae have been found in a projected area of $\sim.0052$ square degrees. A systematic alignment of the major axis of the nebulae in a direction parallel to the local interstellar magnetic field is found. This result suggests that the nebulae are probably related to mass outflow processes in the young Serpens stars, and that the outflows are influenced by the local magnetic field. The new faint Serpens bipolar nebula is associated with the infrared source SVS4. Unlike the other six Serpens nebular objects, this nebula is oriented in a direction almost perpendicular to the magnetic field.

[*] based on observations at the Calar Alto Observatory (Centro Astronomico Hispano-Aleman)
[1] supported by a fellowship of the Comunidad Autonoma de Madrid

FORMATION OF PLANETARY NEBULAE

Arrigo Finzi
Dept of Mathematics, Technion
Haifa, Israel

According to a model presented some years ago (1),(2),(3) the H-rich material that forms a planetary nebula is carried by a supersonic wind blowing from the surface of a ~ 0.4 Mo star of radius ~ 0.02 Ro; the escape energy is provided by partial H burning taking place a few hundred kilometers above the stellar surface, at temperatures ~ 10^8 K; the time scale of the emission stage is ~ 1000 years and the expansion velocity of the nebula ~ 30 Km/s.

The main problem to be solved before the above model of planetaries can be accepted is to explain how stellar evolution leads to a ~ 0.4 Mo star of radius ~ 0.02 Ro with a ~ 0.1 Mo H-rich envelope; one can convince oneself that a central energy sink effective at temperatures ~ 10^6 K is needed. To produce a similar energy sink we assume (4),(5) that the X's - massive elementary components of dark matter - have collision cross-sections with nucleons ~ 10^{-37} cm^2 and that in a collision an appreciable fraction of the center-of-mass energy is lost in the form of unspecified, easily escaping "dark radiation"; X's would steadily accrete onto stars; an energy sink Lx ~ L could result from their presence in the core of a 1 Mo main sequence star.

The apparent ages of globular clusters seem to be older than the age of the Universe deduced from the Hubble shift; the hypothesis of the X's could easily explain this paradox. The ^{71}Ga solar neutrino experiment could provide the best test of the hypothesis since one can easily show (6) that a count significantly larger than ~ 10^6 SNU predicted by the standard solar model, would imply the presence of an energy sink near the solar centre.

References

(1) A.Finzi, R.A.Wolf, 1970, A&A 11, 418
(2) A.Finzi, R.Finzi, G.Shaviv, 1974, A&A 37, 325
(3) A.Finzi, R.Yahel, 1978, A&A 68, 173
(4) A.Finzi, 1986, Nuovo Cimento, 95B, 71
(5) A.Finzi, 1987, Nuovo Cimento B, in press
(6) A.Finzi, 1987, Nuovo Cimento B, in press

BOW SHOCK STRUCTURES NEAR YOUNG STELLAR OBJECTS

T.P. Ray and R. Mundt
Max-Planck-Institut für Astronomie
Königstuhl 17
D-6900 Heidelberg
FRG

ABSTRACT. We have obtained narrow band CCD images of several Herbig-Haro objects resembling bow shocks. As predicted by bow shock theory, we find that the region of highest excitation coincides with the apex of the proposed shock.

1. Introduction

Recent observations suggest that several Herbig-Haro (HH) objects may be extended bow shock structures. Examples include HH1 (Mundt, Brugel and Bührke, 1987), HH34 (Reipurth et al., 1986; Mundt, 1986; Bührke, 1986 and these proceedings), RNO43 North and South (Ray, 1987; Mundt, Brugel and Bührke, 1987) and probably HH39 (Mundt, Brugel and Bührke, 1987). Most of these objects are associated with highly collimated jets - in fact, they may be the working surface of the jet as it ploughs into the surrounding medium.

2. CCD Imaging

Böhm, Raga and Solf (1987) have listed several observational tests of the bow shock hypothesis. One test is to compare the observed and predicted emission line morphologies. With this in mind, we have obtained narrow band (H_α, [OIII] λ 5007 and [SII] $\lambda\lambda$ 6716, 6731) CCD images of several bow shock candidates using the 3.5 m telescope at Calar Alto, Spain. These observations were made in December/January 86/87 under excellent atmospheric conditions (seeing ~ 0.8").

As predicted by the models (see e.g. Raga, 1986), the [OIII] emission is confined to the apex or stagnation point of the shock whereas the [SII] and H_α emission extends to its wings. The wings are in some cases very extensive (e.g. HH1, RNO43 North and South) and this is to be expected if the shock is travelling into a pre-ionised medium. Pre-ionisation may be due to nearby OB associations or UV from the bow shock recombination region.

Simple bow shock models predict a smoothly varying intensity profile whereas our observations show several clumps within the overall bow shock envelope. Such clumpiness may be due to an inhomogeneous pre-shock medium or perhaps thermal instabilities. Recent calculations by Raga and Böhm (1987) have shown how such instabilities could develop behind a bow shock.

ACKNOWLEDGEMENTS. T.P. Ray would like to thank the Alexander von Humboldt Foundation for financial support while in Heidelberg.

REFERENCES

Böhm, K.-H., Raga, A.C. and Solf, J., 1987, Proc. I.A.U. Symp. 122, 'Circumstellar Matter', (Dordrecht, Reidel), in press.
Bührke, T., 1986, Ph.D. Thesis (University of Heidelberg).
Mundt, R., 1986, in 'Jets from Stars and Galaxies', Can.J. Phys, 64, 407
Mundt, R., Brugel, E.W. and Bührke, T., 1987, Ap.J. in press.
Raga, A.C., 1986, A.J., 92, 637.
Raga, A.C. and Böhm, K.-H., 1987, Ap.J., in press.
Ray, T.P., 1987, Astron. Astrophys, 171, 145
Reipurth, B., Bally, J., Graham, J.A., Lane, A. and Zealy, W.J. 1986, Astron. Astrophys., 164, 51

WINDS FROM COLD PRE-MAIN SEQUENCE STARS: IONIZATION STRUCTURE AND LINE INTENSITY

A. Natta[1], C. Giovanardi[2], and F. Palla[2]
[1] Centro Astronomia Infrarossa, C.N.R.
Largo E. Fermi 5, Firenze, Italy.
[2] Osservatorio Astrofisico di Arcetri
Largo E. Fermi 5, Firenze, Italy.

ABSTRACT. Using non-LTE model calculations we compute the ionization structure and hydrogen line intensity in partially ionized winds from cold (T_e =4500 K) PMS stars.

1. INTRODUCTION

Mass loss is a well estabilished property of the early stages of stellar evolution. Observationally, a wealth of data derived from CO, H recombination lines and molecular H emission lines in a large sample of PMS stars is now available (Lada 1985, Levreault 1985, Evans et al. 1987). These observations indicate a large discrepancy between the mass loss rate inferred from molecular lines and H recombination lines, the former being generally higher (sometime by orders of magnitudes) than the latter. Theoretically, various mechanisms to produce the observed mass loss have been presented, none of them fully satisfactory, due to the complexity of the phenomena involved (e.g. Hartmann 1986).

Among various interpretations, it has been suggested (Evans et al. 1987) that the observed discrepancy may be reconciled if the wind is mostly neutral, contrary to the assumption of fully ionized, isothermal wind models on which the interpretation of the data usually rests (Simon et al. 1983). However, there are no quantitative estimates to support the claim.

We report in this paper the first results of an extended set of model calculations of the hydrogen excitation structure and the predicted line intensities in winds from cold PMS stars, that lend support to the interpretation in terms of partially ionized winds.

2. THE MODEL

2.1 Wind structure.

We show here the results for a cold star (T_e=4500 K) loosing mass in the range 10^{-6} - 10^{-8} M_0/yr. The wind is spherically symmetric and the gas density at any given distance from the star is set by the continuity equation.

Both the velocity and the temperature profile of the wind depend on the mechanism which drives the mass loss. We have used a velocity profile given by the analitical expression:

$$V = V_0 + V_1 (1 - 1/R^2) \qquad (1)$$

where R is a radial coordinate in units of the stellar radius, V_0=20 km s^{-1}, V_1=300 km s^{-1}. Eq.(1) is a good approximation of the results of Hartmann and MacGregor (1980) for Alfvén wave driven winds.

For euristic purposes, the models discussed here have constant temperature. However, two cases with more realistic temperature profiles (see Fig.2) have also been included.

2.2 Model calculation.

At each radius, we solve the detailed equilibrium equations for an hydrogen atom consisting of 15 levels (of which the first 4 are split into l-sublevels) and continuum. For each level, we consider collisional and radiative ionizations, collisional and radiative excitations and de-excitations. The radiative transfer is solved by using the escape probability formalism in the large velocity gradient approximation (Giovanardi et al., in preparation).

3. RESULTS

3.1 Ionization structure.

The run of the fractional ionization in the wind is shown in Fig.1 and Fig.2 for several representative cases. From an inspect to the curves in the bottom panel, we notice that ionization from the highest n levels dominates in the innermost regions of the wind, while radiative transitions from n=2 account for the ionization of the outer parts. This occurs due to the large optical depth to the Ly-α photons. The transition radius depends on T and $\dot{m}$.

The exact behaviour of the ionization structure can only be computed via a numerical treatment of the hydrogen atom which includes a large number of levels and a complete set of processes. A more exhausting discussion on this subject can be found in a forthcoming paper (Natta et al., in preparation).

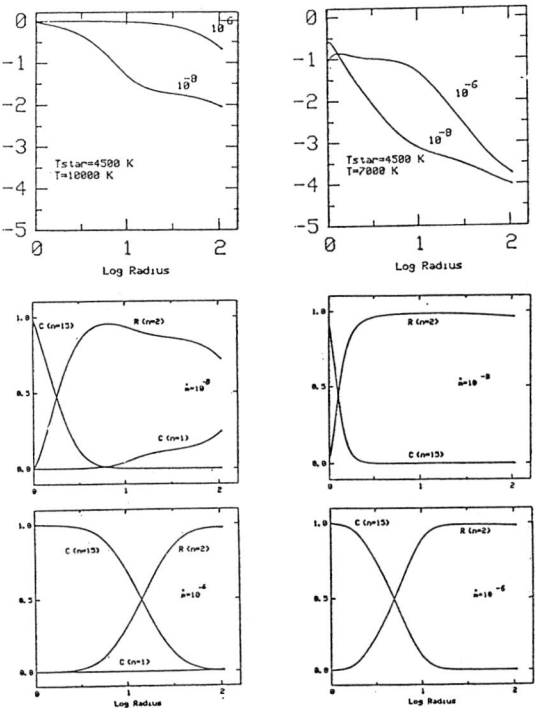

Figure 1. Top: ionization fraction for an isothermal wind as a function of R for two gas temperatures, $T=10^4$ K and T=7000K respectively. The curves are labelled by the rate of mass loss in solar masses per year. Bottom: individual processes contributing to the ionization, normalized to unity. C(n=1) and C(n=15) refer to the ionization rate due to collisions from level 1 and 15 respectively. R(n=2) stands for radiative ionization from n=2.

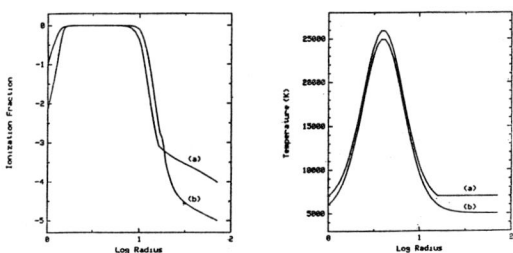

Figure 2. Ionization fraction and temperature profile for two non isothermal winds. Curve (a) refers to a model with T(R=1)=7000K and $\dot{m}=10^{-7}$, curve (b) to a model with $T(R=1)=10^4$ K and $\dot{m}=10^{-6}$.

3.2 Line intensity.

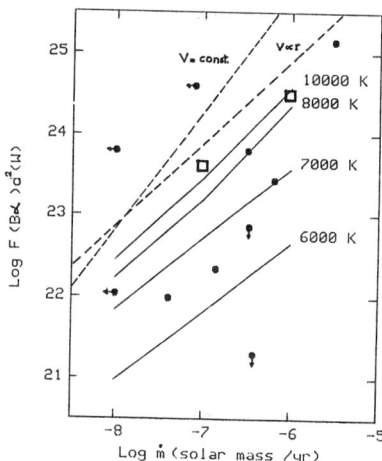

Figure 3. The predicted Bα line flux is plotted versus the mass loss rate for four values of the gas temperature in the wind (solid lines). Filled circles are the observed fluxes in 10 PMS stars by Evans et al. Dashed lines show the predictions of Simon et al. in the case of constant velocity and $v \propto r$ outflows. Squares refer to the non-isothermal model.

The variation of the predicted Bα flux with $\dot{m}$ for various values of the gas temperature is shown in Figure 3. By varying T from 10000 K to 6000 K the line flux decreases by a factor of 30 for $m=10^{-8}$ and 85 for $m=10^{-6}$.

Our computed values cover the entire range of the observed points and are much lower than those expected in conventional wind models for the same value of m, as long as the temperature in the wind remains well below 10^4 K. Therefore, our results provide a quantitative support to the interpretation of line emission from partially ionized winds.

REFERENCES

Evans, N.J. II, Levrealt, R.M., Beckwith, S., and Skrutskie, M. 1987, Astrophys. J., in press.

Hartmann, L. 1986 in 'Fundamentals of Cosmic Physics', Vol. 11, 279.

Hartmann, L., and Mac Gregor, K.B. 1980, Astrophys. J., 242, 260.

Lada, C.J. 1985, Ann. Rev. Astron. Astrophys., 23, 267.

Levreault, R.M. 1985, Ph.D. Thesis, The University of Texas at Austin.

Simon, M., Felli, M., Cassar, L., Fischer, J., and Massi, M. 1983, Astrophys. J., 266, 623.

HAVE THE ENERGETICS OF MOLECULAR LINE FLOWS BEEN EXAGGERATED?

J. E. Dyson[1], J. Cantó[1,2], L. F. Rodriguez[2]
1. Department of Astronomy,
   University of Manchester,
   Manchester M13 9PL,
   U.K.
2. Instituto de Astronomía,
   Universidad Nacional
   Autónomia de México,
   Ap. Postal 70-264,
   04510 México, D.F.,
   México.

ABSTRACT. If flows in molecular clouds are driven by stellar winds, the flow energetics, in principle, reflect the wind energetics. It has been generally accepted that the momentum requirements for the winds driving the flows are so severe that winds driven by conventional mechanisms cannot be responsible for the flows. Various exotic possibilities have therefore been proposed. We suggest that these momentum requirements can be considerably alleviated by the simple expedient of assuming that the flows are stationary. Evaluation of the momentum requirements for various specific molecular flows implies that conventional wind driving mechanisms may suffice.

1. INTRODUCTION

Hypersonic flows of molecular gas are a common phenomenon in regions of active star formation (e.g., Lada 1985) and it is generally accepted that these flows are driven by the interaction of mass loss from recently formed stars with their molecule bearing surroundings. To date, estimates of the required wind momentum fluxes (Lada 1985; Rodriguez et al 1982; Bally and Lada 1983) have led to extremely high values, so high in fact, that conventionally (e.g., radiation or Alfvén wave) driven winds cannot, with present theoretical models, account for these requirements. Various models essentially to tap stellar gravitational binding energy (e.g., Pudritz and Norman 1986) have been put forward. We here briefly report on a re-evaluation of molecular flow energetics which may obviate the need for these more exotic possibilities.

2. FLOW ENERGETICS

The simplest flow resulting from the interaction of a stellar wind with cloud material is the well known two-shock pattern (e.g., Dyson 1984) in which molecular material is accelerated and compressed into a fast moving thin shell which expands into its surroundings. Ultimately, the flow achieves pressure equilibrium with these surroundings and only the inner shock (in the wind) then exists. The canonical interpretation of molecular flows is that the flows are much younger than the timescale to reach this pressure equilibrium. In this interpretation, the shell mass, velocity and radius are identified respectively with the mass of moving molecular gas, the peak gas velocity and the flow scale length. To derive wind energetics, some assumption has to be made regarding the basic interaction process. Flows in which the shocked stellar wind cools adiabatically and not radiatively demand intrinsically the lowest wind momentum output rates (Dyson 1984). However, there are serious momentum problems even in this best possible case. Additionally, it is very hard to see how the mixing of cool molecular gas and hot shocked wind gas, with subsequent strong cooling, can be avoided.

3. STATIONARY FLOWS

Cantó (1985) has argued that the molecular line profiles do not support the premise that the line emitting material is confined to a thin expanding shell. An attractive possibility (Cantó 1985) is that the flows arise as a result of viscous coupling between a cooled shocked stellar wind flowing around the outside of an ovoid cavity excavated by the stellar wind. Since the wind cools radiatively, the molecular flows are driven by wind momentum directly. Such a flow is stationary and the wind momentum is not added at a point, but is added over an extended spatial region (the cavity).

Cantó, Dyson and Rodriguez (1987; in preparation) have examined the dynamics of momentum driven stationary flows and re-evaluated the observational data in terms of simple models. They conclude that wind momentum output rates have been systematically overestimated by factors which can be as great as 200. There are two physical reasons for this. Firstly, the flow lifetimes have been underestimated because in a stationary flow of dimensions R, the flow time is $\sim R/C$ (C is the effective sound speed in the ambient gas), whereas the flow lifetime is generally taken as $\sim R/V$, where V is some characteristic velocity (say the maximum velocity), observed in the molecular flow. Secondly, most of the gas in a stationary flow moves at a velocity about equal to C and not, as generally assumed, about equal to V.

An interesting corollary of a stationary flow pattern is that the flow time ($\sim R/C$) is the absolute minimum age of the flow. Molecular flows could be appreciably (say at least a factor 10) older than previously suspected. If this is so, it is necessary that only 2% of stars more massive than 1 $M_\odot$ (or, alternatively, that all stars more massive than about 4 $M_\odot$) go through an energetic mass loss phase. This contrasts with the need for all stars more massive than 1 $M_\odot$ to go through such a phase if the previous lower age estimates are used (Snell 1987).

Cantó et al (1987) have re-evaluated required wind momentum output rates for a variety of sources and find them in substantial agreement with those derived from Brα and radio continuum data (Persson et al 1984). Unlike those authors, however, Cantó et al deduce that the winds are completely ionized and are not dominated by a neutral component.

4, CONCLUSIONS

The assumption that flows in molecular clouds are directly driven by wind momentum but are in a steady state considerably reduces the required wind momentum rates. In fact, within the frame work of present wind theory, winds from high luminosity stellar sources could be radiatively driven and winds from low luminosity sources could be Alfvén wave driven. Of course more exotic, wind driving mechanisms cannot be ruled out. However the momentum requirements for the winds may be far less than previously supposed.

REFERENCES

Bally, C. J. and Lada, C. J., 1983, Astrophys. J., 265, 824.
Cantó, J., 1985, in Cosmical Gas Dynamics (ed. Kahn, F. D.), VNU Science Press, Utrecht, p.267.
Dyson, J. E., 1984, Astrophys. Space Sci., 106, 181.
Lada, C. J., 1985, Ann. Rev. Astr. Astrophys., 23, 267.
Persson, S. E., Geballe, T. R., McGregor, P. J., Edwards, S. and Lonsdale, C. J., 1984, Astrophys. J., 286, 289.
Pudritz, R. E. and Norman, C. A., 1986, Astrophys. J., 301, 571.
Rodriguez, L. F., Carral, P., Ho, P. T. P. and Moran, J. M., 1982, Astrophys. J., 260, 635.
Snell, R. L., 1987, in Star Forming Regions - IAU Symposium 115 (eds. Peimbert, M. and Jugaku, J.), D. Reidel Publ. Co., Dordrecht, p.213.

Infrared Emission from Outflows Associated with Young Stars

F.O. Clark[1,2], R.J. Laureijs[1], C.Y. Zhang[1,3]
G. Chlewicki[1], P.R. Wesselius[1]

[1]Laboratory for Space Research Groningen, The Netherlands
[2]University of Kentucky, Lexington, KY, USA
[3]Purple Mountain Obs., Academica Sinica, Nanjing, China

Abstract. The extended infrared emission detected near many young stars can be used to divine the luminosity, apparent color temperature, and thermal emitting mass of the surrounding regions. The dominant dust heating mechanism appears to be ultraviolet radiation, either from stellar wind shocks or directly from hotter stars. The infrared luminosities from stellar winds of low mass stars are an appreciable fraction of the luminosities of the driving star, and the mechanical luminosity in the winds must be even higher.

I. Introduction

Extended infrared emission is often detectable around young stars (Clark et al. 1986, Zhang et al. 1987), and can reveal the wind luminosity (Clark et al. 1986), morphology of dust color temperature, and mass of warm dust. Molecular spectral lines confirm the presence of winds and shocks (Snell et al. 1980, Lada 1985, Clark and Turner 1987).

II. Extended IR From Winds

L1551 veils a young pre-main-sequence star (IRS-5) with accompanying extended infrared emission (Figure 1) from

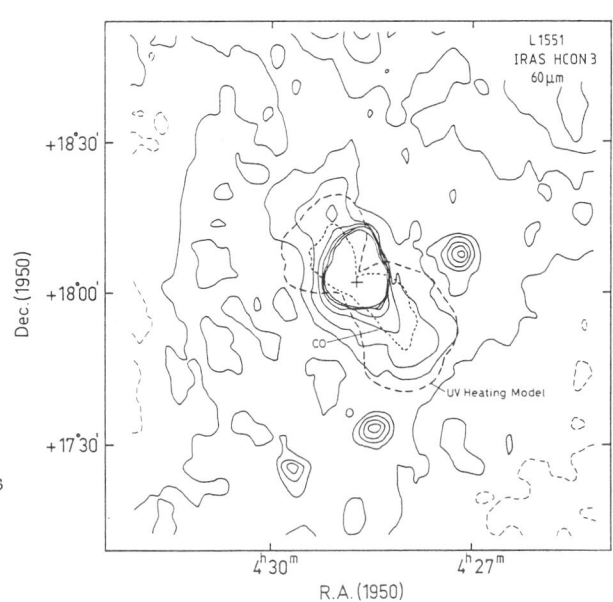

Figure 1. L1551 Extended 60 μm IR

L. Bianchi and R. Gilmozzi (eds.), Mass Outflows from Stars and Galactic Nuclei, 303–307.
© 1988 by Kluwer Academic Publishers.

dust surrounding a bi-polar flow. The dust is heated by shock produced ultraviolet radiation (Edwards et al. 1986, Clark et al. 1986). The luminosity of the central star is ~38 $L_\odot$, while the extended infrared emission is 10 -2 +4. The dust apparent color temperature is 21-24 K (emissivity ~λ^{-1} to 2), so most of the flux associated with the L1551 flow is well beyond the 100 μm IRAS detectors, and a bolometric correction must be made. The bolometric correction yields L> 19-28 $L_\odot$. Even should spectral line emission be important in this extended infrared emission from the flow, such a correction will be necessary, as it is unlikely that all of such conjectured line emission would be covered by the two IRAS filters. The mechanical luminosity in the flow must be even higher, and these enormous wind luminosities allude to the conclusion that the wind derives it's energy from the process of star formation itself, i.e. from the gravitational energy of the cloud, and not by stellar radiation pressure. The uncertainty in the bolometric correction can be relieved by measurements at wavelengths of 200-400 μm.

CED 110 exhibits extended infrared emission detected by the IRAS Chopped Photometric Channel (CPC a high spatial resolution instrument on board IRAS) shown in Figure 2. The visible central star (indicated by +) lies well above the main sequence (Rydgren 1980), and is one of the younger visible objects in the Chameleon Cloud. We estimate an L(bol) = 21 $L_\odot$ for the star (.337 to 100 μm). The extremities of the extended 50 μm emission exhibit apparent color temperature peaks of 34 K. The ambient density of the molecular cloud surrounding the star is 10 times that of L1551, ~3 10^3 cm^{-3} (Toriseva and Mattila 1985). The IRAS pointed observations suffer from a 2 σ pointing uncertainty of about 30", and it is plausible that the star is coincident with the central of the three symmetric peaks.

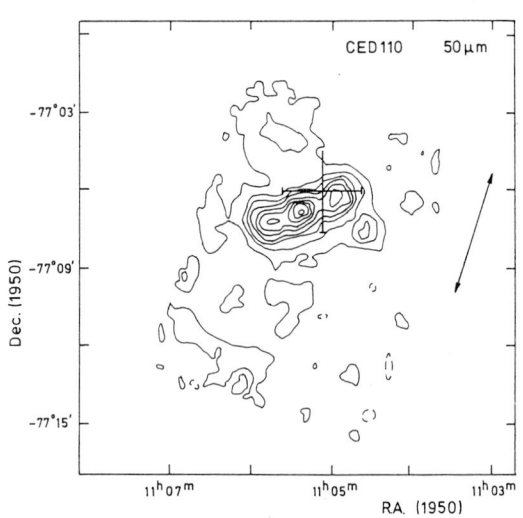

Figure 2. CED 110 50 μm IR

The data of Toriseva and Mattila reveal anomalously strong OH lines and high velocity wings in CED 110, which speak for a stellar wind (Clark & Turner 1987), presumably also responsible for the bipolar infrared emission. The extended infrared luminosity is 2.2 $L_\odot$, ~10 % of that of the star. The dense surrounding cloud results in a short ultraviolet path length from the working surface of the conjectured shock.

A region of warm dust appears near but well away from the bipolar flow source IRAS 05553+1631 (Figure 3). The peak infrared color temperature is high, ~40 K, suggesting local heating, and broad OH emission lines (~50 km/s) are detected on the warm infrared peak, as well as broad CO which peaks towards the star. The CO data indicate that the gas and dust heating are decoupled, as would be expected in this very low density object (with n ~few 100 cm^{-3}). These data are suggestive of the working surface of a shock from a stellar wind (Clark and Turner 1987). This putative working surface is at a considerable distance from the star (2 pc). The dust heating mechanism for the extended infrared emission near IRAS 0553 +1631 appears to be shock produced ultraviolet.

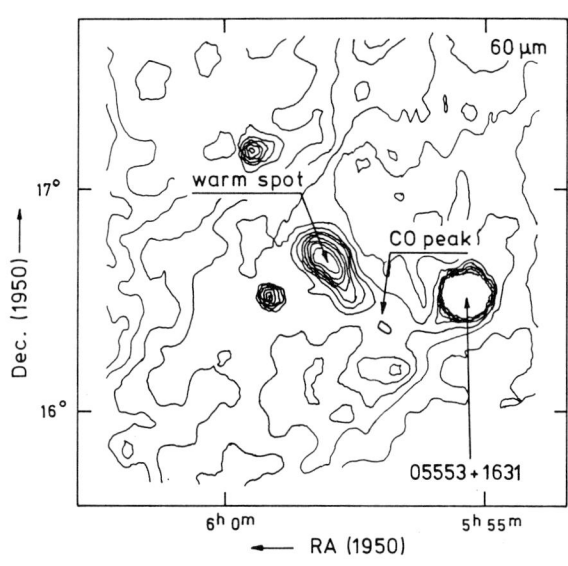

Figure 3. 60 μm emission

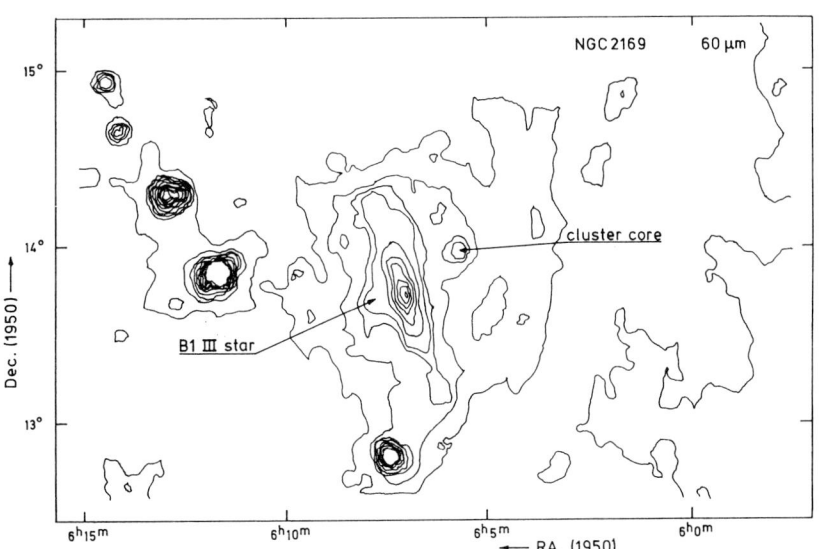

Figure 4. 60 μm IR

Figure 4 shows what is apparently the final step in the destruction of the parent molecular cloud by the young open cluster, NGC 2169. NGC 2169 is considerably younger than the Pleiades in consequence of it's B1 stars still on the main sequence. The infrared luminosity of this object is large, 20,000 L⊙, ~50% of that of the cluster. Such an enormous fractional luminosity requires that a significant fraction of the dust heating derives from the diffuse galactic radiation field, as was also found for extended emission in Serpens (Zhang et al. 1987).

The distance of the cloud is established by extinction in foreground stars, and stars within the cloud which produce extended dust heating. CO has been detected only in the bright IRAS core using the University of Cologne 3 m telescope. Ammonia emission was detected in this core with the MPIfR Effelsberg 100 m telescope.

Very broad OH emission lines with widths of ~40 km/s localized to the IRAS object were detected, indicative of shocks. The average of the stellar velocities is comparable to that of the gas, indicating intimate association. Much work remains to be done on this object. The observed broad OH line widths clearly implicate wind produced shocks from the embedded early B stars, but direct radiative heating is probably the dominant dust heating mechanism in NGC 2169 for two reasons: (1) the infrared luminosity is many orders of magnitude greater than that of measured winds from comparable stars, and (2) the dust temperature is observed to peak on the stars of early B type.

III. Results

Table 1 enumerates the results for extended infrared emission from the three recognized bipolar flows to date: CED 110, L1551, and IRAS 05553+1631, and for the extended emission from the open cluster NGC 2169, which represents an example of radiative heating.

Table 1

Source	spectral class	$L_{*,bol}$ $L_\odot$	L_{IR} $L_\odot$	$V_{heating}$ km/s	V_{OH} km/s	T_{dust} (peak) K	E_{tot} 10^4 yrs ergs
CED 110	G2	21	2.2	56	>20*	34	2 10^{45}
L1551	F-G	38	19	32	>30#	23	1 10^{46}
05553+1631	B3?	>900	>140	46	~50**	33	9 10^{46}
NGC 2169	B1	~40,000	~20,000	--	41	27	(rad. heating)

* maximum extent not determined by observations.
# broad component may be greater.
** preliminary values

A major signature of radiative heating of dust, either by local stars or the galactic diffuse radiation field, is strong emission in all four IRAS bands, 12, 25, 60, and 100 μm, with similar morphology. This apparition is displayed by "normal" diffuse clouds with no internal heating sources. In contrast, the three bipolar flows recognized to date are only clearly detected in 60 and 100 μm emission, with possible accompanying very weak 12 μm emission.

The <u>mechanical</u> luminosities in these flows must be even higher than the infrared luminosities. Clark et al. (1986) made a simple model incorporating shock produced ultraviolet to heat the dust, which indicated that the L1551 mechanical luminosities were 2-7 times the infrared luminosities. Such enormous mechanical luminosities presumably could not be produced by the central star, IRS-5 (Table 1), nor even by the sum of all stars in the immediate vicinity. If the bipolar flow phase lasts for 10^4 years (Lada 1985), then the associated total energy is also uncomfortably large. A plausible driving source for such luminosities and energies is the gravitational energy of the cloud extracted from a small scale size, perhaps 1 au. A mechanism for such extraction, using angular momentum for non-spherical geometry, and magnetic fields as a medium, has been proposed by Draine (1983).

IV. Summary

In summary, extended warm infrared emission is detectable near young embedded stars. The extended infrared emission associated with bipolar flows is surprisingly large, and probably reflects an energy input in excess of the gravitational binding energy of the cloud on the scale of the bipolar flows (Clark et al. 1986). Those objects detected to date all exhibit a dust color temperature which is elevated over their surroundings. Understanding the physical nature of these infrared sources requires supporting spectroscopic observations.

The heating mechanism of extended infrared emission near stars of lower mass seems invariably to be shock produced ultraviolet (CED 110, L1551, IRAS 05553+1631). However some stars of spectral class B (and presumably earlier) seem to heat their surrounding dust primarily by direct radiative heating (like the extended infrared emission in Serpens - Zhang et al. this conference - and NGC 2169). Perhaps this change occurs because the stellar radiative luminosity begins to exceed the wind luminosity by a large fraction in these more massive stars, which may occur as the wind luminosity begins to fade. Extended infrared emission provides a new tool to probe the physical processes of winds from stars interacting with their surroundings.

REFERENCES

Clark, F.O. Laureijs, R.J. 1986 A.&A. Letters 154, L26.
Clark, F.O., et al. 1986a "Space-Bourne Sub-Millimetre Astronomy Mission" ESA SP-260, 173
Clark, F.O., et al. 1986b A.&A. Letters 168, L1.
Clark and Turner 1987 A.&A. 176, 114.
Draine, B.T. 1983 Ap.J. 270, 519.
Edwards, et al. 1986 Ap.J. Letters 307, L65.
Jones, B.F. Herbig, G.H. 1979 A.J. 84, 1872.
Lada, C.J., 1985 Ann. Rev. Ast. & Ap. 23, 267.
Rydgren, A.E. 1980 AJ 85, 444
Snell, R.L., Loren, R. Plambeck, R. 1980 Ap.J. Letters 239, L17.
Toriseva and Mattila 1985 A&A 153, 207.
Zhang, C.Y., et al. 1987 A.& A. "IRAS Study of the Serpens Molecular Cloud: I. Large Scale (submitted)

INDICATION of OUTFLOWS FROM YOUNG STARS IN THE SERPENS MOLECULAR CLOUD

C.Y. Zhang[1,2], F.O. Clark[1,3] and R.J. Laureijs[1]

1. Laboratory for Space Research Groningen, The Netherlands
2. Purple Mountain Obs., Academica Sinica, Nanjing, China
3. University of Kentucky, Lexington, KY, USA

ABSTRACT. The extended infrared emission in the Serpens molecular cloud is associated with three B stars, HD 170634, 170739 and 170784. The dust heating is due to a central stellar and an external interstellar radiation field. The Serpens extended emission exhibits apparent small-scale cavities immediately around the three stars. With the assumption that the cavities have been excavated by stellar winds, mass loss rates for these stars may be estimated within the range of $10^{-13} - 10^{-11}$ $M_\odot$ yr^{-1}. The IR emission from the core of the cloud has a double-peaked morphology, the grains is mainly heated by ultraviolet photons produced by shock waves.

1. INTRODUCTION

The Serpens molecular cloud is visible as a large obscured region on POSS prints (Fig. 1), just north of the T Tauri star VV Ser. The Serpens molecular core (RA~=~$18^h27.4^m$, DEC = $1°12.5'$ [1950]), which contains the well-known Serpens Nebula, has been recognized as a star-forming region and observed extensively at both radio and infrared wavelengths. The possible interaction between stars and their surrounding medium can be explored by analyzing the data of IRAS survey and CPC (Chopped Photometric Channels) pointed observations.

2. EXTENDED IR IN SERPENS

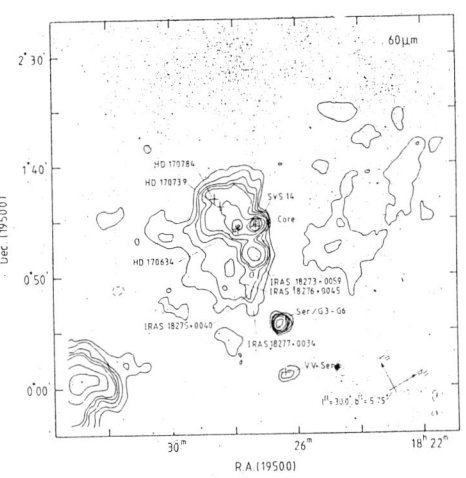

Fig. 1. 60 μm map superimpoed on POSS print

IRAS sky maps revealed a large-scale, $1° \times 1°$, distribution of extended infrared emission at all four IRAS bands (Zhang et al., 1987). The 60 μm map superimposed on the POSS print is shown in Fig. 1. The extended emission appears as a "fan" (Serpens Fan) spreading out from the Core in a symmetric manner with respect to the axis defined by the three stars: HD 170634, 170739 and 170784. For the purpose of modelling the surface brightness distribution in the Fan, we have taken cuts in SE-NW directions perpendicular to the axis of symmetry, with one through HD 170634, and another through HD 170739/784. The surface brightness profiles in the SE-NW directions are shown in Fig. 2 for all four IRAS wavelengths. The Fan has the flux densities of 210, 200, 1000 and 4100 Jy respectively for 12, 25, 60 and 100 μm. The 12/25 μm and 60/100 μm color temperature of the Fan are about 310 K and 25 K respectively. The 60/100 μm color temperature distribution in the Fan peaks (27-29 K) near the stars and remains high over the outer regions, as high as 22-23 K even at edges about 17' to 28' away from the stars. The equilibrium heating of grains by both the central stellar radiation field and the interstellar radiation field (ISRF) is required to explain the 60 and 100 μm extended emission. The total bolometric infrared luminosity of the Fan is estimated as 1300 $L_\odot$.

The apparent central minimum in the surface brightness profiles for HD 170739/784 is suggestive of a cavity around the stars (Fig. 2). In order to understand the infrared extended emission around these B type stars, an extended dust shell model is constructed to match the 60 and 100 μm surface brightness profiles and the 60/100 μm intensity ratio profiles of the extended emission associated with HD 170634 and HD 170739/784. The best fits show that the dusty clouds have inhomogeneous density distribution and cavities around the stars. The radii of the cavities are $6.8 \; 10^{17}$ cm and $2 \; 10^{18}$ cm, the larger of which is around the two stars HD 170739/784 and visible in the IRAS sky maps. A B7 star (HD 170634) can produce only $1.05 \; 10^{41}$ phtons sec^{-1}

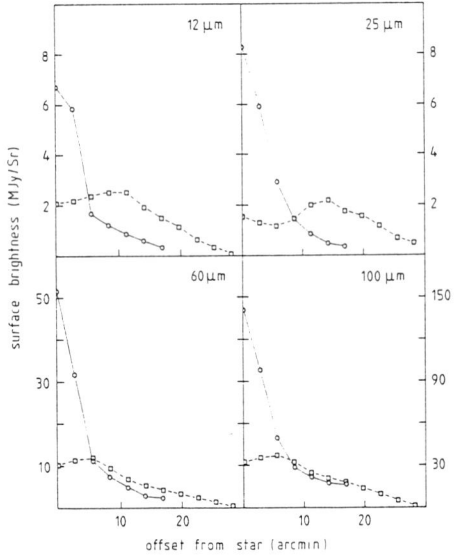

Fig. 2. Surface brightness profiles in the Serpens Fan

(Thompson, 1984) and a Strömgren radius of about $1.9\ 10^{16}$ cm, much smaller than the radius of the cavity derived from the IRAS data. With an assumption that the cavities are excavated by stellar winds, using the formalism of Königl (1982), mass loss rates for these stars may be estimated within the range of $10^{-13} - 10^{-11}\ M_\odot yr^{-1}$.

3. THE SERPENS CORE

The IRAS Chopped Photometric Channel (CPC) pointed observations with higher spatial resolution reveal that the 50 and 100 μm infrared emission of the core of the Serpens molecular cloud, containing the well-known Serpens Nebula, has a double-peaked morphology similar to that of NH_3 spectral line emission (Fig. 3; Ungerechts and Güsten, 1984).

Comparison of the 50/100 μm color temperature map with the map of CO high velocity outflow arising from the embedded low-mass young stars (Bally and Lada, 1983) in this core region (Fig. 4) suggests that the majority of the dust appears to be heated by shock produced ultraviolet photons (Clark and Laureijs, 1986; Clark et al., 1986a; Clark et al., 1986b; Zhang et al., 1987)

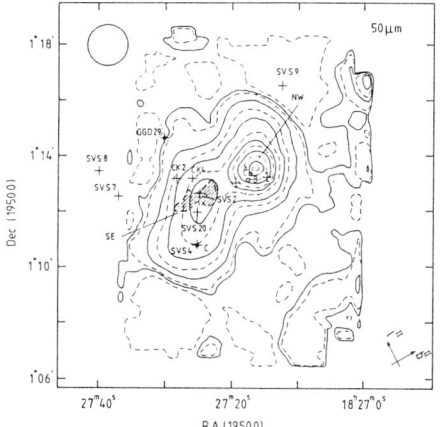

Fig. 3. IRAS CPC 50 μm map of the Serpens Core

4. CONCLUSIONS

The three B stars, HD 170634, 170739 and 170784 are surrounded by large dust and gas complexes indicated by the IRAS extended emission, which may be remnants of the parental clouds of these stars. A model indicates that the emitting material forms two extended shells with cavities required around the star HD 170634 and the stars HD 170739 and 170784 respectively. These two cavities could have been excavated by stellar winds, with $\dot{M}_w = 10^{-13}\ 10^{-11}\ M_\odot\ yr^{-1}$. The heating due to both the stellar radiation field and ISRF is needed to reproduce the 60 and 100 μm extended emission. Shock produced ultra-violet photons is the main heating source for the grains in the Core.

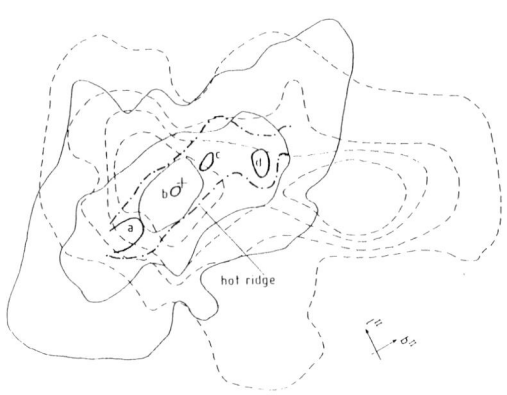

Fig. 4. A few 50/100 μm color temperature contours (thick lines) superimposed on the CO high velocity flow map (Bally and Lada, 1983)

REFRENCES

Bally, J., Lada, C.J.: 1983, Ap.J. 265, 824
Clark, F.O., Laureijs, R.J.: 1986 A.&A. Letters 154, L26
Clark, F.O., et al.: 1986a "Space-bourne Sub-Millimetre Astronomy Mission" ESA SP-260, 173
Clark, F.O., et al.: 1986b A.&A. Letters 168, L1
Königl, A.: 1982, Ap. J. 261, 115
Thompson, R.I.: 1984, Ap.J. 236, 598
Ungerechts, H. and Gusten, R.: 1984, A.&A. 131, 177
Zhang, C.Y., et al.: 1987, A.&A. "IRAS Study of the Serpens Molecular Cloud: I. Large Scale" (submitted)
Zhang, C.Y., et al.: 1987, in preperation

THE JET OF HH 34: NEW RESULTS

Th. Bührke
Max-Planck-Institut für Astronomie
Königstuhl
6900 Heidelberg
FRG

ABSTRACT. Narrow band filter imaging of the HH 34 jet system reveals an inhomogeneous excitation structure in HH 34, in particular a compact [OIII]-emission region near the apex of the bow shock. About 2' N of the jet source a counter bow shock was discovered of very similar appearance to HH 34. Long-slit spectroscopy with high resolution shows a two velocity component flow in the jet and an inhomogeneous velocity structure in HH 34. These phenomena are briefly discussed in terms of interactions of the flow with the surrounding medium and/or thermal instabilities in the cooling regions of a bow shock.

1. OBSERVATIONS
1.1 Narrow Band Filter Imaging

CCD images through narrow band filters had been obtained by R. Mundt using an RCA-CCD camera mounted at the prime focus of the 3.5 m telescope on Calar Alto, Spain. The following filters were used: [SII] (λ_c = 6740 Å, $\Delta\lambda$ = 70 Å), [OIII] (λ_c = 4990 Å, $\Delta\lambda$ = 100 Å) and H_α (λ_c = 6580 Å, $\Delta\lambda$ = 100 Å). All observations were made in January 1987 during excellent seeing of about 0.8".

Fig. 1 shows a montage of two CCD-images obtained through the SII -filter. The jet is pointing towards HH 34, having the shape of a bow shock. Due to the good atmospheric conditions about 6 extra knots in the jet became visible in addition to the five bright knots already known (see also Mundt 1986, Reipurth et al. 1986, Bührke and Mundt 1987). At the opposite side about 2' away from the source the northern bow shock is seen very similar to HH 34. Buth there is no sign of a counter jet. While HH 34 appears to be an extended bow shock in [SII] and in particular in H_α, in [OIII] only a compact region is seen near the apex. Fig. 2 shows an isointensity contour plot of the [SII]- (light line) and OIII -emisson (heavy line). The maximum of the [OIII]-emission region does not coincide exactly with the SII maximum but is shifted some 0.5" ($\triangleq 10^{15}$ cm) towards the outer edge of the bow shock. The jet is not visible in the [OIII]-line, a fact already noted by Bührke and Mundt (1987).

Fig. 1: A montage of two CCD images taken through an [SII] filter. It shows the knotty jet (1) pointing towards HH34 (3). About 100" N of the jet source (2) the counter bow shock (4) appears very similar to HH34.

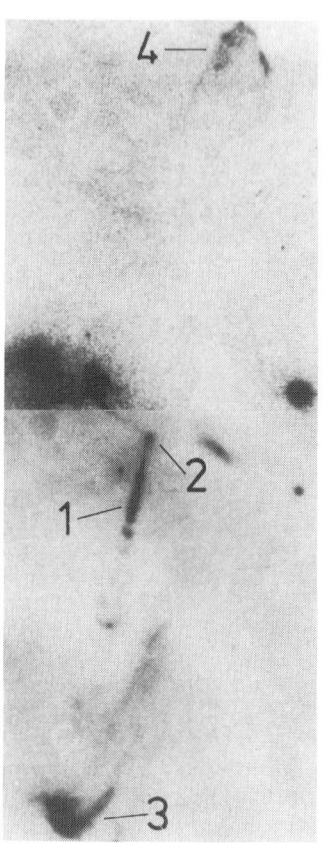

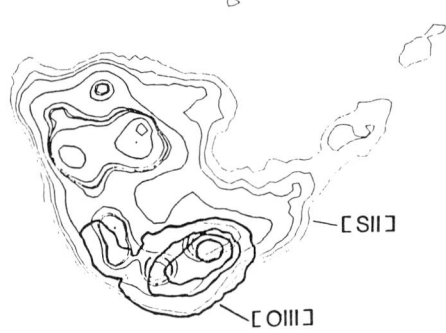

Fig. 2: Isointensity contour plot of HH34 showing the knotty structure in [SII] (light lines) and the compact [OIII] region (heavy lines) near the apex.

1.2 Long-slit Spectroscopy

High resolution spectroscopy was carried out at the Coudé-spectrograph of the 2.2 m telescope on Calar Alto with a spectral resolution of about 20 km/s (FWHM). The spectral range covered the region from 6500 Å to 6780 Å, including H_α, the [NII] $\lambda\lambda$6548/6583 Å- and [SII] $\lambda\lambda$6716/6731 Å lines. The slit width was about 3" the accuracy in velocity is in general ±2 km/s.

a) Jet

The jet emission comes from the brightest knots denoted as A - E by Reipurth et al. (1986). As fig. 3 (a space-velocity-diagram of the [SII] λ 6731 Å-line) shows, there is a main component with a nearly constant velocity of about -80 km/s. Another component with a lower velocity (v $\approx$ -50 km/s) arises in the region of knots B and C and becomes accelerated with increasing distance from the jet source. A very similar behaviour was also seen in the jet of HH24 (Solf 1987).

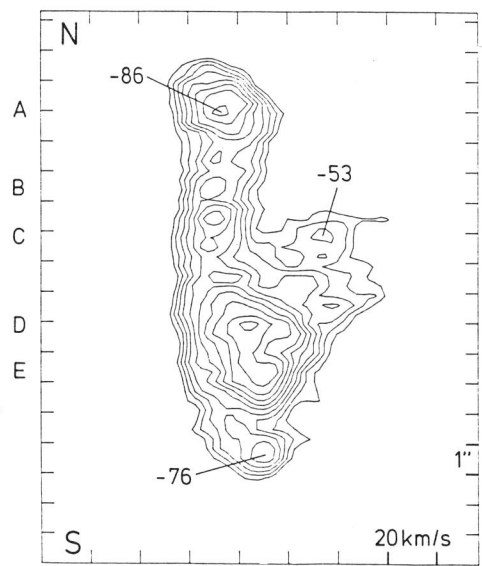

Fig. 3: Isointensity contour plot of the [SII] λ6731 Å-line in the jet. The designation of the knots is marked on the left side. A nearly constant velocity component is visible at v ~ - 85 km/s. In the region of knots B/C a second component with lower velocity arises and becomes accelerated with increasing distance from the star.

The main component remains spectroscopically unresolved, i.e. it has a FWHM ≤ 20 km/s at least in the upper part of the flow, where the line width is measurable.

b) HH34

In order to obtain a velocity map of HH34 four slit positions at different position angles were chosen. The final result is given in fig. 4. We see an inhomogeneous and asymmetric flow structure in HH34: in the western wing of the suspected bow shock a continuous decrease in velocity from -130 km/s to -35 km/s takes place away from the apex. Such a trend in radial velocity is expected from theoretical models of bow shocks (e.g. Raga and Böhm 1986). But the eastern part behaves differently: the velocity actually increases slightly from the apex to the brightest knots in the wings. From measurements of the [NII]/H$_\alpha$ -ratio one can also derive an inhomogeneous excitation structure. From the bright knots in the eastern part of HH34 only the one near the apex is detectable in the [NII] -line.

2. DISCUSSION

We interpret the jet as a continuous stream of material flowing in a narrow channel with a small opening angle of about 5. The series of knots may represent emission from the cooling regions behind incident shocks in the jet, induced by the interaction with the surrounding medium. The low decrease in velocity, small line widths and low excitation fit the model of Falle et al. (1987) very well. However, the small observed knot spacing of about 2" is lower than predicted by the models. This fact cannot be explained only by projection effects. The reason for the second flow component which appears to accelerate with increasing distances from the source might be entrained ambient material, which can become accelerated at the rim of the jet boundary. Entrainment of interstellar material might be due to Kelvin-Helmholtz instabilities (e.g. Payne and Cohn, 1985).

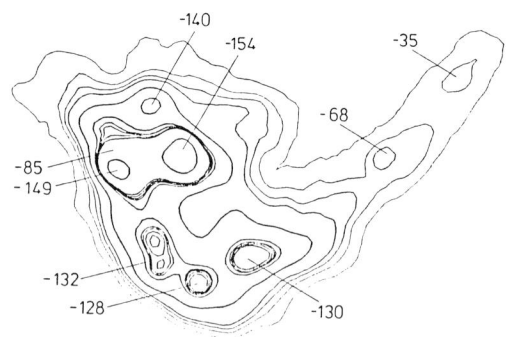

Fig. 4: SII isointensity contour plot of HH 34 revealing the inhomogeneous velocity structure derived from the Coudé-spectra.

The general shape of HH 34 suggests that it is the bow shock of the jet. The compact [OIII]-emission region near the apex and the velocity gradient in the western wing are in agreement with the bow shock models of Raga and Böhm (1986). The clumpy structure and the inhomogeneous velocity field, particularly in the eastern part of HH 34, might be due to the inhomogeneous interstellar medium and/or to thermal instabilities in the cooling region of the bow shock. This idea was recently pointed out in a model of time dependent bow shocks by Raga and Böhm (1987).

REFERENCES

Bührke, Th. and Mundt, R. 1987, in IAU Symposium 122 on "Circumstellar Matter", Heidelberg 1986, eds. I Appenzeller and C. Jordan, Reidel, Dordrecht, in press
Falle, S.A.E.G., Innes, D. and Wilson, M.J. 1987, Mon. Not. Roy. Astron. Soc., in press
Mundt, R. 1986, Can. J. of Phys. 64, 407
Payne, D.G. and Cohn, H. 1985, Astrophys. Journ., 291, 655
Raga, A.C. and Böhm, K.-H. 1986, Astrophys. Journ. 308, 829
Raga, A.C. and Böhm, K.-H. 1987, in press
Reipurth, B., Bally, J., Graham, J.A., Lane, A.P., Zealey, W.J. 1986, Astron. Astrophys. 164, 51
Solf, J. 1987, "High-Resolution Spectroscopy of Jets from Young Stars", this issue

HIGH-RESOLUTION SPECTROSCOPY OF JETS FROM YOUNG STARS

J. Solf
Max-Planck-Institut fuer Astronomie
Koenigstuhl
D-6900 Heidelberg 1
F.R.G.

ABSTRACT. High-resolution long-slit spectra of the HH 24 complex are presented. The data reveal the presence of a bipolar system of jet-type mass flows emanating from the infrared source SSV 63 considered to be the "exciting" star of HH 24. The detailed position-velocity (PV) structure derived for the jet-like component HH 24C is compared with a kinematical model of highly collimated mass flow consisting of a cylindric inner core and a co-axial envelope. In the course of propagation along the flow axis, the core presents a nearly constant velocity, whereas significant deceleration is found in the envelope.

1. OBSERVATIONS

Long-slit spectra (7.8 Å/mm) of HH 24 were obtained at the Calar Alto Observatory (Spain) using the 2.2-m coude spectrograph and a two-stage image-intensifier tube. The various slit positions are indicated on the isophotic contour map of HH 24 in Fig. 1. The main components of HH 24 (A,B,C,D in the notation of Herbig, 1974) and the position of the infrared source SSV 63 (Strom, Strom and Vrba, 1976) are marked as well. The obtained spectrograms show spatially resolved line features of Hα, [NII] and [SII] due to the various nebular condensations. Fig. 2 presents a position-velocity diagram of the [SII]λ6716 line structure observed at slit position (a) intersecting the components HH 24 A and C and the position of SSV 63 (see Fig. 1). The "cross" inserted near the center of the PV diagram indicates the position of SSV 63 (ordinate) and the velocity of the associated molecular cloud (abscissa) assumed to be the same as the (unknown) velocity of SSV 63. On the PV diagram, the compact HH 24A is represented by a prominent roundish feature at small redshift, whereas the jet-like HH 24C is shown as an arc-like feature at high negative velocities. In addition, the PV diagram presents an elongated, highly redshifted feature (labelled "E") due to a nebular condensation (designated HH 24E hereafter) which is rather inconspicuous on the direct map of HH 24 (Fig. 1). Slit position (b) intersects HH 24E at a different position angle. The obtained spectrogram shows a compact line feature due to HH 24E barely resolved along that slit direction.

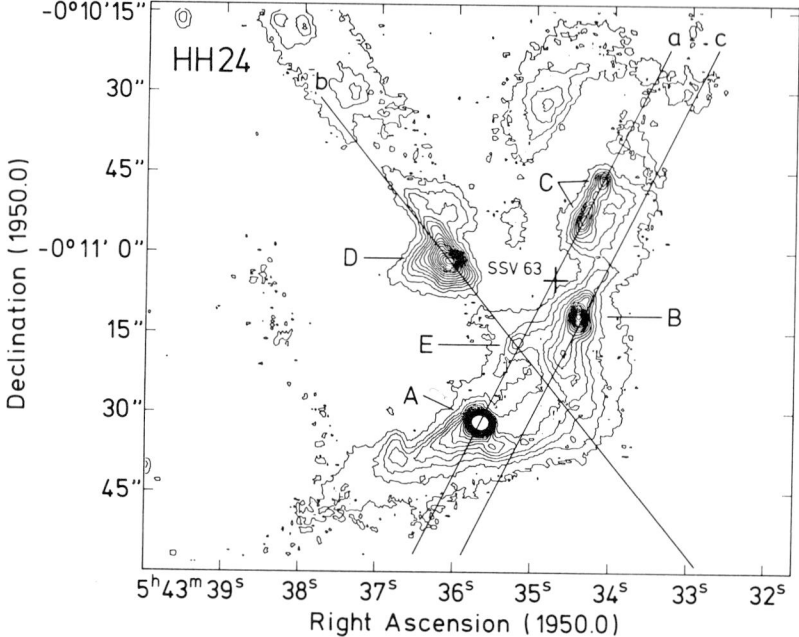

Figure 1. Isophotic contour map of HH 24 derived from a narrow-band red CCD image (kindly provided by R. Mundt). The contour spacing corresponds to a factor of $\sqrt{2}$ in the intensity. The positions of the spectrograph slit (a to c) are indicated.

2. A BIPOLAR SYSTEM OF JETS

The data show that HH 24C and HH 24E form mirror images of one to each other. Both components are rather elongated structures located on opposite sides of and pointing towards SSV 63. Their radial velocities are of nearly the same amount but of the opposite sign. Hence both the morphology and the kinematics suggest that HH 24C and HH 24E represent a bipolar system of highly collimated, jet-like mass flows of high velocity (>200 km/s) emanating from SSV 63 considered to be the "exciting" source of HH 24.

3. KINEMATICAL MODEL OF COLLIMATED MASS FLOW

The line features due to the jet-like component HH 24C observed with the spectrograph slit parallel to the flow axis (Fig. 2) resembles the shape of capital "lambda" (Λ). Its apex corresponds to the head of the jet. The high-velocity sub-component (left) of the feature presents a small downstream decrease ($\sim$5 percent) of the velocity, while the low-velocity sub-component (right) shows a large upstream decrease ($\sim$50 percent).

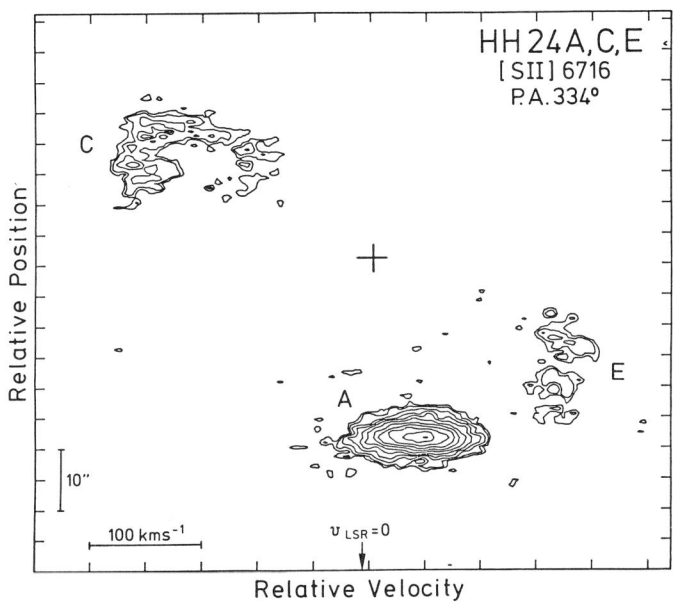

Figure 2. Position-velocity contour diagram of the [SII]λ6716 line observed at slit position (a) (Fig. 1). The contours indicate a factor of $\sqrt{2}$ in intensity. Line features due to HH 24A, C and E are labelled accordingly.

The line shape observed in HH 24C seems to deviate in several respects from those calculated on the basis of models of a bow shock formed around a "bullet" of dense gas plowing into the ambient medium (Boehm and Raga, 1986). These models predict an asymmetric line shape with two peaks (one of them near the velocity of the powering star), quite different to the $\bigwedge$ shape shown in Fig. 2.

A simple kinematic model for the collimated mass flow is proposed as sketched in Fig. 3 (left) which in a qualitative way explains some of the characteristics of the lines observed in HH 24C. The highly collimated flow is assumed to be confined within two separate cylindric regions: an inner core around the flow axis and a co-axial envelope surrounding the core. In each of the regions, the amount of the flow velocity varies in a specific way as indicated by the length of the arrows in Fig. 3. In the core region, the velocity is nearly constant presenting only a slight downstream decrease, whereas in the envelope, the velocity presents a large upstream decrease. Near the head of the jet flow, the velocities of the core and envelope are the same, but become progressively different along the upstream direction. Adopting an inclination of the flow axis towards the observer (as it is the case in HH 24C) the model predicts a PV distribution of the lines as shown in Fig 3 (right), if the slit is oriented parallel to the flow axis.

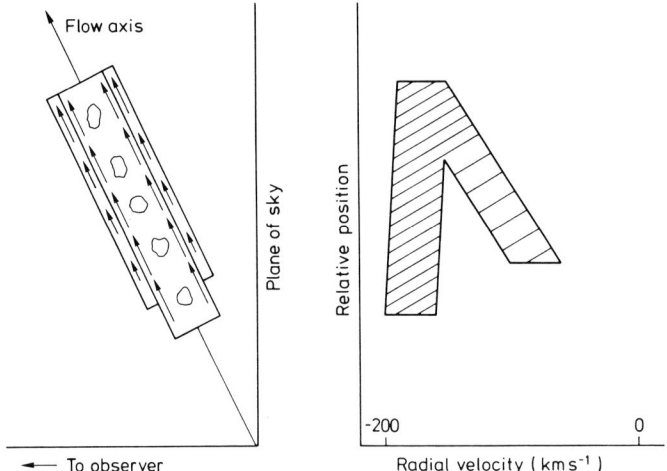

Figure 3. Schematic representation of the kinematical model of highly collimated mass flow as discussed in the text. The left hand panel shows the plane defined by the direction of the slit and the vector of the mass flow (inclined with respect to the line of sight). The right hand panel sketches the resulting position-velocity structure of an emission line originating from the two co-axial regions of the model.

A comparison of the PV diagrams in Fig. 2 and 3 shows that the observed line shape is similar to that predicted by the model, suggesting that the basic assumptions of the model may be correct. It seems that (in the course of propagation along the flow axis) the gas of the core region is only marginally decelerated, whereas the material of the envelope is subject to strong decelerating forces in the ambient gas and hence becomes progressively decoupled from the high-velocity motion of the jet core. It should be noted that recent high-resolution observations of the HH 34 jet (Buehrke and Mundt, this workshop) present line shapes of striking similarity to those found in HH 24C. This suggests that the core/envelope structure may be a characteristic property of jets from young stars. The core/envelope structure of the kinematical model in a qualitative way resembles the gross morphology of the "beam/cocoon" structure predicted in hydrodynamic model calculations of jets by Norman, Winkler and Snarr (1983).

4. REFERENCES

Herbig, G.H.: 1974, Lick Obs. Bull. No 658
Norman, L.M., Winkler, K.H., Snarr,L.: 1983, in Astrophysical Jets, ed. A, Ferrari and A.G. Pacholczyck, Reidel, p. 227
Raga, A.C. and Boehm, K.H.: 1986, Astrophys. J. 308, 829
Strom, K.M, Strom, S.E, Vrba, F.J.: 1976, Astron. J. 81, 308.

A THIN-SHELL MODEL FOR MOLECULAR OUTFLOWS

M. Robberto[1,2] and G. Silvestro[2]

(1) Osservatorio Astronomico di Torino, Pino Torinese, Italy.
(2) Istituto di Fisica Generale, Università di Torino, Italy.

1. INTRODUCTION

The morphology and emissivity of shell-shaped bipolar outflows, driven by a spherically-symmetric stellar wind which interacts with a non-isotropically distributed circumstellar gas within a dense molecular cloud, was estimated previously (Silvestro and Robberto, 1987) in the framework of the shell model of collimated outflows suggested by Barral and Cantò (1981). Our computation allowed to obtain a spatial map of the profiles of the molecular lines emitted in the region of interaction between the wind and the circumstellar material. The position-velocity map of the rotational CO lines from the source Mon R2 turns out to be in very good agreement with the assumed thin-shell model for the outflowing lobe.

Recent observations of optical and radio jets and chains of Herbig-Haro objects moving with high velocities show that the collimation must take place very close to the source (Rodriguez, 1987, Silvestro et al, 1987). Our previous computations, which assumed an emissivity proportional to the shock luminosity $L_{shock} \propto \rho V^3$, afforded a value for the luminosity of the high-velocity molecular component that, compared with observations, increases too steeply in the vicinity of the driving source.

We consider here a model with the following new features:

(1) an intrinsically collimated stellar wind,

(2) molecular emission limited to that part of the post-shock cooling region where the temperature has dropped below 10^4 K.

2. THE OUTFLOW GEOMETRY

We assume that, at a distance from the driving source which is small compared with the angular resolution of observations, the stellar wind has an angular structure described by the equation:

$$\dot{M}_*(\theta) = \dot{M}_{*0} F(\theta)$$

where:
$$F(\theta) = \left[\frac{\sin^2(\theta)}{5} + \cos^2(\theta)\right]^{-1/2}$$

The shock surface locus is given by an equilibrium condition between the wind pressure and the pressure of the surrounding material, taking into account the centrifugal correction (Barral and Cantò, 1981). Fig.1 shows a typical lobe profile which we obtain by using the new model, compared with the case of an isotropic stellar wind.

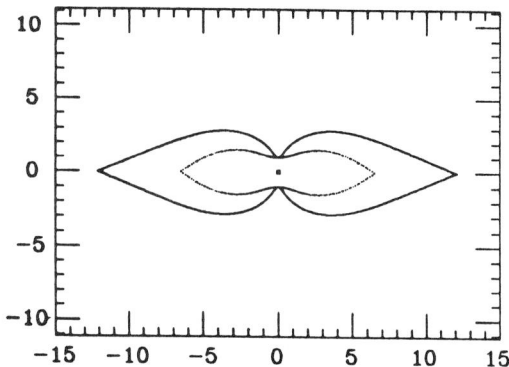

Figure 1. The locus of the shock surface for an intrinsically collimated stellar wind (solid line) and an isotropic stellar wind (dotted line).

3. EVALUATION OF CO EMISSIVITY ON THE SHOCK FRONT

We follow the classical formalism by Kaplan (1966), which makes use of a parameter η defined by
$$\rho = (J^2/P\eta)$$
where $J = \rho V$ is the momentum flux and $P = \rho\left(\frac{RT}{\mu} + V^2\right)$ is the total pressure (J and P are conserved across the shock surface). The parameter η has a value $(1/4)$ on the shock front and decreases to zero on the outer edge of the cooling region.

By using the equation of energy conservation we obtain the total emissivity across the cooling region:
$$W_{tot} = \frac{P^2}{2J}\left[4\eta^2 - 5\eta + 1\right] \ .$$

The CO emissivity can be written as
$$W_{CO} = \alpha_{CO}\left[W_{tot} - W(T > 10^4 K)\right] \ ;$$

we then obtain
$$W_{CO} = \alpha_{CO} W_{tot}\left[4\frac{RT}{\mu}\frac{1}{V_s^2} + \frac{1 - \left(1 - 4\frac{RT}{\mu V_s^2}\right)^{1/2}}{2}\right]$$

where α_{CO} is the fraction of the molecular emission in CO lines, and $T = 10^4$ K according to our assumption. We notice that the term $1/V_s^2$ in the parenthesis would decrease the fractional emission in the molecular component when the strenght of the shock is increased.

4. CONCLUSIONS

As one can see by comparing Figures 2a and 2b, we obtain, by restricting molecular emission to temperatures $T < 10^4$ K, a brightness distribution in CO lines that is spatially more extended and which displays a clear bipolar morphology. The new model then appears to give a better description of the observational features of molecular outflow sources.

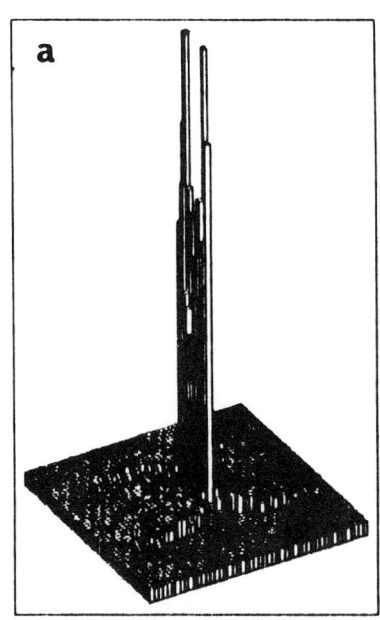 

Figure 2. The two-dimensional brightness distribution in CO lines: (a) CO emissivity $\propto \rho V_s^3$; (b) CO emission restricted to the cooling region ($T < 10^4$ K).

REFERENCES

Barral, J.F. and Cantò, J., 1981, *Rev. Mexicana Astron. y Astrof.*, **5**, 101.
Kaplan, S.A., 1966, *Interstellar Gas Dynamics*, Pergamon Press, Oxford, p. 83.
Rodriguez, L.F., 1987, in *Star Forming Regions*, IAU Symposium No. 115, p. 239.
Silvestro, G., Ferrari, A., Trussoni, E., Rosner, R., and Tsinganos, K., 1987, *Nature* **325**, 228.
Silvestro, G., and Robberto, M., 1987, in *Circumstellar Matter*, IAU Symposium No. 122, in press.

EVIDENCE OF THIN DUST SHELLS IN SOME RS CVN STARS

F.Scaltriti[1], M.Busso[1], M.Robberto[1], P.Persi[2], G.Silvestro[3]

1) Osservat. Astronomico Torino-Pino Torinese-Italia
2) Istituto Astrofisica Spaziale CNR-Frascati-Italia
3) Istit. Fisica Generale Univ. Torino-Torino-Italia

ABSTRACT. UBVRIJHK light curves of four active RS CVn binaries together with coordinated spectrophotometric observations in the range 1.4-2.5 μm are presented. From the reconstruction of the whole spectral distributions (from U to K) we may infer the presence of infrared excesses which cannot be simply explained by a combination of stellar spectra (even changing the accepted spectral types) and of spot-related photospheric perturbations. Our analysis shows that thermal radiation from a thin dust shell surrounding the system may explain the data.

The RS CVn-type binaries belong to a class of active objects that show solar-like phenomena at a level much larger than in the Sun owing to the high angular velocity enforced by the tidal synchronization between the components. From a photometric point of view, the stellar activity introduces a wave-like distortion, due to large areas covered by dark spots.
Berriman et al. (1983) showed that IR observations of RS CVn stars are significant, a) for spectral classification of the components when also visual data are available, b) for studying IR excesses and the underlying mechanisms. Though, on evolutionary grounds, one does not expect appreciable IR excesses neither from external layers nor from circumstellar envelopes, past observations provided contradictory results. While Berriman et al. (1983), Antonopoulou (1983) and Antonopoulou and Williams (1984) did not find significant IR excess from circumstellar emission, Atkins and Hall (1972), Hall et al. (1975), and Milone (1976) inferred the presence of an excess in several binaries. Moreover, analyzing MgII lines, Glebocki et al.

(1986) found traces of circumstellar matter in the long-period RS CVn system λ Andromedae.

In order to clarify whether excess emission in the near infrared is observable, we present here the results of symultaneous UBVRIJHK photometry of the RS CVn-type stars RU Cancri, VV Monocerotis, UV Piscium, and TY Pyxidis performed at ESO (La Silla) in 1984-1985. In the same campaign CVF spectrophotometry in the range 1.4-2.5 μm was carried out.

The V-filter observations (Busso et al. 1987) show clearly that a perturbing wave is affecting the light curve; however, due to photometric uncertainty in IR, it is hard to recognize trends in the infrared matching the visual ones. We tried to perform CVF spectrophotometry near the phases of maximum and minimum spot visibility, as close as allowed by weather conditions. Figure 1 shows the infrared spectra

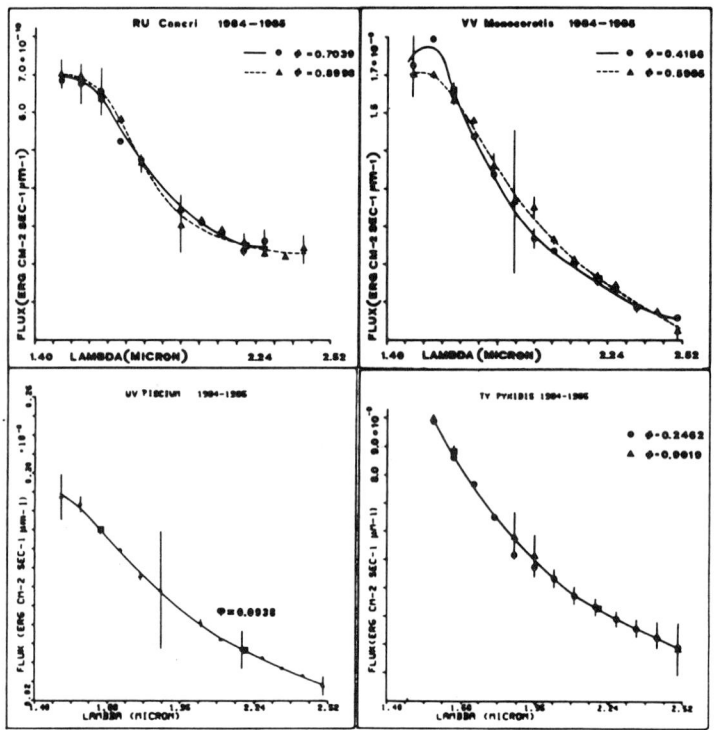

Figure 1 - The near-IR fluxes of RU Cnc, VV Mon, UV Psc and TY Pyx resulting from spectrophotometry at different phases. Squares indicate the fluxes from H and K photometry.

obtained, together with observational errors, when they are appreciable on the plots. Cloudy weather prevented from getting both spectra for UV Psc; for the other binaries observational accuracy and/or non exact matching with the maximum and minimum of the wave-like distortion do not allow firm conclusions about spot-related changes in the spectral distribution and/or in the level of emission.

Since the flux variations with phase (even at maximum and minimum spot visibility) are within the limits of detectability, the energy distributions of the system studied could be reconstructed, from our UBVRIJHK photometry, regardless to phase information.

The observed fluxes were then compared to the ones expected on the basis of the spectra of G and K stars, after combining them according to the known dimensions and spectral types of the components of the binaries studied. The fluxes of normal stars from U to K were taken from Johnson (1966), Bessell (1979), Koornneef (1983) and Straizys and Sviderskiene (1972) (see Figure 2).

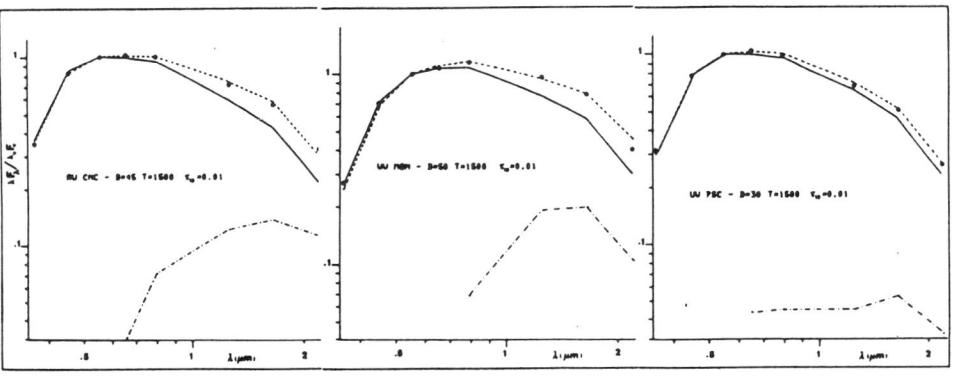

Figure 2 - The flux distribution of RU Cnc, VV Mon and UV Psc as resulting from our UBVRIJHK observations. The solid line is the expected flux on the basis of the known spectral types. The dash-dotted line shows the shape of the IR excess. The dashed line indicates the fit obtained adding a dust shell characterized by the given parameters.

Generally, the optical part of the spectrum is well reproduced; as far as the red and infrared parts of the spectrum are concerned, only for TY Pyx we obtained a good agreement between the calculated flux and the observations; all the other stars show remarkable IR excesses, clearly observable already in the I band.

We first tried to interpret the above excesses as a result of errors in the spectral types. Actually, as Berriman et al. (1983) pointed out, visual versus IR colours of G-K stars change very quickly passing from one subclass to the adjacent one. For the above reason, we allowed the spectral types of the components to vary over a large range, spanning over several subclasses, but still no agreement was possible for three out of four systems.

The observed IR excesses cannot be attributed to free-free emission from a disk or from mass loss, since neither the spectrum nor the excess show the theoretically predicted power law (Felli and Panagia 1974).

An alternative explanation is the presence of a thin dust shell surrounding the system. Assuming that at a distance D from the binary a dust shell exists characterized by a certain temperature T and optical depth $\tau(\nu)$, the effect of the cloud is to absorb the incident radiation $(F(\nu)/D^2)$ from the binary and to emit as a black-body at temperature T, according to the formula (Lang 1974):

$$S(\nu) = (F(\nu)/D^2) \exp(-\tau(\nu)) + B(\nu,T)(1-\exp(-\tau(\nu)))$$

Assuming T=1500 K, τ as given by the model of Jones and Merril (1975) for their "dirty" silicates and allowing D to vary in the interval 70-100 (D in units of the radius of the larger component), the dashed lines of Figure 2 show that the fits are strongly improved, despite the strong oversimplifications of the model, in which no dependence of the dust density on distance is introduced.

We are now studying, with techniques similar to the ones presented here a wide sample of RS CVn stars for which optical and near-IR photometry is available, in order to ascertain if thin dust shells can be considered a common property of those systems.

REFERENCES
- Antonopoulou,E.: 1983,Astron. Astrophys.120,85.
- Antonopoulou,E.,Williams,P.M.:1984,Astron.Astroph.135,61.
- Atkins,H.L.,Hall,D.S.:1972,Publ.Astron.Soc.Pacific 84,638.
- Berriman,G.,De Campli,W.M.,Werner,M.W.,and Hatchett,S.P.: 1983, Mon. Not. R. Astron. Soc. 205,859.
- Bessell,M.S.: 1979, Publ. Astr. Soc. Pac. 91, 589.
- Busso,M.,Scaltriti,F.,Persi,P.,Robberto,M., Silvestro,G.: 1987, Astron. Astrophys., in press.
- Felli,M.,and Panagia,N.: 1975, Astron. Astrophys. 39,1.
- Glebocki,R.,Sikorski,J.,Blielioz,E.,and Krogulee,M.: 1986, Astron. Astrophys. 158,392.
- Hall,D.S.,Montle,R.G.,Atkins,H.L.:1975,Acta Astron.25,125.
- Johnson,H.L.: 1966, Ann. Rev. Astron. Astrophys. 3, 193.
- Jones,T.W.,and Merril,K.M.: 1976, Astrophys. J. 209,508.
- Koornneef,J.: 1983, Astron. Astrophys. 128,84.
- Lang,J.: 1974,in "Astrophysical Formulae",Springer-Verlag.
- Milone,E.F.: 1976, Astrophys. J. 31,93.
- Straizys,V.,and Sviderskiene,Z.: 1972, Bull. Vilnius Astron. Obs., no. 35.

NEAR INFRARED-IRAS CANDIDATES FOR MOLECULAR OUTFLOWS IN LYNDS CLOUDS

A. Mampaso[1], J.P. Phillips[2], R. Gomez-Reñasco[1], R. Carballo.[1]
1. Instituto de Astrofisica de Canarias,
38200 - La Laguna, Tenerife, Spain.
2. Queen Mary College, London University, Mile End Road,
London E1 4NS, UK.

ABSTRACT. Near infrared observations of cold IRAS sources associated with Lynds dark clouds allowed the detection of 13 objects which, in some cases, could represent young embedded stars energizing molecular outflows.

1. INTRODUCTION

Cold IRAS sources have been shown to be frequently associated with CO radio emission and could represent places of molecular outflows (Casoli et al., 1986). Furthermore, some of them, associated with dense molecular cloud cores, could be low-mass protostars still accreting material from their placental cloud (Beichman et al., 1986).

In this work the IRAS Point Source Catalogue and the Lynds Catalogue of Dark Clouds were cross-correlated to find bright infrared sources with 100 μm fluxes higher than their corresponding fluxes at 60μm. Our final aim was the detection of new candidates for bipolar molecular outflow sources and/or young embedded PMS stars associated with those clouds.

In the following we report the preliminary results of a near-infrared study of seven of these fields, which yielded the detection of thirteen new NIR sources. In the following section we describe the observations and in Section 3 we discuss the nature of the detected sources.

2. OBSERVATIONS

The observations were taken at the 1.5m SM telescope at the Observatorio del Teide (Tenerife, Spain) equipped with a solid Nitrogen InSb detector system, using standard broad-band J H K L and M filters. We used a 20 arcsec circular aperture and 20 arcsec chopper throw in declination.

An area of approximately 90x90 arcsec around the IRAS nominal position was scanned at 2.2 μm to a limit of ~10 mag. A set of standard stars from Engels et al (1981) was used for calibration. Table 1 lists

TABLE 1

IRAS source	COORDINATES [a] α(1950) δ(1950)	NIR source	INFRARED PHOTOMETRY [b]				
			J	H	K	L	M
17547 - 1832	17^h 54^m 46^s.4 -18° 32' 11"	IRS 1	11.14 (3)	8.82 (1)	7.81 (1)	6.99 (40)	
		IRS 2	9.52 (1)	6.70 (1)	5.50 (1)	4.56 (1)	>4.1
		IRS 3	11.47 (14)	9.34 (2)	8.51 (1)		
18195 - 1407	18 19 34.0 -14 07 05	IRS 1	10.93 (1)	9.00 (4)	8.18 (2)	7.90 (40)	
		IRS 2	10.03 (4)	8.62 (1)	7.90 (1)	7.08 (30)	
18232 - 1154	18 23 15.2 -11 54 22	IRS	11.05 (10)	8.34 (5)	7.36 (1)	6.34 (24)	>4.0
18277 - 0844	18 27 42.8 -08 44 46	IRS 1	10.74 (10)	8.23 (4)	7.10 (2)	5.77 (5)	5.88 (30)
		IRS 2 [c]	10.55 (3)	10.03 (2)	9.39 (9)		
18342 - 0655	18 34 16.2 -06 55 51	IRS 1	10.78 (5)	8.82 (1)	7.86 (4)	6.80 (4)	>5.9
		IRS 2 [c]	8.30 (1)	7.10 (1)	6.72 (1)	6.54 (1)	6.8 (20)
19295 + 1637	19 29 35.6 16 37 03	IRS 1		9.51 (3)	8.93 (5)	7.99 (3)	
		IRS 2	11.61 (2)	9.35 (1)	8.37 (1)	7.6 (1)	>5.6
21023 + 5002	21 02 19.6 50 02 40	IRS	10.32 (1)	7.98 (1)	6.32 (1)	2.97 (4)	3.26 (20)

NOTES: (a) Position from the IRAS Point Source Catalogue.
(b) Infrared magnitudes (with 1σ statistical error in hundredths) Lower limits are at a 3σ level.
(c) Source coincident with a visible star in the TV-guiding camera.

the NIR photometry of the thirteen sources detected, together with the IRAS positions for the centres of scanned boxes. The 1-100 μm fluxes of the sources are shown in Fig. 1.

3. DISCUSSION

Figures 2 and 3 show the loci of the different sources found in our survey (represented by solid squares) in NIR colour-colour diagrams, together with the position of main-sequence stars (+), giants (x), T Tauri stars (o) (Rydgren and Vrba, 1983) and Herbig Ae-Be stars (*) (Finkenzeller and Mundt, 1984). In the J-H/H-K diagram, our sources occupy a rather restricted area suggesting that, at these wavelengths, they could be explained as heavily reddened normal (or giant) stars. In the H-K/K-L diagram however, they are well mixed with the other PMS stars discussed above, showing substantial excesses towards longer wavelengths and indicating dust temperatures of around 1500 K. The IRAS data, on the other hand, show a strong temperature gradient (as deduced from the 25/12, 60/25 and 100/60 μm fluxes ratios) with means of 170, 60 and 30 K respectively.

Although the precise nature of these sources is by no means clear, their near and far infrared properties and their association with dark clouds makes them good candidates for embedded young stars possibly associated with molecular outflows.

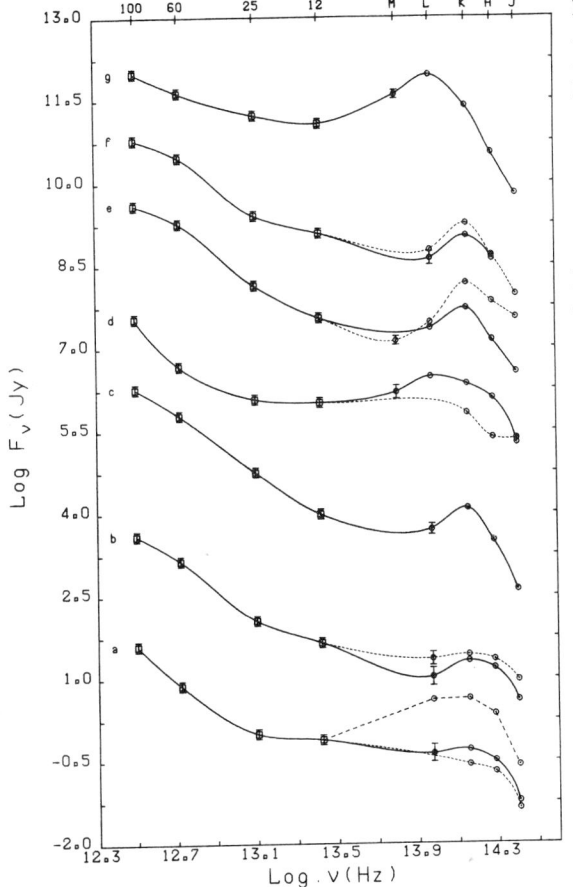

Fig. 1: Infrared photometric spectra for the observed sources. Different factors have been added to the log F_ν for clarity in the plot.

a) 17547-1832:IRS1,2,3,
b) 18195-1407:IRS1,2 (+1.7)
c) 18232-1154:IRS1 (+3.8)
d) 18377-0844:IRS1,2 (+6.3)
e) 18342-0655:IRS1,2 (+7.6)
f) 19295+1637:IRS1,2 (+9.4)
g) 21023+5002:IRS1 (+10.7)

References.

Beichman, C.A., Myers, P.C., Emerson, J.P., Harris, S., Matheu, R., Benson, P.J., Jennings, R.E.: 1986, Astrophys. J. **307**, 337.
Casoli, F., Dupraz, C., Gerin, M., Combes, F., Boulanger, F.: 1986, Astron. Astrophys. **169**, 281.
Engels, D., Sheerwood, W.A., Wamsteker, W., Shultz, G.V.: 1981, Astron. Astrophys. Suppl. Ser. **45**, 5.
Finkenzeller, U., Mundt, R.: 1984, Astron. Astrophys. Suppl. Ser. **55**, 109.
Rydgren, A.E., Vrba, F.J.: 1981, Astron. J. **86**, 1069.

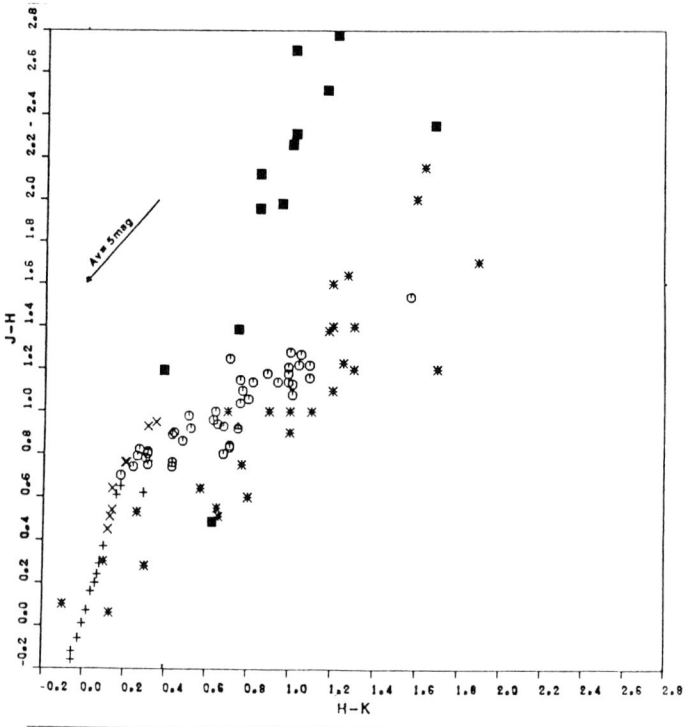

Fig. 2: Near-infrared colour-colour diagram for the observed sources (solid squares) together with normal and PMS Stars (see text for details)

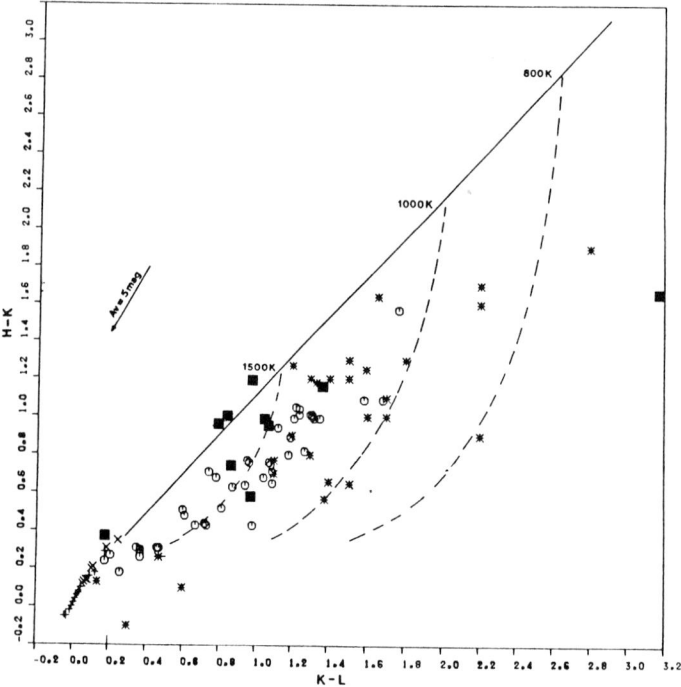

Fig. 3: As in Fig. 2 for H-K vs. K-L. The solid line represents the blackbody loci and the dashed line indicates the loci of a normal MS photosphere surrounded by a dust layer of constant temperature with increasing optical depth.

MASS OUTFLOWS AND JETS IN SYMBIOTIC STARS

R. Viotti[1], G.B. Baratta[2], L. Piro[3]
A. Altamore[4], A. Cassatella[5]

1. Istituto Astrofisica Spaziale, Frascati, Italy
2. Osservatorio Astronomico, Roma, Italy
3. Istituto TESRE, CNR, Bologna, Italy
4. Istituto Astronomico, Universita' La Sapienza, Roma, Italy
5. IUE Observatory, European Space Agency, Madrid Spain

ABSTRACT. We discuss the origin and structure of jet components in symbiotic stars, such as R Aqr, CH Cyg and AG Dra. It is shown that the sudden appearance of jets is associated with the increase of the ionizing flux from the central source followed by ionization of circumstellar material, rather than with violent ejection of matter. Recent X-ray observations are discussed.

Symbiotic stars are known for their peculiar composite spectrum characterized by the superposition of a late type absorption spectrum and prominent emission lines which give evidence of ionized circumstellar nebulae. A major feature of these objects is the large and irregular variability, including nova-like outbursts. Many symbiotic stars are radio sources and recent high resolution radio imagery has disclosed the presence of small nebulae around them, with a variety of structures, such as shells, jets and bipolar nebulae (e.g. Hjellming and Gibson 1984; Viotti 1987). In a number of cases the structure appears variable in time. In the following we shall discuss the possible origin of these nebulae also in the light of the most recent UV and X-ray observations with IUE and EXOSAT.

The symbiotic Mira R Aqr can be considered as the prototype of jet activity in symbiotic stars. In 1977 Wallerstein and Greenstein (1980) discovered a jet-like feature about 7" from the star which was not present in 1970 (Herbig 1980). VLA radio maps not only confirmed the jet, but also disclosed the presence of five separate sources nearly placed on the same line (Hollis et al. 1985, 1986; Kafatos et al. 1986). The appearance of the jet might suggest violent expulsion of matter from R Aqr before 1977, possibly associated with the recurrent anomalies of the Mira light curve with an almost complete disappearance of the pulsation (Mattei and Allen 1979). This could be due to tydal effects during the periastron passage of a close binary system moving in an eccentric orbit, as suggested by Willson et al. (1981). The last anomaly occurred in 1974-77, and if associated to the last close approach of the R Aqr companion, this could have caused an increase of mass transfer in the system, followed by mass ejection. However, the radio observations of Hollis et al. (1985) give no evidence of radial expansion of the radio features during 1982-84. Therefore the appearance of the jet features in R Aqr should be rather ascribed to ionization of pre-existing circumstellar matter by the increased flux of ionizing photons from the central hot source. X-rays from R Aqr were recently detected with EXOSAT in 1985 (Baratta et al. 1985, Viotti et al. 1987) which are probably emitted from the jet. The absence or weakness of the X-ray flux in the previous HEAO-2 observations of June 1980 (Jura and Helfand 1984, see also Viotti et al. 1987) might suggest a recent increase of the X-ray flux from R Aqr associated with the appearance of high ionization emission lines in the UV spectrum of the jet (Kafatos et al. 1986; Viotti et al. 1987).

Another interesting object is CH Cyg which recently underwent a radio outburst followed by the appearence of ejecta (Taylor et al. 1986). This is the brightest symbiotic star which in July 1984 declined by about 1.5 mag (Tomov 1984). In the same time Selvelli and Hack (1985) observed a dramatic increase of the line excitation. Taylor et al. (1986) reported a strong rise of the radio flux during April 1984 to May 1985 of about a factor 35. Moreover they mapped CH Cyg at 2 cm with VLA and found two radio components on 8 November 1984 separated by 0.18", and three components 75 days later with a total separation of 0.4". They concluded that the jets are expanding at a rate of 1.1 arcsec/yr, corresponding to a projected velocity of about 1000 km/s. This conclusion seems also to be supported by the large width of the hydrogen lines. Thus CH Cyg could be considered as the best example of jet activity. There is however an alternative and simpler explanation of the observations: the two sources observed in November 1984 more likely correspond to the central and South-East radio components in the January 1985 VLA map, while the North-West source is a new one. In such a hypothesis the apparent motion of the radio structures is much lower than that claimed by Taylor et al. (1986) and probably close to zero. Also the broadness of the Ly-a emission is easily explained by very large optical thickness without the need of Doppler broadening. We think that the appearance of the jets in CH

Cyg is associated with a sudden increase of the flux of the ionizing photons from the central source in mid 1984, followed by the propagation of a ionizing front through the pre-existing circumstellar matter. In this case the linear structure is probably caused by the presence of dense material in the orbital plane of a central binary system as in the case of R Aqr discussed above. It should also be noted that CH Cyg was detected with EXOSAT in May 1985 (Leahy and Taylor 1987), while only an upper limit was obtained with HEAO-2 in October 1979 suggesting a long-term variability probably associated with the 1984 radio outburst.

AG Dra is the strongest X-ray source among symbiotic stars. It was observed with HEAO-2 in April 1980, when the star was at minimum (Anderson et al. 1981), and with EXOSAT during two recent minor outbursts (in March 1985 and February 1986) and in the minimum phase between these outbursts. Cassatella et al. (1987) discovered a large fading of the X-ray flux in correspondence of the two maxima. Recent radio observations at 6 cm indicate an increase of the radio flux from this star, and the presence of a faint satellite component about one arcsec South-West (Torbett and Campbell 1987). If this behaviour is associated to the recent outbursts, we should expect a rapid radio evolution of AG Dra in the following years.

X-rays were also observed in the symbiotic stars RR Tel, V1016 Cyg and HM Sge which are known to have undergone a nova-like explosion in recent years (Allen 1981). Allen noted that in the more recently exploded objects V1016 Cyg and HM Sge the flux was larger than in the older symbiotic nova RR Tel. This may suggest a secular decrease of the X-ray luminosity after the outburst. This behaviour was confirmed by Willson et al. (1984) who reanalyzed the data and added a new observation of HM Sge obtained in 1981. Radio imagery of these stars disclosed the presence of small nebulae with complex structure (see e.g. Hjellming and Gibson 1985) again associated with the nova-like outbursts as for instance suggested by the radio brightening of HM Sge (Kwok 1982).

In summary, in symbiotic stars the jet structures seem to be associated with phases of enhanced activity, and with a presence of high temperature circumstellar regions, but there is so far no clear evidence of violent ejection of matter. Certainly, processes such as propagation of ionization fronts and/or shock waves play an important role in their jet activity. The recent occurrence of this phenomenon in the above described symbiotic objects is a key to understand the formation and time evolution of astrophysical jets.

REFERENCES.

Allen, D.A.: 1981, Mon. Not. R. astr. Soc. 197, 739.
Anderson, C.M., Cassinelli, J.P., Sanders, W.T.: 1981, Astrophys. J. 247, L127.
Baratta, G.B., Piro, L., Viotti, R., Cassatella, A., Altamore, A., Ricciardi, O., Friedjung, M.: 1985, Cosmic X-Ray Spectroscopy Mission, ESA SP-239, p.95.
Cassatella, A., Cordova, F.A., Friedjung, M., Kenyon, S.J., Piro, L., Viotti, R.: 1987, IAU Colloquium 93, Cataclysmic Variables, Astrophys. Space Sci. Vol.131.
Herbig, G.: 1980, IAU Circular No.3535.
Hollis, J.M., Kafatos, M., Michalitsianos, A.G., McAlister, H.A.: 1985, Astrophys. J. 289, 765.
Hollis, J.M., Michalitsianos, A.G., Kafatos, M., Wright, M., Welch, W.J.: 1986, Astrophys. J. 309, L53.
Hjellming, R.M., Gibson, P.M. eds.: 1985, Radio Stars, Reidel, Dordrecht.
Jura, M., Helfand, D.J.: 1984, Astrophys. J. 287, 785.
Kafatos, M., Michalitsianos, A.G., Hollis, J.M.: 1986, Astrophys. J. Supplem. Ser. 62, 4.
Kwok, S.: 1982, The Nature of Symbiotic Stars, M. Friedjung and R. Viotti eds., Reidel, Dordrecht, p.17.
Leahy, D.A., Taylor, A.R.: 1987, Astr. Astrophys. in press.
Mattei, J. A., Allen, J.: 1979, J. Roy. astr. Soc. Canada, 73, 173.
Selvelli, P.L., Hack, M.: 1985, Astronomy Express 1, 115.
Taylor, A.R., Seaquist, E.R., Mattei, J.A.: 1986, Nature Vol. 319, 38.
Tomov, T.: 1984, Inf. Bull. Var. Stars. No. 2610.
Torbett, M.V., Campbell, B.: 1987, Astrophys.J. submitted.
Viotti,R.: 1987, in Planetary and Protoplan. Nebulae from IRAS to ISO, A. Preite-Martinez ed., Reidel, Dordrecht.
Viotti, R., Piro, L., Friedjung, M., Cassatella, A.: 1987, Astrophys. J. Letters, in press.
Wallerstein, G., Greenstein, J.L.: 1980, Pub. Astr. Soc. Pacific 42, 275.
Willson, L.A., Garnavich, P., Mattei, J.: 1981, Inf. Bull. Var. Stars no. 1963.
Willson, L.A., Wallerstein, G., Brugel, E., Stencel, R.E.: 1984, Astron. Astrophys. 133, 154.

IRAS AND NEAR INFRARED OBSERVATIONS OF PECULIAR NEBULOSITIES

P. Persi[1], M. Busso[2], M. Ferrari-Toniolo[1], L. Origlia[3], M. Robberto[2,3], F. Scaltriti[2] and G. Silvestro[3]

(1) Istituto di Astrofisica Spaziale del CNR, CP 67, 00044 Frascati, Italy.
(2) Osservatorio Astronomico di Torino, Pino Torinese, Italy.
(3) Istituto di Fisica Generale, Università di Torino, Italy.

ABSTRACTS. We used the IRAS survey to search for IR emission from 95 peculiar nebulosities with CO emission. Fifty-two IRAS sources are associated with them. We discuss the nature of the sources on the basis of their IRAS color-color diagram and of near-IR observations.

1. INTRODUCTION

Recent observations have revealed that many young stellar objects undergo an energetic phase of mass outflow. This activity occurs presumably when the star is deeply embedded in its parental molecular cloud. The interaction of the collimated stellar wind with the ambient gas may explain the Herbig-Haro phenomenon and the presence of peculiar nebulosities associated with star forming regions.

Torrelles et al. (1983) searched for CO outflows from 73 "cometary nebulae" listed by Parsamian and Petrosian (1979) (PP objects), and 22 new Herbig-Haro like objects found in the Palomar Sky Survey Plates by Gyulbudaghian (1982) (GY objects). CO emission was observed in almost all regions, but no evidence of high-velocity carbon monoxide emission was found.

Using the IRAS survey, we searched for newly forming stars associated with these peculiar nebulosities. IRAS color-color diagrams, and JHKL photometry obtained at the 1.5 m Italian Infrared Telescope (TIRGO) are presented in this paper and used for discussing the nature of the sources.

2. IRAS

Areas $6 \times 6\,arcmin^2$ centered on 95 peculiar nebulosities and H-H like objects observed by Torrelles et al. (1983), were searched for sources in the IRAS Point Source Catalog. To ensure association with the nebulae, the sources had to lie within $1-2\,arcmin$ of the nebular position, and had to be detected either at $25\mu m$ or at both 60 and $100\mu m$.

We have found that nearly half the peculiar nebulosities have IRAS sources associated with them. The IRAS flux densities, color corrected according to the prescription given in the Catalog Supplement (Beichman et al., 1985), were used to obtain $12/25/60\mu m$ and $25/60/100\mu m$ color-color diagrams (Figures 1,2).

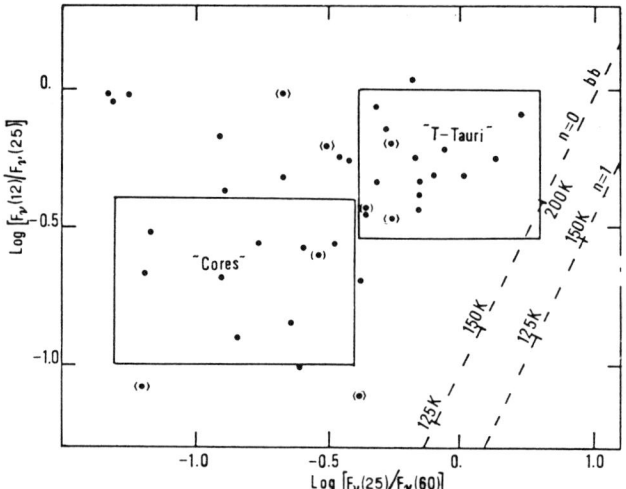

Figure 1. IRAS $12/25/60\mu m$ color-color diagram. (•) GY objects; • PP objects.

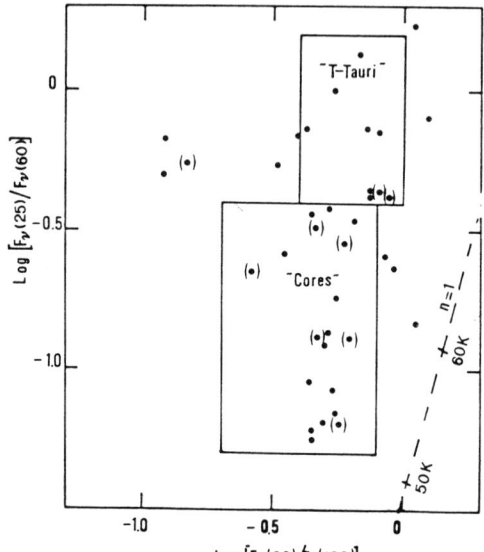

Figure 2. IRAS $25/60/100\mu m$ color-color diagram. Symbols as in Figure 1.

From the location of the sources in the color-color plots, we have deduced that 21 objects could be classical T-Tauri stars, while 19 sources have IRAS colors similar to that of IR sources associated with small and dense molecular clouds indicated as "cores" by Beichman et al. (1986). Besides, most of the IRAS sources in the T-Tauri area are associated with visible stars. In addition, a small fraction of our sources lie in the upper left corner of the diagrams, far from all other sources in the sample. The apparent 60 and 100μm excess for these sources might be due to the contamination by cirrus.

In Tables 1 and 2 we give a list of the T-Tauri and "core" type sources found in our survey.

Table 1
Peculiar Nebulosities: T-Tauri type

Nebula	IRAS	Near-IR	Opt.Id.*
PP1	00044+6521	JHKL	star,V=17.3
PP2	00087+5833	JHKL	V376 Cas
GY2	00338+6312	JHKL	star,V=14.1,RNO1
GY5	03134+5959	JHKL	star
PP8	03220+3035	JHK	star,V=19.6,RNO13
PP10	03257+3034	JHK	LH 325,V=14
PP11	03507+3801	JHKL	-
PP13	04073+3800	-	Lynds 1473
PP16	04188+2819	-	RY Tau
GY13	04591-0856	JHK	-
PP37	05358-0704	-	V883 Ori
PP38	05363+2620	JHKL	RR Tau
PP43	05426+0903	-	FU Ori
PP49	05591+1630	JHKL	B9e,V=9.7
PP81	15420-3408	-	Ge,V=10.5
PP82	16239-2438	-	star,V=15
GY20	17554-2606	-	red star
PP87	18585-3701	-	A5e,V=13.5
PP97	20453+6746	JHKL	PV Cep
PP99	20595+5009	-	V1331 Cyg
GY21	21004+7811	JHKL	Ge,V=15.8,RNO129

*RNO,Cohen (1980)

Table 2
Peculiar Nebulosities:"Cores" type

Nebula	IRAS	Near-Ir	Id.*
PP9	03245+3002	HK	Lynds 1455
GY10	04073+3800	JHKL	Lynds 1473
GY14	05173-0555	-	H-H,RNO40
PP28	05280+3408	JHK	RNO41
PP33	05338-0624	JHKL	RNO47
PP36	05355+3039	JHKL	AFGL5158
PP39	05369-0728	-	-
PP40	05375+3540	JHKL	S235,RNO52
GY18	05439+3035	JHKL	-
PP50	05598-0906	JHK	-
PP51	06010-0943	-	RNO61
PP52	06013+3030	JHKL	CED061
PP84	16316-1540	-	Lynds0043
PP88	19266+0932	JHK	-
PP89	19403+2258	JHKL	NGC6820
PP90	20193+3700	JHKL	-
PP93	20222+3541	-	-
PP101	21407+5441	H	Lynds1084
PP106	23561+6609	JHKL	LH 259

* RNO,Cohen (1980)

3. NEAR INFRARED

During two different observing runs at the 1.5 μm Italian Infrared Telescope (TIRGO) on November 1986 and January 1987, 36 peculiar nebulosities associated with IRAS sources were photometrically observed in JHK and sometimes L broad-bands.

We have analyzed our photometric data by studying the conventional J-H vs. H-K and H-K vs. K-L color-color diagrams (Figure 3). These diagrams are effective in separating stars with IR excesses due to thermal dust or free-free emission, from reddened stars.

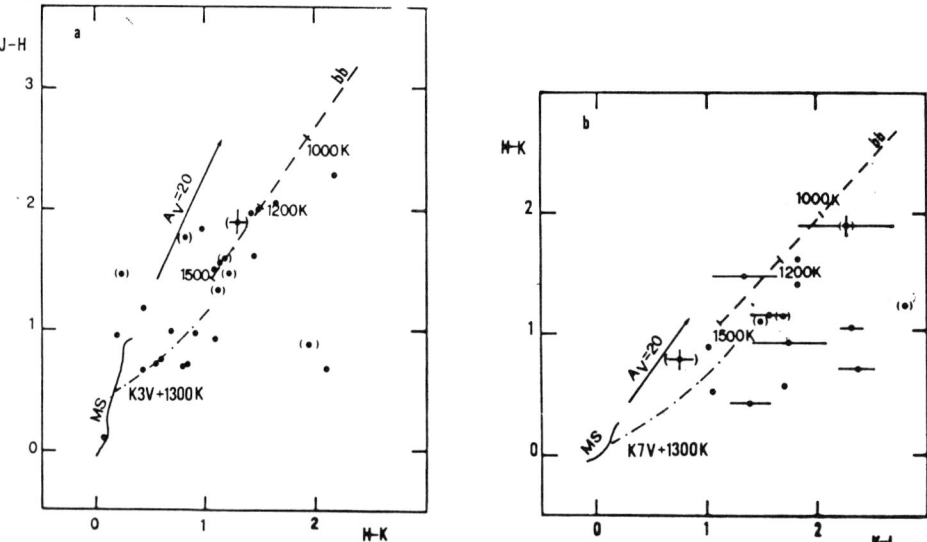

Figure 3. (a) J-H vs. H-K two-color diagram. (b) H-K vs. K-L two-color diagram. Symbols as in previous figures.

Most of the sources reported in the J-H vs. H-K diagram have colors similar to those of visible T-Tauri stars. Their IR excess is explained by the presence of a circumstellar shell dust at T = 1300 K surrounding a late K star (Rydgren et al., 1982). Sources with IRAS colors similar to "cores" show higher IR excesses than visible T-Tauri stars. These sources, in agreement with Beichman et al. (1986), could correspond to either pre-main sequence stars embedded in their parental clouds, or accreting protostars. Their $1-100\mu m$ energy distributions, that will be presented in a forthcoming paper, are of Class I according to the classification given for young stellar objects by Lada (1987). Class I objects may be protostars.

In summary, 52 IRAS sources were found associated with peculiar nebulosities: 21 sources have beeen classified as T-Tauri stars, while 19 sources show emission similar to IR sources in dense molecular clouds. The remaining sources have anomalous IRAS and near-IR colors, that make their identification difficult.

REFERENCES

Beichman, C.A., Neugebauer, G., Habing, H.J., Clegg, P.E., Chester, T.J., 1985, *Explanatory Supplement to the IRAS Catalogs and Atlases* (Washington:

GPO).

Beichman, C.A., Myers, P.C., Emerson, J.P., Harris, S., Mathieu, R., Benson, P.J., Jennings, R.E., 1986, *Astrophys. J.*, **307**, 337.

Gyulbudaghian, A.L., 1982, *Pis'ma Astron. Zh.*, **8**, 232.

Lada, C.J., 1987, *Star Forming Regions*, M. Peimbert, J. Jugaku eds, Reidel Publ. Co., p. 1.

Parsamian, E.S., Petrosian, V.M., 1979, *Soobshenia Biurakanskai Observatori*, Akad. Nauk. Armianskoi S.S.R., No. **51**.

Rydgren, A.E., Schmeltz, J.T., Vrba, F.J., 1982, *Astrophys. J.*, **256**,168.

Torrelles, J.M., Rodriguez, L.F., Cantò, J., Marcaide, J., Gyulbudaghian, A.L., 1983, *Rev. Mexicana Astron. y Astrof.*, **8**,147.

RADIO MAPPING OF TYPE I POST-MAIN-SEQUENCE NEBULAE

J.P. Phillips[1] and A. Mampaso[2]

[1]Physics Department
Queen Mary College
London E1 4NS

[2]Instituto de Astrofisica de
Canarias
La Laguna
Tenerife

Type I post-main-sequence nebulae appear to be characterised by a range of unusual, and in certain cases extreme physical characteristics, including high velocity (and possibly high mass) outflows, and anomalous He, N, and C abundances; a feature which is almost certainly indicative of high mass nebular projenitors (cf. Renzini and Voli, 1981). The nature of the flows in these sources is not fully understood, although our own extensive NIR mapping and optical spectroscopy reveals that many contain extremely compact ionised cores, wherein high excitation winds extend to velocities of $\sim 10^3$ km.sec^{-1} (and in certain cases very much greater). The primary nebular mass appears to expand at very much lower velocities, although even here it is clear that dense, mass condensations are being accelerated to velocities of order $\sim 10^2$ km.sec^{-1} or greater; an order of magnitude larger than characterising the broad range of PMS nebulae. These condensations appear to display a lower range of excitations than the high velocity flows, perhaps reflecting the inclusion of dense HI condensations (and corresponding UV shadow zones), and dominate the lower excitation nebular structures [OI], [NII] and [OII]. The result, as noted in figure 1 for M2-9, is that [NII] and Hα line profiles differ appreciably. Infrared spectroscopy of these nebulae also reveals that the compact NIR cores are in certain cases associated with H_2 S(1) V = 1-0 λ2.1 μm quadrupole emission (Phillips et al 1983, 1985); presumably arising from the transmission of shock and ionisation fronts into similarly compact, enveloping HI cores, or reflecting the prevalence of radiation pumping mechanisms in the presence of hard U.V. radiation fields.

In either case, it seems likely that the cores of many of these structures contain dense and, in all probability, massive HI shells, responsible for HI absorption in NGC 6302 (Rodriguez et al, 1985), the confinement of extremely restricted, high density plasmas, and the subsequent collimation of this material to form high velocity bipolar winds. With this in mind, therefore, we have acquired high resolution VLA mapping of five type I nebulae, with the aim of investigating the inner core structures of these sources. The results were taken in snapshot mode at λ 2, 6 and 20 cm, and with the VLA in A and C configurations, whilst nearby calibrators were bootstrapped to 3C286, resulting in fluxes accurate to within 10%. Certain of the results are briefly summarised in figure 2, and described below:

HB5

This unusual and irregular optical structure appears extremely bright in the FIR (Phillips and Mampaso, 1987), and shows evidence for two jets emerging at high velocity from the central core (Phillips et al; in preparation). Our 6 cm map of this source reveals a somewhat complex structure, with a predominantly box-shaped configuration framing the central star location. The brightest regions of the nebula, north and south of the core source, almost certainly delineate the inner edges of a collimating toroid, whilst lower resolution 2 cm and 20 cm images reveal the inner segments of the E-W gas outflow. The emission measure and density of the core take the respective mean values $\sim 2.10^7$ pc.cm^{-6} and $4.8.10^4$ cm^{-3}, although these figures increase to $\sim 3.3.10^7$ pc.cm^{-6} and $8.6.10^4$ cm^{-3} for the more northerly bright condensation. The source appears to be optically thick at 20 cm, and possesses a continuum turn-over frequency close to 2.0 GHz, whilst up to 86% of the flux is found to be contained in the core.

NGC 6537

Another irregular, bipolar outflow source, with an exceptionally strong and dominating NIR core of size ~ 6 arcsecs. Our high resolution spectroscopy on the INT reveals the presence of substantial clumps of high velocity material near the core - clumps which are not present within the NII source structure. The 6 cm radio map indicates a complex and tortuous arrangement of filaments, with local densities approaching 10^5 cm^{-3} in the primary bright filaments, and emission measures of order $9.4.10^7$ pc.cm^{-6}. The turn-over frequency takes a value ~ 2.6 GHz, and the ionising flux is $\sim 3.2.10^{47}$ photons. sec^{-1}, yielding a central star temperature $3.0.10^5$ K for an (extinction corrected) visual magnitude $M_V = 15.1$ mags. Up to 96% of the radio flux is contained in the core.

NGC 6741

We show, in figure 2, the 2 cm map of this source. This appears to represent a classical bipolar structure with size ~ 0.1 pc. The core emission measure is of order $3.1.10^6$ pc.cm^{-6}, implying a high mean density 8.10^3 cm^{-3}.

M3-28

This is a relatively weak radio continuum source, with a 6 cm core

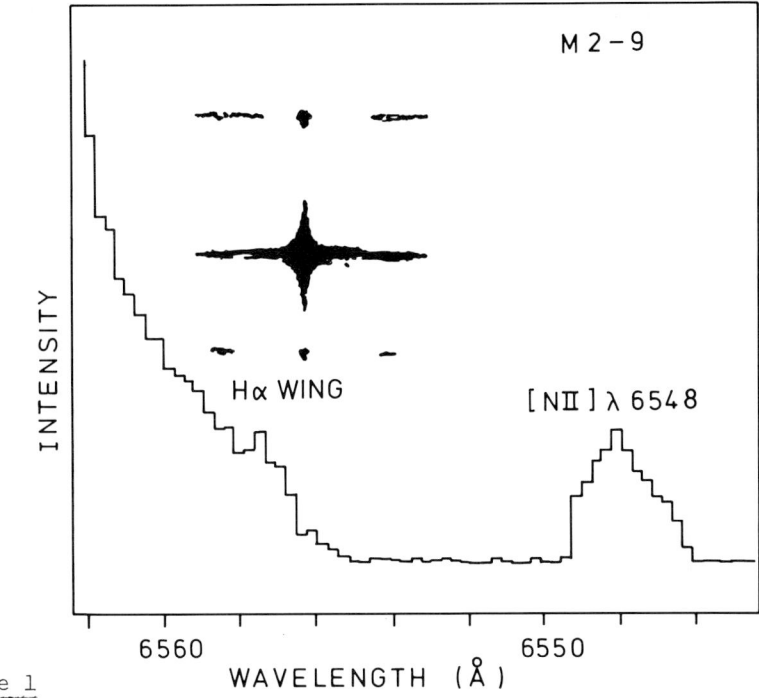

Figure 1

INT spectroscopy of M2-9, showing Hα wing and [NII] λ 6548. The superimposed image shows the Hα λ 6654 + [NII] λλ 6548/6583 lines as recorded by the CCD array.

flux of .04 Jy; the emission appears optically thin at $\lambda \sim 20$ cm, with a mean emission measure near $\sim 2.4.10^6$ cm^{-3}, and a density $6.6.10^3$ cm^{-3}. The structure is again evidently bipolar, a trend which is also replicated in the lower resolution 2 and 20 cm maps.

NGC 6445

This is the most extended of the sources observed here, and our map at 20 cm reveals a rather irregular bipolar structure, with mean emission measure $1.5.10^5$ pc.cm^{-6}, and density $\sim 8.2.10^2$ cm^{-3}. These figures increase to $9.2.10^5$ pc.cm^{-6} and $3.8.10^3$ cm^{-3} for the bright outer lobes alone. The source appears to be optically thin out to λ 20 cm, and we estimate a central star temperature $3.5.10^5$ K.

An interesting feature of this source is that, unlike most other comparable bipolar configurations, the brightest nebulosities are located at the extremities of the major axis; a trend which may indicate an intrinisically prolate structure, or alternatively, a toroidal formation viewed from the side, the remnants (perhaps) of the inner collimating toroid, now fully ionised, expanded and of size $\sim .45$ pc.

346

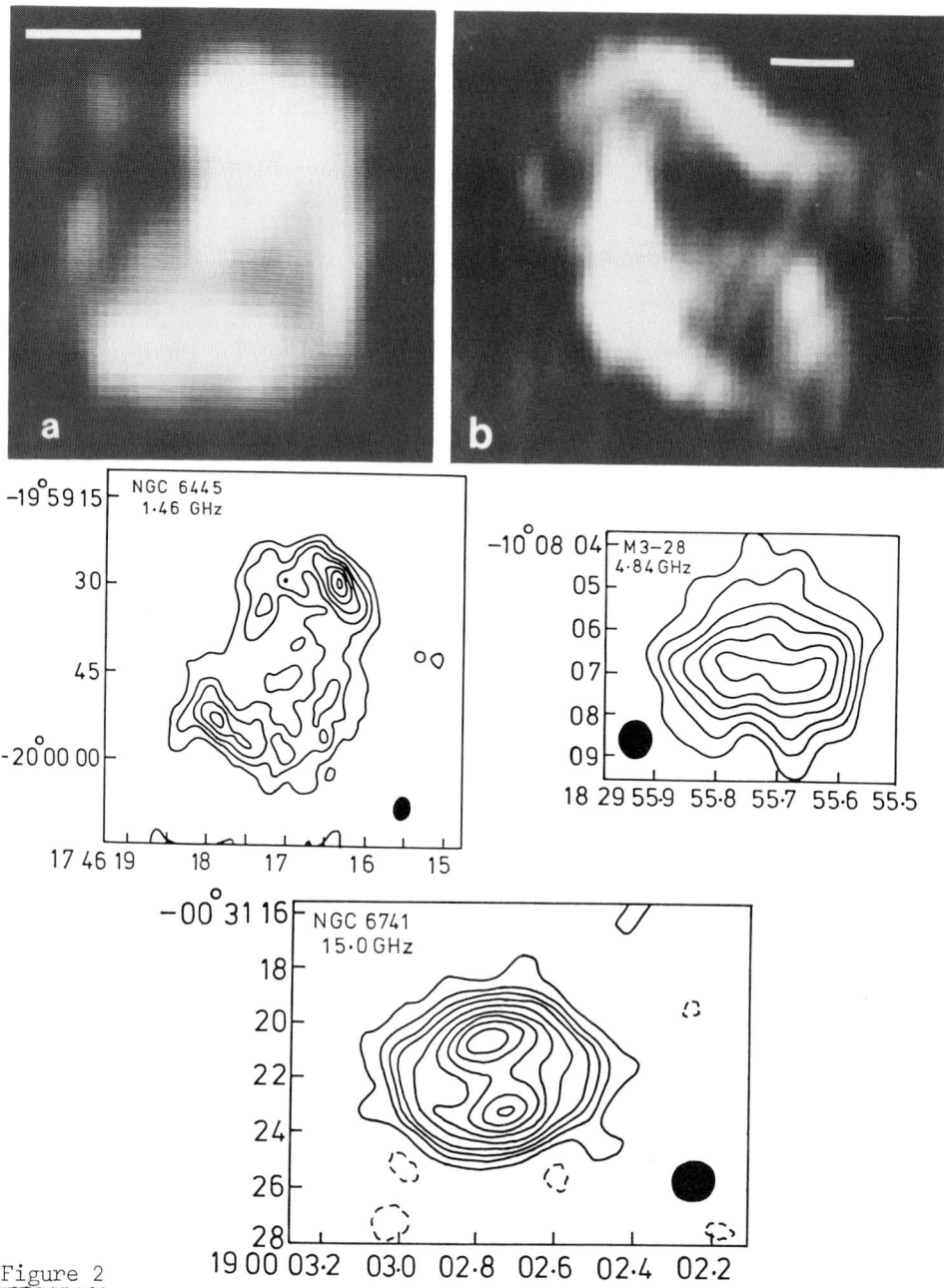

Figure 2

VLA maps of five type I nebulae. The top left scale image
corresponds to NGC 6537 at 6 cm, and the top right image corresponds
to HB5 at 6 cm; in both cases, the horizontal bar indicates 1 arcsec.

REFERENCES

Phillips, J.P., Reay, N.K. and White, G.J., 1983 *Mon. Not. Roy. Astron. Soc* 203, 977.

Phillips, J.P., White, G.J. and Harten, R., 1985. *Astron. Astrophys.* 145, 118.

Phillips, J.P. and Mampaso, A., 1987. Submitted for publication.

Renzini, A. and Voli, M., 1981. *Astron. Astrophys.* 94, 175.

J = 3-2 AND J = 2-1 CO MAPPING OF HIGH VELOCITY OUTFLOW SOURCES

J. P. Phillips and Glenn J. White
Physics Department
Queen Mary College
Mile End Road
London E1 4NS

High velocity molecular outflow sources have been studied extensively over the last decade or so, although it must be confessed that there is, as yet, relatively little direct knowledge concerning the mechanisms of mass acceleration, or the processes leading to outflow collimation, and the formation of distinctively bipolar source morphologies. In terms of the outflow flux of mechanical energy, for instance, it is clear that the transference of a relatively modest fraction (.002 - .02) of the radiant flux is sufficient to power many of these zones, although the corresponding outflow momenta are less readily explained; the typical observed values $\sim 10^{-3}$ $M_\odot.yr.^{-1}$ being some $\sim 10 - 10^3$ times greater than the momentum flux L/c carried by the outward radiation field. One way out of this problem was addressed by Phillips and Beckman (1980), whereby multiple scattering within an optically thick dusty thermosphere may lead to high momentum coupling to the outflowing mass, and material outflow momenta >> L/c. Such a process would not, however, explain the high momentum outputs of certain low optical depth sources (such as T Tauri and MWC 1080).

In an attempt to gain a clearer understanding of the kinematics and structures of a broad sample of these sources, we have therefore undertaken an extensive programme of J = 2-1 and J = 3-2 CO mapping of K3-50, CRL 2591, W3, S88, NGC 2264, NGC 2024, S140, G35.2 - 0.74, NGC 1333 (HH7-11) and NGC 1333 (IRAS 1). The results were taken with the CFHT and UKIRT facilities at Mauna Kea, Hawaii, and with the typical resolutions 57 arcsecs at J = 3-2, and 85 arcsecs at J = 2-1. We provide, below, a brief summary of some of the more important conclusions, together with a reprise of certain mapping and spectral data.

1. One of the more surprising aspects of our results, and a feature which we do not, as yet, fully comprehend, is a tendency for the J = 3-2 spectral line data to be *considerably* weaker than either the J = 2-1 or J = 1-0 peak radiation temperatures T_R. A summary of these ratios is provided in figure 1a, whilst a typical set of J = 2-1, J = 1-0 and J = 3-2 spectra is indicated in figure 1b for S140 IRS. In figure 1a, we have also drawn a curve indicating the maximum value of $\Phi_{2/3} = T_R(J = 2-1)/T_R(J = 3-2)$ as a function of $T_R(J = 2-1)$, the

J = 2-1 radiation temperature; a locus which is essentially invariant with kinetic temperature T_k for the homogeneous LVG modelling adopted here. It is clear, from this, that at least 3 of the sources observed at J = 3-2 are not comprehensible in terms of standard LVG modelling, nor do we find it possible to explain this trend in terms of source clumping (frequency invariant or otherwise), non-linear (Sobolevian) velocity gradient models, self-absorption, infrared line pumping and so forth. It may be necessary, therefore, to invoke a non-Sobolevian model for the line cores, in which local line excitation is a function of the overall source mass, kinematic, and thermal structures.

2. A further interesting characteristic of these sources is a tendency to extremely low values of X(CO)/dv/dr, and correspondingly large densities $n(H_2)$ for the outflow wings; a trend clearly demonstrated in figure 2, where it is seen that the LVG solutions follow the approximate relation:

$$\frac{dv}{dr} \underset{\sim}{\sim} 2.5 \cdot 10^{-4} \, n^{3/2}(H_2) \text{ km.sec}^{-1}.\text{pc}^{-1}$$

$$(\text{for } X(CO) = 5 \cdot 10^{-5})$$

Whilst clumping may again somewhat alter these solutions, we find very approximately that the ratio X(CO)/dv/dr is proportional to the reciprocal beam dilution W^{-1}. Even therefore for a value $W \sim 0.2$, say, corresponding to an appreciable level of clumping, it is unlikely that the high density solutions will fall remotely near the stability/free-fall line in figure 2, corresponding to sources which are undergoing either free-fall collapse, or are in a state bordering on turbulent, thermal, or rotational stability.

This, of course, is hardly surprising, considering the clear evidence of high velocity outflow in these sources. We also find that the velocity range in the source wings ΔV, divided by the observed widths of the outflow lobes Δr, gives a maximum gradient which is very close to the value dv/dr in the low density solutions of figure 2 (supposing a relative CO abundance $X(CO) \sim 5.10^{-5}$); a correspondence which suggests a reasonably smooth and continuous variation of velocity throughout these zones. Rather more surprising, however, are the disparities between $\Delta V/\Delta r$ determined from the observed source and line structures, and the values of X(CO)/dv/dr in the regime of high density solutions ($n(H_2) \gtrsim 10^4$ cm^{-3}); a difference which is most readily understood in terms of extremely high velocity gradients dv/dr ≈ 5.10^2 km.sec^{-1}, and an emission zone width Δr_e which is very much less than the observed dimensions of the outflow lobes.

We believe, in short, that the present results provide clear evidence for shock refraction and compression at the edges of the outflow cavities, of kind noted by Uchida (1987), and investigated theoretically by Uchida and Shibata (1984, 1985). An interesting characteristic of this plot is that several of the sources are observed to contain both low and high density solutions, corresponding to either the red or blue wings - a trend which may suggest differing degrees of shock deceleration and collimation in the immediate source environment.

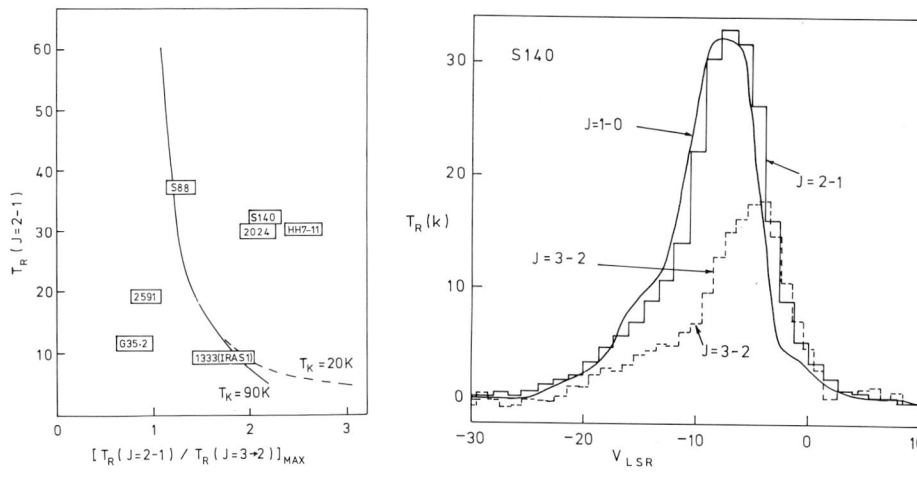

Figure 1

a) Observed values of T_R (J = 2-1) and $\Phi_{2/3} = T_R$ (J = 2-1)/ T_R (J = 3-2), together with a curve indicating the maximum value of $\Phi_{2/3}$ for homogenous LVG models.

b) CO J = 1-0, J = 2-1 and J = 3-2 spectra for S140.

3. Three of the high velocity sources (S88, NGC 1333 (IRAS 1) and NGC 2264) are revealed, for the first time, to possess bipolar outflow structures, whilst the bipolarity in one further source (NGC 2024) is confirmed at J = 3-2 CO. Two of the outflow maps are illustrated in figure 3, where we see that S88 appears to possess a very restricted red wing, size $\sim$ 1 arcminute, and a very much more broadly distributed blue wing. The full extent of the outflow in NGC 1333 (IRAS 1) is not fully determined in our present mapping, although it is clear that the large ratio T_R (J = 2-1)/T_R(J = 3-2) for the line core is in this case attributable to low gas kinetic temperatures $T_K \sim 15K$. The final nebula for which we have discovered a bipolar flow is NGC 2264, a source which is apparently located at the edge of a dense rotating toroid. Schwartz et al (1985) attributed broad velocity wings in this region to the formation of a blister zone, although our present results appear, on the contrary, to indicate the presence of directed red and blue shifted velocity wings - presumably indicative of interaction between collimated stellar outflows and the ambient gas.

4. A conformity is noted between the LVG wing solutions of S88 and CRL 2591, and the apparent morphology of these sources on the sky. In particular, the less confined, and apparently more broadly distributed jets appear to be associated with low density solutions, whilst the very much more restricted (and perhaps retarded) outflows appear

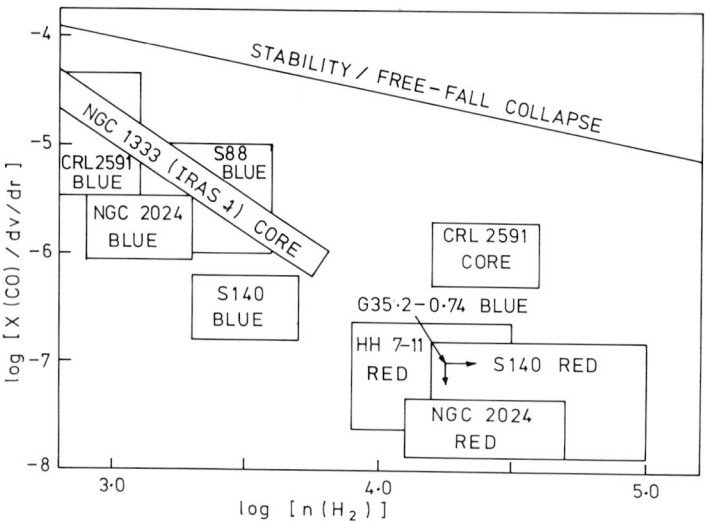

Figure 2

LVG solutions for the line wings and cores of high velocity outflow sources.

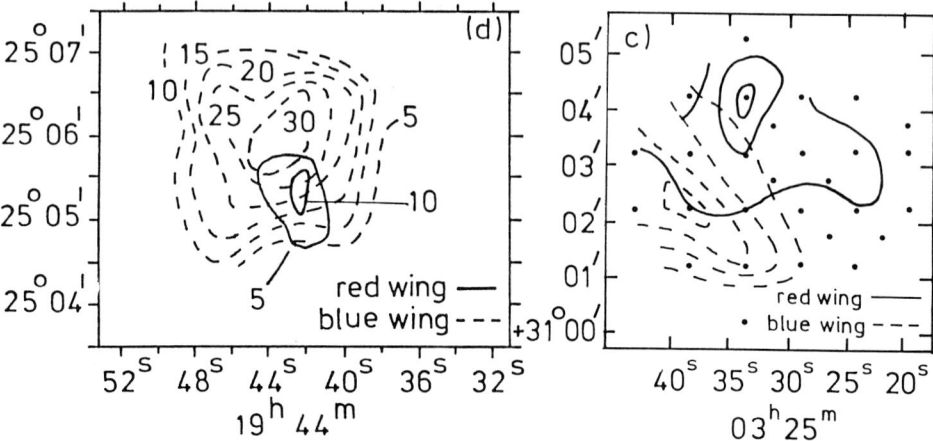

Figure 3

Distributions of integrated wing emission in a) S88, and b) NGC 1333 (IRAS 1).

to correspond to high density environments. It is not unlikely, therefore, that the observed morphologies of these sources are very strongly determined by interaction with the ambient clouds - and that the nature of the interaction may differ appreciably between differing sources and, within the same source, between the opposing jets.

5. Finally, we note that a large proportion of the sources appear to have co-spatial red and blue flows (cf. S140, G35.2 - 0.74, and K3-50), a feature which may be attributed to a jet inclination angle $i \sim \pi/2$ (i.e. perpendicular to the line of sight), and/or large jet opening angles θ, resulting in appreciable transverse jet expansion along the line of sight. Several of the sources (e.g. S140, CRL 2591, and probably G35.2 - 0.74) are also found to possess significantly weaker counter jets, leading to the appearance of monopolar flows. The nature of these structures is not entirely clear, although it is possible that we are again witnessing interaction with a non-symmetric distribution of ambient material, or star formation in clouds which contain appreciable density gradients; one jet entering the denser material, and building up high post shock densities, the other jet entering low density cloud material (or even exiting the cloud), resulting in a very much reduced interaction with surrounding material, and low antennae temperatures T_R^*.

REFERENCES

Phillips, J.P., and Beckman, J.E., 1980. *Mon. Not. R. Astron. Soc.* 193, 245.

Schwartz, P.R., Thronson, H.A., Odenwald, S.F., Glaccum, W., Lowenstein, R.F., and Wolf. G., 1985. *Astrophys. J.* 292, 231.

Uchida, Y., 1987. *Proc. of I.A.U. Symp. 115, D. Reidel Publishing Co., Dordrecht, Holland.*

Uchida, Y., and Shibata, K., 1984. *Publ. Astron. Soc. Japan* 36, 105.

Uchida, Y., and Shibata, K., 1985. *Publ. Astron. Soc. Japan* 37, 515.

STELLAR MASS LOSS BY TURBULENT ALFVÉN WAVES

Reuven Opher and Vera J.S. Pereira
Instituto Astronômico e Geofísico - USP
Caixa Postal 30627
01051 São Paulo SP
Brasil

ABSTRACT. In the present study, we analyze the effect of the change in the opening angle of the magnetic field on mass loss and terminal velocity of the "typical" K5 supergiant (Model 6 of Hartmann and MacGregor (1980)). We use a flux of turbulent Alfvén waves as the acceleration mechanism of the wind, with non-linear and surface Alfvén wave absorption. We show for a range of initial damping lengths $L_0 \sim 0.1 - 0.2\ r_0$ and an isothermal atmosphere that a divergent geometry can produce the observed large mass loss rates and the small ratios u_∞/v_{eo} ($\sim 1/2$) of the supergiants, where u_∞ is the terminal velocity and v_{eo} is the initial escape velocity.

1. INTRODUCTION

Several observations indicate the presence of stellar mass loss in almost all regions of the HR diagram. For late-type giant and supergiant stars, the observations are interpreted as indicating the presence of cool and massive winds (Reimers 1975; Goldberg 1979; Hagen 1980; Castor 1981; Hartmann 1981; Cassinelli and MacGregor 1982), with terminal velocities (u_∞) lower than the surface escape velocity (v_{eo}). In order to explain the observed $u_\infty < v_{eo}$ several mechanisms of wind acceleration have been proposed, and one of the most promising involves the mass loss by an outward-directed flux of Alfvén waves.

2. THE MODEL

We examine the effect of the change in the opening angle of the magnetic field on (i) the mass loss, and (ii) terminal velocity of the "typical" K5 supergiant star studied by Holzer, Flå and Leer (1983) (which is the model 6 of Hartmann and MacGregor (1980)). The geometry studied is shown in Figure 1.

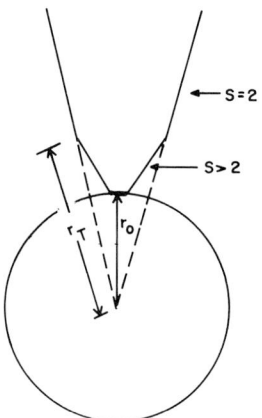

Figure 1. Diverging Geometry Model

We use the diverging geometry $A(r) = A(r_o) (r/r_o)^S$ suggested by Kuin and Hearn (1982) and Parker (1963) up to the radius r_T, where $A(r)$ is the cross sectional area of the geometry at a radial distance r, r_o is the initial radius and S is a parameter that determines the divergence of the geometry. The radius r_T is defined by the relation $(r_T/r_o)^{S-2} = A(r_T)/A_S(r_T) \equiv F$, where $A_S(r) \equiv A_S(r_o)(r/r_o)^2$. In the present study we use $F = 10$. Initially we have a rapidly diverging geometry with $S > 2$ for $r < r_T$. The geometry is radial for $r > r_T$.

We use a flux of turbulent Alfvén waves as the acceleration mechanism of the wind. We assume that the dominant mechanism for the absorption of the Alfvén waves is: 1) nonlinear absorption with damping length (e.g. Lagage and Cesarsky 1983) given by

$$L_1 = L_{10} \left(\frac{V_A}{V_{AO}}\right)^4 \frac{<\delta v^2>_o}{<\delta v^2>} \qquad (1)$$

where: $\rho_o <\delta v^2>_o$ is the initial energy density of the Alfvén waves; V_A is the Alfvén velocity; and L_{10} is the initial damping length; and 2) resonance absorption of surface Alfvén waves with damping length (Lee and Roberts 1986) given by

$$L_2 = L_{20} \left(\frac{r_o}{r}\right)^{S/2} \left(\frac{V_A}{V_{AO}}\right)^2 \qquad (2)$$

where L_{20} is the initial damping length.

The origin of turbulent Alfvén waves was suggested by Opher and Pereira (1986) to be the annhilation of twisted magnetic fields near the stellar surface. We suggest initial diverging geometries having

$S \gtrsim 10$, which do not allow mass loss for L_{10}, $L_{20} \sim 0.1 - 0.2\ r_o$. The pressure of the outflowing matter reduces S until mass flow begins.

We solved the equation of motion (eq. 52, Holzer, Flå and Leer 1983) for an isothermal atmosphere with $T = 10^4$ K to obtain the dependence of the ratio u/v_{eo} as a function of r, S, L_{10} and L_{20}, using the parameters given by the model 6 of Hartmann and MacGregor (1980). The mass loss $\dot{M}$ is defined as $4\pi r_o^2 u_o \rho_o$, with $\dot{M}_6 \equiv \dot{M}/10^{-6}\ M_\odot/yr$.

3. RESULTS

In Figure 2 we show the dependence of u/v_{eo} as a function of r, for $L_{10} = 0.1\ r_o$ (S = 4, 4.5) and $L_{20} = 0.2\ r_o$ (S = 5, 5.5). For all the curves we have $\dot{M}_6 > 1$.

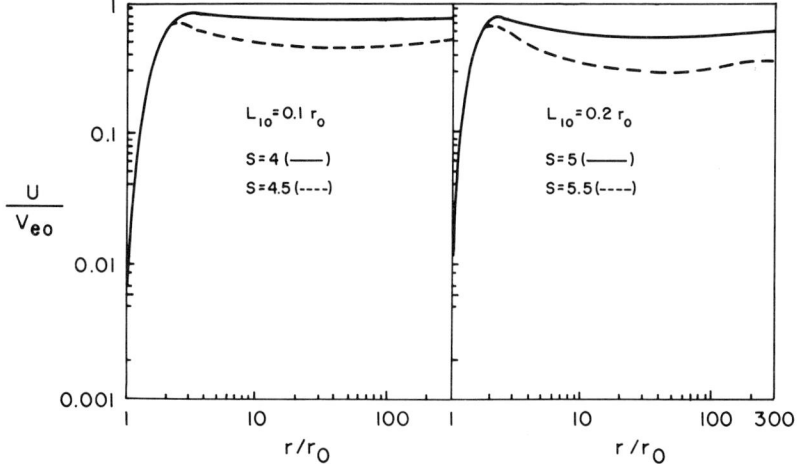

Figure 2. Radial profiles of the flow speed as a function of the distance from the star for $L_{10} = 0.1\ r_o$ (S = 4, 4.5) and $L_{20} = 0.2\ r_o$ (S = 5, 5.5).

The curves are characterized by a rapid rise of u/v_{eo} in the region $1 < r/r_o \lesssim 2$. For a given L_{10}, L_{20}, we find that u/v_{eo} after reaching a peak at $r \sim 2\ r_o$, appreciably decreases for larger r for the higher S. The higher S values are also characterized by peak values at smaller r/r_o. The value u_∞ is defined as the value of u at $r = 300\ r_o$.

4. CONCLUSIONS

Our results indicate that diverging geometries $S \sim 4 - 5.5$ and initial damping lengths $L_0 \sim 0.1 - 0.2\ r_o$ for nonlinear and resonance surface wave absorption can produce the observed large mass loss rates and small terminal velocities of cold supergiants.

ACKNOWLEDGEMENTS

R.O. would like to thank the Brazilian Agency CNPq for partial support.

REFERENCES

Cassinelli, J.P., and MacGregor, K.B. 1982, In Physics of the Sun, ed. P.A. Sturroch, T.E. Holzer, D. Mihalas, and R.K. Ulrich (Chicago: University of Chicago Press).
Castor, J.J. 1981, In Physical Processes in Red Giants, ed. I. Iben, Jr. and A. Renzini (Dordrecht: Reidel).
Goldberg, L. 1979, Quart. J.R.A.S., 20, 361.
Hagen, W. 1980, In Cool Stars, Stellar Systems and the Sun, ed. A.K. Dupree (Cambridge: Smithsonian Astrophysical Observatory).
Hartmann, L. 1981, In Solar Phenomena in Stars and Stellar Systems, ed. R.M. Bonnet and A.K. Dupree (Dordrecht: Reidel).
Hartmann, L., and MacGregor, K.B. 1980, Ap. J. 242, 260.
Holzer, T.E., Fla, T., and Leer, E. 1983, Ap. J. 275, 808.
Kuin, N.P.M., and Hearn, A.G. 1982, Astron. Ap. 114, 303.
Lagage, P.O., and Cesarsky, C.J. 1983, Astron. Ap. 125, 249.
Lee, M.A., and Roberts, B. 1986, Ap. J. 301, 430.
Opher, R., and Pereira, V.J.S. 1986, Astrophysical Letters 25, 107.
Parker, E.N. 1963, In Interplanetary Dynamical Processes, ed. Wiley, New York.
Reimers, D. 1975, In Problems in Stellar Atmospheres and Envelopes, ed. B. Bascheck, W.H. Kegel, and G. Traring (Berlin: Springer).

MASS LOSS FROM THE PROGENITOR OF KEPLER'S SUPERNOVA:
CHARACTERISTICS AND EVOLUTIONARY CONSTRAINTS

R. Bandiera
Osservatorio Astrofisico di Arcetri
Largo E. Fermi 5
50125 Firenze
Italy

ABSTRACT. A scenario for the origin of Kepler's supernova remnant is presented. According to this scenario the progenitor was a massive runaway star, subject to strong mass loss. Due to the interaction of the wind with the interstellar medium, a bow shock has formed in the direction of the stellar motion. Using the available data on the remnant, various constraints on the progenitor's wind can be set. A search has been carried out, among classes of known runaway objects to determine the nature of the progenitor. Similarities have been found with runaway Wolf-Rayet stars; some of them are also surrounded by nebulae, similar to that surrounding Kepler's supernova progenitor. Also these nebulae, commonly interpreted as wind blown bubbles, could actually consist of matter collected on a bow shock.

1. KEPLER'S SUPERNOVA REMNANT

1.1. Morphology of the remnant

The structure of Kepler's SNR is very asymmetric both in radio and in X-ray (Matsui et al. 1984): the northern limb is brighter than the rest of the remnant, while the southern side is blurred. At optical wavelengths the remnant appears very clumpy: the bulk of the emission comes from compact knots. Most of the optical knots lie on the northern limb; other knots are confined to the central region.

1.2. Kinematics of the optical knots

Van den Bergh and Kamper (1977) present astrometric and spectroscopic measurements of the motions of knots. The data do not show the radial expansion expected for stellar ejecta (nearly 6000 km/s), but are consistent with no expansion (or, at most, with an expansion not faster than 200 km/s). The velocities of knots look like random motions added to a common translation: its longitudinal component, determined spectroscopically, is -230 km/s; while its transverse component, derived from an astrometric velocity of 0.0115 arcsec/yr, is 250 km/s (for a distance of 4.5 kpc; see Section 2.3): then the total velocity of the knots system is nearly 340 km/s.

1.3. The origin of the knots

The knots cannot be stellar ejecta, because they expand too slowly. They cannot even be interstellar matter, because their high density and clumpiness is inconsistent with their large distance (530 pc) from the galactic plane. Therefore, they must be composed of circumstellar matter.

2. A MODEL FOR THE REMNANT

2.1. A stellar comet

If the knots have a circumstellar origin, their average motion should reflect the motion of the SN progenitor: consequently it is a runaway star. The stellar wind is eventually deflected, due to the interaction with the interstellar medium, and a bow shock appears: matter gets denser near the bow shock apex, which points to the direction of the stellar motion.

If the star evolved to SN, information on the circumstellar density distribution can be extracted from the passage of the blast wave: the denser the region crossed, the brighter the shock front becomes. This view is confirmed, in Kepler's SNR, by the fact that the brighter limb is in the direction of the average knots motion.

2.2. Constraints on the progenitor's wind

The strong asymmetry observed in the emission from Kepler's SNR indicates that the blast wave is presently moving through the bow shock apex. Therefore the bow shock size is nearly equal to the blast wave radius, namely 2.2 pc.

At the bow shock apex the stellar wind ram pressure is balanced by the interstellar ram pressure. If the interstellar density outside the galactic plane is $10^{-2.5}$ cm^{-3} (McKee and Ostriker 1977), one derives $\dot{M}w = 5.4 \cdot 10^{-4}$ ($M_\odot$/yr)(km/s).

The absence, in X-ray, of a sharp limb on the southern side (where the density is likely to follow an r^{-2} profile), brings one to the conclusion that the blast wave is in Sedov phase. Assuming that the SN released 10^{51} erg, one derives $\dot{M}/w = 4.5 \cdot 10^{-6}$ ($M_\odot$/yr)/(km/s).

Therefore the wind parameters can be separated as $\dot{M} = 5 \cdot 10^{-5}$ $M_\odot$/yr and $w = 10$ km/s (Bandiera 1987). To form a bow shock the wind should have lasted for more than $2 \cdot 10^5$ yrs; therefore more than 10 $M_\odot$ have been lost: this implies a rather massive progenitor.

2.3. The distance of Kepler's supernova remnant

Density peaks, namely knots, are confined to the bow shock; this can be idealized as a thin, pseudo-paraboloidal layer. Optical emission comes only from knots that interact with the blast wave, idealized as a spherical layer. Those knots lie at the intersection of the two surfaces: this intersection has the shape of an annulus. An observer will then see an ellipse, whose axis ratio gives the cosine of the angle between the bow shock direction and that of the observer: this is the ratio between longitudinal and total velocity. Since the astrometric velocity is known, one can derive the distance of the object. An appropriate fit of the ellipse to the knots pattern can be carried out, using a model for the bow shock (Huang and Weigert 1982) and the constraints derived from the velocity measurements. From the fitted ellipse (Fig.1), a distance of 4.5 kpc is derived.

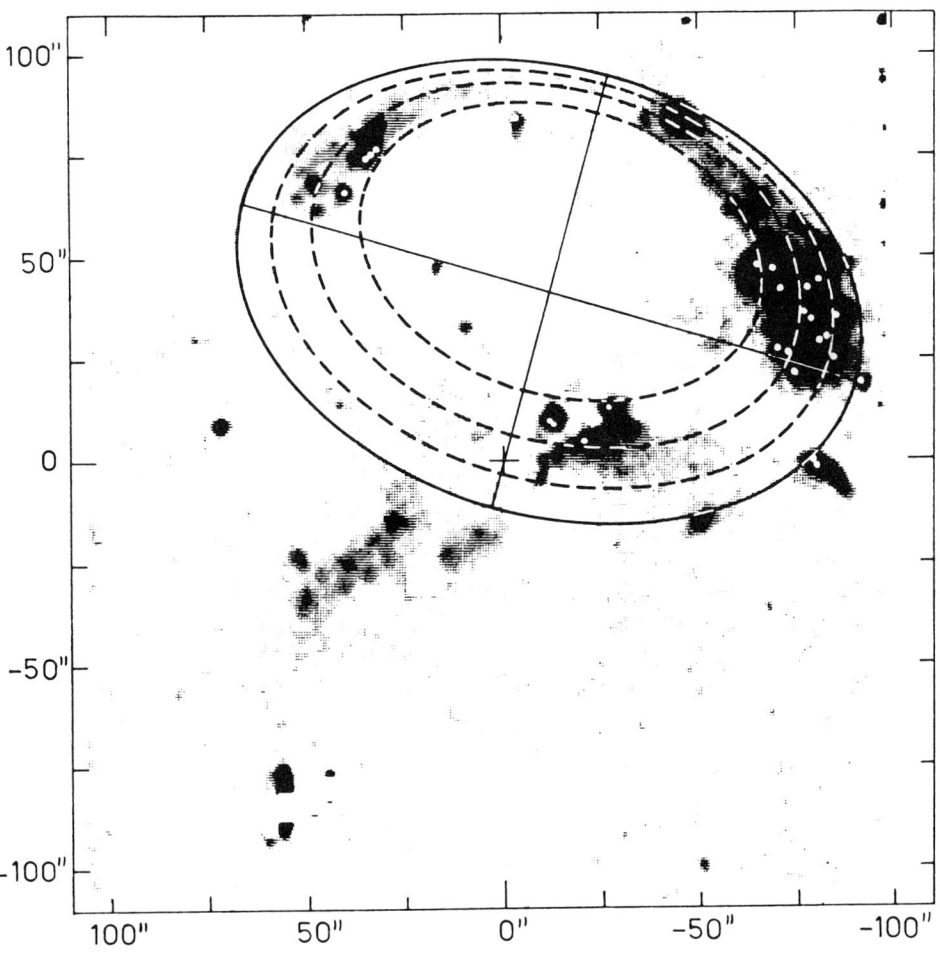

Figure 1. Overlay of the best fitted ellipses to a deep optical image of the knots pattern (d'Odorico et al. 1986); the solid line ellipse gives the present location of the annulus, while the dashed line ellipses give its position 20, 40 and 60 years ago.

3. KEPLER'S SUPERNOVA PROGENITOR

3.1. Constraints on the progenitor

The progenitor was a massive, young population star with strong mass loss; its high velocity suggests a runaway nature. Astrometric measurements confirm that it left the galactic plane $3 \cdot 10^6$ yrs ago: during this time it evolved into a SN.
  Since Kepler's SN exploded only 380 yrs ago, its progenitor should not be too exotic: objects like this must exist in our Galaxy. They have the following characteristics:
i) runaway velocity;
ii) high mass;
iii) strong mass loss;
iv) possibly an associated nebula, made of circumstellar matter.

3.2. Was the progenitor a runaway Wolf-Rayet star?

Runaway Wolf-Rayet stars satisfy all these requirements. They are far from the galactic plane, and possess velocities up to 200 km/s. They represent a phase in the evolution of massive close binary systems (see e.g. van den Heuvel 1976). After the explosion of the primary star, a runaway system is formed: if it does not disrupt, it will consist of a massive star with a compact companion. Many runaway Wolf-Rayet stars have been reported to be spectroscopic binaries, with an unseen companion. Possibly a runaway Wolf-Rayet star produced Kepler's SN.
  However, the wind responsible for the formation of knots has the same parameters as that of a typical red supergiant wind, rather than a Wolf-Rayet wind. It is possible that knots have been formed before the star evolved into a Wolf-Rayet. After that, the Wolf-Rayet wind leaked through the knots, without appreciably accelerating them.

3.3. Nebulae surrounding runaway Wolf-Rayet stars

Some Wolf-Rayet stars are associated with nebulae, for which the generic name of "ring nebulae" is used. However this class is not homogeneous; Chu (1981) introduced a further subdivision into: radiatively excited amorphous HII regions; radiatively excited shell-structured HII regions; stellar ejecta; wind blown bubble.
  If a runaway Wolf-Rayet star was actually the progenitor of Kepler's SN, one would expect that some ring nebulae, associated with these stars, present characteristics similar to those of the matter surrounding Kepler's SN progenitor. Objects with the following characteristics have been selected:
i) a limb brighter in the direction of the stellar motion; otherwise, if the motions are not known, a limb pointing away from the galactic plane;
ii) an associated runaway Wolf-Rayet star;
iii) a displacement of the star towards the brighter limb, and with nearly the same radial velocity of that limb;
iv) orderly internal motions, with velocities of a few tens km/s;
v) a clumpy nebula, with density up to 10^3 cm^{-3}.
Four galactic candidates have been found: all except one have been classified

as wind blown bubbles.

3.3.1. S308. This nebula is almost spherical, with a brighter limb on the north-west side: the bow shock is probably not sufficiently developed. The star is slightly displaced towards the direction of the brighter limb, and the nebula is filamentary. In the opposite direction to the brighter limb there is a young cluster, NGC 2362. If the star originated from this cluster, its velocity must be higher than 150 km/s (Chu et al. 1982). The star has been reported to be a single-line spectroscopic binary, which is typical for runaway Wolf-Rayet stars.

3.3.2. NGC3199. This nebula has an almost paraboloidal shape, as expected for a well developed bow shock; the star is close to the apex, this is also the brightest part of the nebula. In the opposite direction there is NGC 3293, an open cluster from which the star possibly originated.

3.3.3. NGC6888. This nebula has an elliptical shape, with a developed filamentary structure. The star is located near the brighter side, pointing to the north-west. If this is the direction of the stellar motion, the star is moving away from the galactic plane. Also this star is a spectroscopic binary.

3.3.4. M1-67. This nebula has a slightly elliptical shape, and is very clumpy. It is associated with the 209 BAC star, famous for being the fastest known runaway Wolf-Rayet star: a lower limit to its peculiar velocity is 178 km/s.

3.4. Bow shock versus expanding shell

The flow pattern of the matter in a bow shock is very different from that in an expanding shell: in the former case the bow shock is steady, and the material flows away along the bow shock surface, while in the latter there are only radial motions, with spherical symmetry. How does the internal kinematics affect the observed spectra? Can the right kind of flow be recognized by spectral observations? It can actually be done, if detailed spectral information is available.

This is the case of M1-67, of which Solf and Carsenty (1982) took various long slit spectra. A comparison between real and simulated spectra can be performed (Fig. 2). In the bow shock model the only free parameter is the direction of the apex; for an appropriate direction, the bow shock model successfully reproduces two features observed in the real spectra, namely the following properties:
i) the radial velocity of the star is closer to that of the brighter limb (close to the bow shock apex);
ii) spectra taken symmetrically with respect to the nebular center do not necessarily show the same separation between blue- and red-shifted edges.

However a totally developed bow shock cannot easily explain the almost circular shape of the nebula. A more general model is in preparation, simulating the effects of the gas flow in a partially developed bow shock.

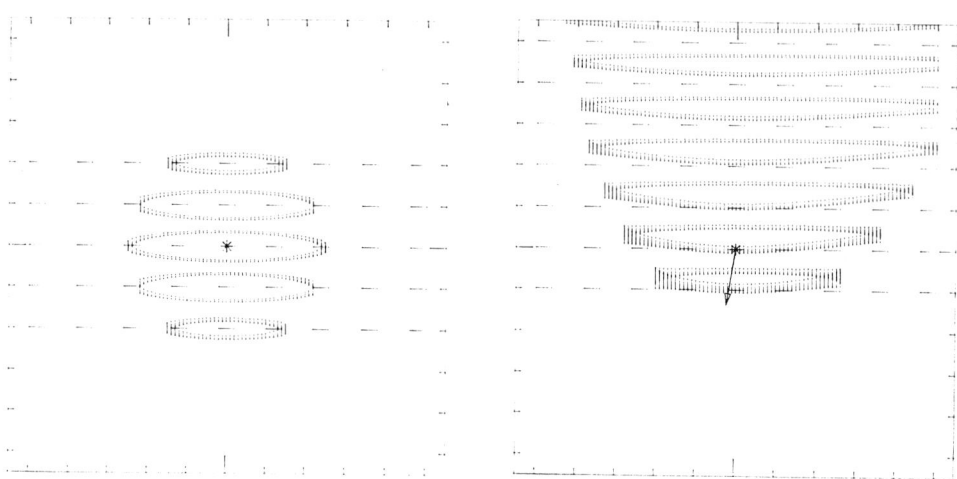

Figure 2. Simulations of long slit spectra observable in the case of an expanding shell (left), and in the case of a bow shock (right): these solutions should be compared with the results presented by Solf and Carsenty (1982) (see Fig.2 in their paper). The star is located at the center; the dashed horizontal lines indicate the positions of the various long slits. For each long slit horizontally one has the spatial coordinate, while the spectrum is dispersed vertically (blue upwards); each dashed line indicates now the stellar redshift for that specific spectrum. The shaded curves represent the simulated shape of an emission line: the width is proportional to the column density, while the various vertical displacements from the dashed line are proportional to the radial velocities relative to the stellar one. In the bow shock model, the direction of the apex has been adapted to fit the case of M1-67; in the simulation presented here the angle between the direction of the apex and that of the observer is 160°, while the polar angle of the projection of the apex on the sky is 170° (see arrow).

4. REFERENCES

Bandiera, R. 1987, Ap. J. (in press)
Chu, Y.H. 1981, Ap. J. **249**, 195
Chu, Y.H., Gull, T.R., Treffers, R.R., Kwitter, K.B., and Troland, T.H. 1982, Ap. J. **254**, 562
d'Odorico, S., Bandiera, R., Danziger, I.J., and Focardi, P. 1986, Astr. J. **91**, 1382
Huang, R.Q., and Weigert, A., 1982, Astr. Ap. **239**, 348
Matsui, Y., Long, K.S., Dickel, J.R., and Greisen, E.W. 1984, Ap. J. **287**, 295
McKee, C.F., and Ostriker, J.P., 1977, Ap. J. **218**, 148
Solf, J., and Carsenty, U., 1982, Astr. Ap. **116**, 54
van den Bergh, S., and Kamper, K.W., 1977, Ap. J. **218**, 617
van den Heuvel, E.P.J., 1976, I.A.U. Symposium No.73, "Structure and Evolution of Close Binary Systems", p.35

OBSERVATIONS OF SN1987A WITH I.U.E.

Roberto Gilmozzi
ESA IUE Observatory
Apartado 54065, Madrid, Spain

An overview of the IUE observations obtained at Vilspa by the European Target of Opportunity Team for supernovae was presented at the workshop. Since most of the results shown have already appeared in the literature, here I only briefly outline the most relevant conclusions derived from the early IUE observations of the LMC supernova, together with the relevant bibliographical references.

The most striking aspect of SN1987a in the ultraviolet has been its unprecedently fast fading (Wamsteker et al, 1987): no Type II supernova previously studied with IUE has shown such a behaviour. From this fact, it has been possible to place some constraints both on the initial configuration of the progenitor star, which had to be rather compact (Panagia et al, 1987), and on the physics of the absorption mechanisms, identifying for the first time the species responsible for the fast drop of the ultraviolet flux: singly and doubly ionized metals of the iron group (Cassatella et al, 1987)

Peculiar in many ways, SN1987a has displayed very unusual light curves, both in the optical and in the ultraviolet, the former never reaching the "predicted" initial maximum, but rather, after a rapid rise and a short stop, slowly increasing for several weeks (it was still rising at the time of the workshop); the latter decreasing since the first IUE observations (Kirshner et al, 1987; Panagia et al, 1987) at all wavelengths, with a rate inversely proportional to wavelength, to then flatten on some asymptotic value (although the in the range 2700-3200 A it started slowly increasing again around the time of the workshop).

However, while in the long wavelength range of IUE this asymptotic value was determined by the supernova itself, in the short wavelength (1200-1900A) it appeared to be the result of a superimposed residual stellar spectrum of an early B type. Since the SN coincides positionally with Sk-69 202, a blue LMC supergiant, this was initially interpreted as the spectrum of the supergiant (Gry et al, 1987). However, an analysis of the image of Sk-69 202 before outburst showed that there were two more stars close by: a reanalysis of the spatially resolved IUE spectra showed indeed two spectra which were interpreted as those of Sk-69 202 itself plus the one of the brighter of the two companions (Sonneborn and Kirshner, 1987). Panagia et al (1987) suggested, on the basis of the separation between the two spectra and their intensity, that Sk-69 202 had instead disappeared and that only the two companions were visible in the ultraviolet. A more accurate analysis of the spatially resolved spectra, plus a direct measurement of the flux at the position of Sk-69 202 (Gilmozzi et al, 1987) proved this to be the case, definitively identifying Sk-69 202 as the progenitor of the LMC supernova.

References

Cassatella,A, et al, 1987, Astron Astrophys 177,L29

Gilmozzi,R, et al, 1987, Nature 328,318

Gry,C, et al, 1987, IAU Circ 4327

Kirshner,R, et al, 1987, Astrophys J, in press

Panagia,N, et al, 1987, Astron Astrophys 177,L25

Sonneborn,G, and Kirshner,R, 1987, IAU Circ 4366

Wamsteker,W, et al, 1987, Astron Astrophys 177,L21

NEUTRINO OUTFLOW FROM SUPERNOVA 1987A DETECTED IN THE MONT BLANC
OBSERVATORY

M.Aglietta[a]   G.Badino[a]   G.Bologna[a]   C.Castagnoli[a]
A.Castellina[a]   V.L.Dadykin[b]   W.Fulgione[a]   P.Galeotti[a]
F.F.Kalchukov[b]   V.B.Kortchaguin[b]   P.V.Kortchaguin[b]
A.S.Malguin[b]   V.G.Ryassny[b]   O.G.Ryazhkaya[a]   O.Saavedra[a]
V.P.Talockin[b]   G.Trinchero[a]   S.Vernetto[a]   G.T.Zatsepin[b]
V.F.Yakushev[b]

a) Istituto di Cosmogeofisica del CNR, Torino, Italy, and
   Istituto di Fisica Generale, Università di Torino, Italy
b) Institute of Nuclear Research, Academy of Sciences of
   USSR, Moscow, USSR

(presented by P.Galeotti)

ABSTRACT. We discuss here the neutrino outflow connected with the event detected in the Mont Blanc Underground Neutrino Observatory (UNO) on February 23, 1987, and consisting of 5 interactions recorded during 7 seconds. The measured energies of the 5 pulses, the duration of the burst, and the advance of the detection time in comparison with the first optical observations give evidence that the event can be explained in terms of detection of neutrinos emitted during the stellar collapse in the Large Magellanic Cloud, which originated supernova SN 1987A.

1. INTRODUCTION

An Underground Neutrino Observatory (UNO) has been built[1] by our two Insitutes with the main aim to search for bursts of low energy neutrinos from stellar collapses. The UNO has been running[2] since October 1984 in the Mont Blanc Laboratory, at a depth of 5200 hg/cm^2 of standard rock underground. The very large coverage of rock, and an additional shielding, allows us to operate the UNO at a very low energy threshold.
An event, considered as a candidate of a neutrino burst, was detec-

ted$^{(3)}$ on February 23.12 ($2^h52^m36^s$ UT). On February 24.23$^{(4)}$ Shelton in Las Campanas Observatory (Chile) reported observation of an optical supernova (SN 1987A) in the Large Magellanic Cloud, 50 kpc faraway. Optical data indicate that no star brighter than magnitude 12 was present at February 23.08, and that the supernova was of magnitude 6.1 at February 23.44.

We discuss here the characteristics of the event observed in the UNO, and the connected neutrino outflow from the stellar collapse which originated SN 1987A.

2. THE MONT BLANC UNDERGROUND NEUTRINO OBSERVATORY

The Mont Blanc neutrino telescope is a 90 tons Liquid Scintillation Detector (LSD) consisting of 72 counters (1.5 m^3 each) in 3 layers, arranged in a parallelepiped shape with 6x7 m^2 area and 4.5 m height. The low energy local radioactivity background from the surrounding rock has been reduced by shielding each counter and the whole detector with more than 200 tons of Fe slabs. In underground experiments, the main source of background is due to cosmic ray muons and their interactions in the rock surrounding the detector, which may induce contained pulses from secondary neutrons or gamma rays. This source of background is very low at the depth of the Mont Blanc Laboratory, where 3.5 muons per hour have been recorded on the average in the whole LSD detector after several months of running time.

The liquid scintillator is watched from the top of each counter by 3 photomultipliers, in a 3-fold coincidence within 150 ns. Our calibrations$^{(1)}$, both from muons and with a ^{252}Cf source, show that a 1 MeV energy loss yields on the average 15 photoelectrons in 1 scintillation counter.

The electronic system consists of 2 levels of discriminators for each scintillation counter. A high-level discriminator for pulses above the energy threshold $\sim$ (6-7) MeV for the 56 surface counters, and $\sim$ 5 MeV for the 16 internal ones, with a total trigger rate of 0.012 Hz. A low-level discriminator for pulses above the energy threshold 0.8 MeV, is active only during a 500 μs wide gate, opened for all the 72 counters by the main high-level trigger. Two ADCs per counter measure the energy deposition in the scintillator in 2 overlapping energy ranges. A TDC gives the time with a resolution of 100 ns. Three memory buffers, for the 2 ADCs and the TDC of each scintillation counter, allow us to record all pulses without dead time. On-line software prints any burst of pulses satisfying our operational definition of a neutrino burst, namely a burst of pulses above a given multiplicity in

a given time.
This recording system allows us to detect both products of $\bar{\nu}_e$ interactions with the free protons of the scintillator (namely, positrons and gammas in a delayed coincidence within 500 μs), through the capture reaction:

$$\bar{\nu}_e + p \rightarrow n + e^+ \qquad (1)$$
$$\hookrightarrow n + p \rightarrow d + \gamma$$

which gives the main signal in detecting neutrinos from collapsing stars. In the LSD, the total number of target free protons is 8.4 10^{30}. In addition, also electrons from elastic scattering of neutrinos of other species with the electrons of the scintillator can be detected in the LSD. For positron detection, the pulse amplitude is given by the sum of the positron kinetic energy and the annihilation energy ($\sim$ 1 MeV) of gammas. The efficiency to detect gammas from the (np,dγ) capture reaction, in the same counter where the neutron was produced is $\sim$ 40% on the average.

The absolute time in the LSD is recorded by using the signal broadcasted by the Italian Standard Time Service (IEN Galileo Ferraris). The accuracy of the absolute time is better than 2 msec.

3. THE NEUTRINO BURST DETECTED IN THE UNO ON FEB. 23.12

Since January 1, 1986, the LSD experiment has been running with an average efficiency of 90% (and almost 99% since October 1986). Recently, the detector shielding has been partly increased for test purposes, with paraffin and lead in order to further decrease the low energy background from the surrounding rock. Trigger pulses are analysed in order to have a long term statistics and to search for bursts. The experimental distributions of these pulses, recorded over several months, show that the counting rate is stable and agrees very well with a Poisson distribution.

For monitoring purposes the detector counting rate is also checked on-line every 100 triggers. By analysing these data, and from the tape analysis several days before and after the event discussed here, we have been able to verify that the apparatus was running properly throughout the entire period.

On February 23.12, 1987, ($2^h 52^m 36^s$ UT), an event, consisting of a burst of 5 pulses and printed in real time during its occurence, was recorded in 5 different counters (3 of them internal) during 7 seconds. Table I gives the event number, the counter number, the

Table I - Characteristics of the pulses in the burst detected on February 23rd, 1987

Event no.	Counter no.	Time (UT)	E_{vis} (MeV)
994	31	$2^h 52^m 36^s.79$	6.2
995	14	40.65	5.8
996	25	41.01	7.8
997	35	42.70	7.0
998	33	43.80	6.8

absolute universal time (with an accuracy better than 2 msec), and the visible energy of the detected pulses (with an average accuracy of ~ 10%). A low energy pulse, with energy E = 1 MeV, accompanying the 3rd interaction, was recorded 278 μs after the main pulse in the same scintillation counter. From the measured efficiency to detect γ's from neutron capture, we expect on the average 2 such pulses in the 5 counters involved in the burst.

Fig 1 shows the distribution of bursts of pulses, with multiplicity n ≥ 5, n ≥ 10, and n ≥ 15, recorded during 2 days of measurements encompassing the event discussed here. The full lines are computed according to a Poisson distribution of the trigger counting rate, with a binning of 10 seconds, and mean value given by our average trigger rate, which has the value of 0.012 Hz during this run as for the previous ones. From fig.1, excellent agreement between the expected and measured distributions is found, except for the point corresponding to the event considered here. The imitation rate from the background is 0.7 per year for this burst, or ~ 4 10^{-4} in the time interval corresponding to the

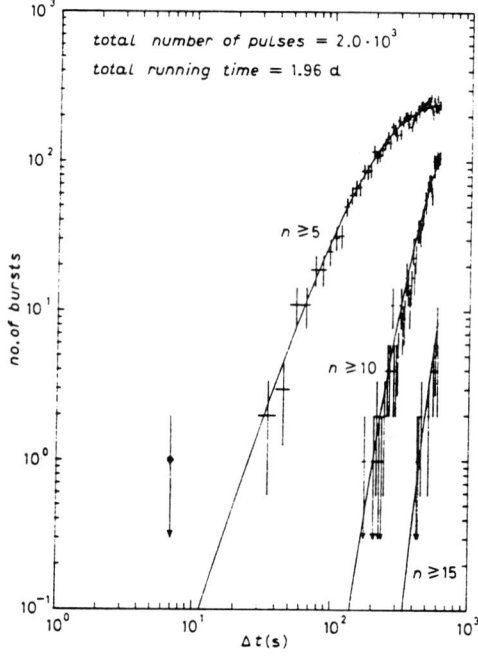

Fig.1 - Experimental distributions of bursts of pulses as a function of their duration.

uncertainty (~ 5 hours) of the instant of collapse, as suggested by optical data.

4. DISCUSSION

During the stellar collapse, the difference of binding energy of the collapsed and initial core is available energy, that goes mainly into neutrinos while only a small fraction (of order of 1%) produces other observable effects. Only a negligible fraction of the total energy output is emitted as ν_e during the neutronization peak; the bulk of the neutrino luminosity is emitted as neutrino pairs during the cooling phase of the new born neutron star, for times as long as several seconds.

The energy spectrum of neutrinos escaping from the collapsing core can be approximated by a distribution similar to a Fermi Dirac one, namely:

$$\Phi \ (\bar{\nu}_e/\text{sec MeV}) \propto \frac{\varepsilon^2 \exp[-\alpha \varepsilon^2]}{1 + e^\varepsilon} \qquad (\varepsilon = E_\nu/KT)$$

where E_ν is the $\bar{\nu}_e$ energy (in MeV), and T the temperature of the neutrinosphere. The correction factor $\exp(-\alpha \varepsilon^2)$ takes into account neutrino absorption in the stellar envelope above the neutrinosphere.

The central temperature KT of a type II presupernova is supposed to be less than 1 MeV, and the temperature of the neutrinosphere, after neutrino trapping, is of the order of a few MeV. Depending on the values of the absorption parameter α, the energy spectrum of the neutrinos emitted from the collapsing stellar core can be different from the spectrum produced during the collapse: high energy neutrinos are indeed more absorbed than the low energy ones, and bigger the mass of the star envelope

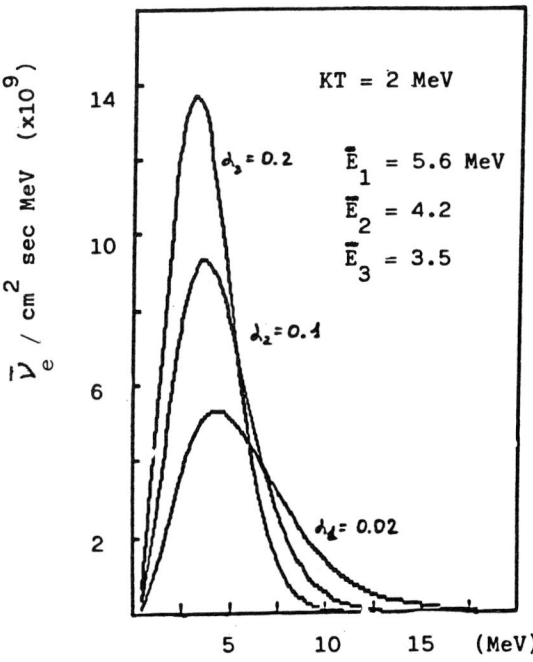

Fig.2 - $\bar{\nu}_e$ flux at the earth for some α values, assuming a collapse at d = 52 kpc emitting 10^{53} erg in $\bar{\nu}_e$

stronger is the shift of the spectrum to low energies. Fig.2 shows the effect of the absorption parameter α, in the case of a neutrinosphere at the temperature KT = 2 MeV; without absorption in the envelope (α = 0), the average neutrino energy would be 3.15 KT, characteristics of a Fermi Dirac gas. The neutrino interactions recorded in the LSD experiment agree with a temperature of the neutrinosphere KT $\simeq$ 2 MeV and absorption parameter $\alpha \simeq 0.1$.

The total energy involved in the burst can be estimated assuming the 5 pulses recorded in the LSD are due to $\bar{\nu}_e$ capture processes (1). In this case the neutrino energies are connected to the visible energies of the detected pulses by $E_\nu = E_{vis} + 1.8 - 1.0$ MeV, 1 MeV being the energy released in the scintillator by the positron annihilation, and 1.8 MeV is the energy threshold of the reaction. Adopting a distance of 52 kpc for SN 1987A, and using the cross section:

$$\sigma(E_\nu) = 9.45 \cdot 10^{-44} (E_\nu - \Delta M) \left[(E_\nu - \Delta M)^2 - m_e^2 \right]^{1/2} \text{ cm}^2$$

where $\Delta M = 1.293$ is the neutron to proton mass difference, and $m_e = 0.511$ MeV is the electron rest mass, the total energy outflow in $\bar{\nu}_e$ is $6.7 \cdot 10^{53}$ erg.

5. CONCLUSIONS

The neutrino burst recorded on real time in the Mont Blanc Underground Neutrino Observatory during the occurence of supernova 1987A, indicate that the collapse had a duration of several seconds, during which a high luminosity burst of neutrinos was emitted by a low temperature neutrinosphere. Even if at large distance, this stellar collapse produced a significant number of interactions in the LSD experiment because of its very low energy threshold and high efficiency to detect low energy pulses.

REFERENCES
1. G.Badino et al.,Nuovo Cim.7C,573,1984
2. M.Aglietta et al., Nuovo Cim.9C,185,1986
3. M.Aglietta et al.,IAU Circ.no.4323, and Europhys.Lett.3,1315,1987
4. I.Shelton, IAU Circ.no.4316

GEOMETRICAL SIMULATIONS OF ASTROPHYSICAL JETS

L.Zaninetti

Istituto di Fisica Generale , Torino , Italy

ABSTRACT.The observed morphological features of extragalactic radio jets and H.H objects include wiggles and pinches . To gain understanding of these properties , we review the model of helical and pinching instabilities . This model enable us to reproduce these features and to deduce a number of important physical parameters, such as the Mach number and the density contrast .

1. INTRODUCTION

In this paper , we apply both geometrical and hydrodynamical computations to address the following questions about some of the observed morphological features of astrophysical jets :

Is it possible to reproduce observed features such as wiggling, pinching, and lateral displacement of a jet from the results of hydrodynamical calculations ?,

Is the observed bending in some jets produced by ram pressure effects, or is it the result of a decrease in the average flow velocity of the jet ?,

What is the effect of precession on the shape of a jet ? ,

What are the projection effects on various jet trajectories ?.

To explore these questions, we examine in sec. 2 the hydrodynamical behavior of a straigth jet subject to cylindrical instabilities.

2. INSTABILITIES

The Kelvin-Helmholtz instabilities of an axisymmetric flow have recently been reinvestigated for cases in which the wavelengths of the perturbations, λ, are of the order of, or greater than, the radius a of the jet (see Zaninetti and Van Horn 1987). This work contains a new derivation of the time t_{min} and of the distance l_e, over which the most unstable mode grows by a factor e . The corresponding wavelength λ_{max} also has been obtained for the ordinary mode of instability. The best fitting curves are given by:

$$l_e = \left(\frac{-7.68 + 13.08M}{1 + 12.40m}\right)\nu_0^{0.03} a, \tag{1}$$

$$\lambda_{max} = \left(\frac{-1.83 + 3.25M}{1 + 0.8m}\right)\nu_0^{-0.20} a, \tag{2}$$

$$t_{min} = \left(\frac{0.18}{log(1+0.31m)} + \frac{-0.34}{-0.46+M}\right)\nu_0^{-0.15}\frac{a}{s_i}. \tag{3}$$

where M is the Mach number, ν_0 is the ratio of the density inside the jet to that outside, m is the azimuthal number, a is the radius of the jet and s_i is the internal sound speed

The range over which we have allowed these parameters to extend is : $3 \leq M \leq 50$, $1 \leq m \leq 9$ and $0.1 \leq \nu_0 \leq 100$.

It must be stressed that these results apply to classical fluid jets with different azimuthal numbers, but we have not considered jets with constant opening angles or with gradients of the density ratio ν_0 (see Hardee 1982, 1984). The reason is that terms involving jet expansion affect the growth rate and the wavenumber of the maximally unstable mode at the 10 % level, when the expansion angle is small, and thus are not essential for our present purposes.

To apply these results to interpret the observed morphological shapes of astrophysical jets, we consider motion in the x-y plane, taking the axis of the unperturbed jet to lie along the x-axis. The oscillatory displacement of the perturbed jet in the direction perpendicular to the axis is given by :

$$y = A_0 sin(2\pi x/\lambda_{max}^1), \tag{4}$$

where the superscript 1 corresponds to azimuthal number $m = 1$. The jet is also subject to pinching instabilities, which correspond to modes with $m = 0$. The wavelength of the most unstable pinching mode is

$$\lambda_{max}^0 = 2.18 M^{0.07} \nu^{0.18} a. \tag{5}$$

The quantity λ_{max}^0 is the predicted distance between " knots" along the jet, as deduced from the linear analysis ; the real structure, of course ,develops from the nonlinear evolution of the instabilities.

Another important observed parameter is the rate of increase of the oscillation amplitude. From the observations, this amplitude appears to increase exponentially with angular separation from the nucleus. This phenomena is connected with the growth of the internal perturbed pressure $\delta p \ exp(\frac{2t}{t_{min}^1})$ and we obtain for the lateral displacement , in place of (4), the result :

$$y = A_0 exp\left(\frac{2t}{t_{min}^1}\right) sin\left(2\pi \frac{x}{\lambda_{max}^1}\right). \tag{6}$$

Of course, in order for the linear analysis to be applicable, the offset y $(x=L) = \Delta L_{obs}$ of the center of the jet must be small compared to the total length L of the jet.

Fig. 1 - The best fitting model of M87 at 2 cm ; the shadowed circles indicate the positions of the theoretical "knots", and the filled circle represent the parent galaxy. The physical parameters are written on the figure. The opening angle, θ, of the jet and the angle of rotation of coordinates, α, are expressed in degrees.

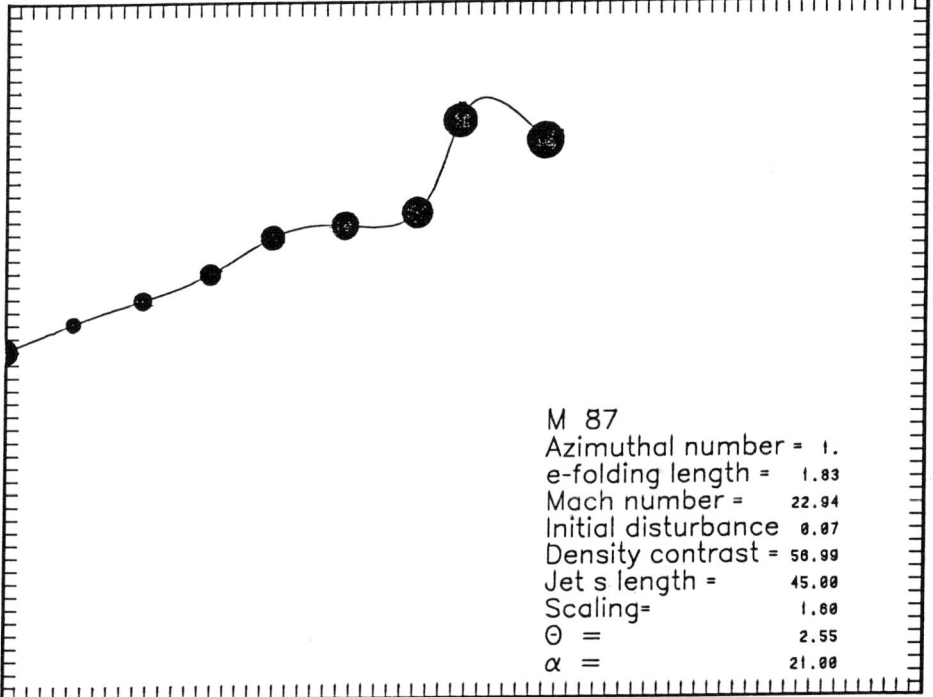

At this point we must supply particular values of the parameters A_0, ν_0, and M in order to reproduce the observed features of the radio maps for particular sources. These parameters can actually be obtained by relating the theoretical parameters to the oserved structures :

$$\lambda^1_{obs} = \lambda^1_{max}, \qquad (7a)$$

$$\lambda^0_{obs} = \lambda^0_{max}, \qquad (7b)$$

$$\Delta L_{obs} = A_0 \, exp(\frac{L}{M t_{ad} a}). \qquad (7c)$$

were $t_{ad} = t_{min} \cdot s_i/a$. In addition, we take the observed length of the jet to be given by :

$$L_{obs} = n l_e , \qquad (8)$$

where n is assumed to be of the order of unity. Here the observed quantities are the total length L_{obs}, the wavelength λ^1_{obs} of the wiggles along the jet, the distance λ^0_{obs} between the knots, and the final offset ΔL_{obs} of the center of the jet. We report the observed parameters of M87 given in units of the jet radius a in fig. 1 where we plot our best fitting model.

References

Hardee, P.E. : 1982, *Astrophys. J.* **257**, 509
Hardee, P.E. : 1984, *Astrophys. J.* **287**, 523
Zaninetti, L. ,Van Horn ,H. : 1987 , *Astron. Astrophys. submitted*

ON THE STABILITY OF ROTATING SHEAR FLOWS

W. Glatzel
Max - Planck - Institut für Astrophysik
Karl - Schwarzschild - Straße 1
D - 8046 Garching bei München
FRG

ABSTRACT. In connection with the stability of accretion tori and accretion disks the stability properties of a simplified model for a rotating shear flow are investigated analytically. Two types of instabilities can be identified: One of them is due to energy loss from the system, in particular to acoustic radiation. The second can be understood as a resonant interaction of two modes whose energy has opposite sign.

1. INTRODUCTION

Accretion tori are discussed as a major ingredient in the models for the central engine of active galactic nuclei and quasars (cf. Rees (1984)). In particular they can exhibit steep funnels in which jets may be formed and collimated, where tori having constant angular momentum distribution provide the most attractive models for jet production. However, crucial for the viability of these models are their stability properties, which have been first investigated by Papaloizou & Pringle (1984). They found that accretion tori are unstable to global nonaxisymmetric perturbations on dynamical timescales. Meanwhile this instability has been investigated by many authors under various aspects (for a list of references see e.g. Papaloizou & Pringle (1987)).

Here we examine the stability of a simplified model for the shear flow in an accretion torus, which allows for an analytical treatment and is still in qualitative agreement with more realistic models as described by Glatzel (1987a). A detailed discussion of this investigation is given in Glatzel (1987b).

2. BASIC ASSUMPTIONS AND EQUATIONS

We consider a cylindrical rotating shear flow having constant angular momentum distribution. The fluid obeys a polytropic equation of state and for simplicity the sound speed is assumed to be constant. The vertical component of the velocity is required to vanish and the vertical dependence of both the unperturbed and the perturbed quantities is ignored. In the limit of thin shear layers, high azimuthal wavenumbers m and high Mach numbers M the linear perturbation equation reduces to Whittaker's equation, where $Q = \bar{\sigma}^{-1/2}\tilde{p}$ is the dependent and $\varsigma = i\frac{m}{2}M\bar{\sigma}^2$ is the independent variable. $\tilde{p}$ denotes the Eulerian pressure perturbation and $\bar{\sigma}$ is defined as $\bar{\sigma} = \omega + \Omega$, where $\sigma = \omega m$ is the complex eigenfrequency and $\Omega = 1/r^2$ is the angular velocity of the unperturbed flow. The parameters κ and μ of Whittaker's equation turn out to be $\mu = 1/4$ and $\kappa = \frac{i}{4}\frac{m/2}{M}$.

3. LINEAR SHEAR LAYER VERSUS ROTATING SHEAR LAYER

A comparison with the analysis of the linear shear layer (see Glatzel, this volume) shows that the structure of the stability problem is essentially the same, if we identify the wavenumber $m/2$ with k and the unperturbed angular velocity Ω with $\bar{V}$ (normalized by half of the thickness of the shear layer). The results obtained for the linear shear layer therefore also apply to rotating flows. This means in particular that rotation, i.e. centrifugal and Coriolis forces are not responsible for the instabilities.

A hint on the possible cause of the instabilities is given by the evolution equation for the perturbation energy $\tilde{E}$ defined as the positive definite sum of the kinetic and potential energy of the perturbation (see Glatzel (1987a)):

$$\frac{\partial}{\partial t}r\langle\tilde{E}\rangle = r^2\bar{\rho}\langle-\tilde{v}_r\tilde{v}_\varphi\rangle\frac{d\Omega}{dr} - \frac{d}{dr}(r\langle\tilde{v}_r\tilde{p}\rangle) \qquad (3.1)$$

$\tilde{v}_r$ and $\tilde{v}_\varphi$ are the components of the velocity perturbation, $\bar{\rho}$ denotes the unperturbed density and $\langle\rangle$ is an angular average. The second term on the r.h.s. of equation (3.1) represents the acoustic energy flux and the first describes the production of perturbation energy due to the Reynolds stress. Since it is proportional to the shear we may conclude that the physical origin of the instabilities is the shearing motion and surface or sound waves are only necessary as a mechanism to produce the actual instability from the unstable tendency.

4. THE DISPERSION RELATIONS AND A MECHANISM FOR THE INSTABILITY

The dispersion relations are obtained in the same way as for the linear shear flow (see Glatzel, this volume). If the pressure is required to vanish at one boundary and a radiation condition is applied at the second, we find a set of radiation unstable modes given by

$$\bar{\sigma} = \pm\left(\frac{\pi}{Mm/2}\right)^{1/2}\left(2n - 1/2 \pm i\frac{\ln 2}{2\pi}\right)^{1/2} \quad ; \quad n = 1, 2, 3, \ldots \qquad (4.1)$$

where the sign in front of the root has to be chosen to ensure a corotation point of the mode within the flow, the sign of the imaginary term to ensure an outgoing wave. $\bar{\sigma}$ is to be evaluated at the boundary where the pressure is required to vanish. If we require zero energy flux instead of a radiation condition the imaginary term in equation (4.1) vanishes and we get a set of neutral modes. Their pattern speed is plotted in Fig. 1 as a function of q for two boundary positions at $r = 1 \pm q/m$ where q is defined as $q = (\frac{2}{\gamma-1})^{1/2} \frac{m}{M}$. We have adopted $m = 10$ and $\gamma = 1.4$ for the adiabatic index in Fig. 1.

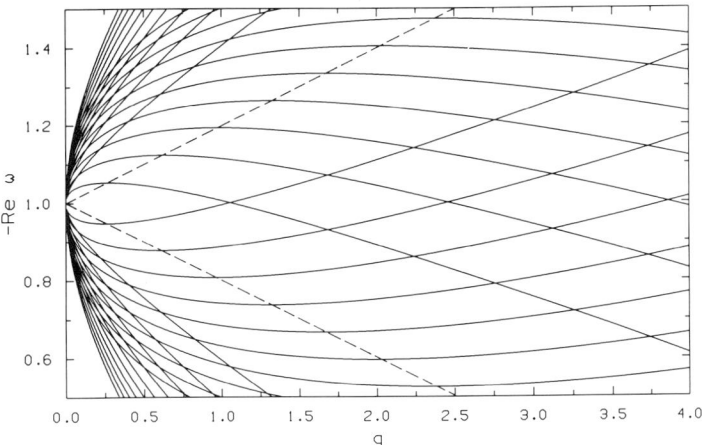

Figure 1. The pattern speed of the neutral modes (see text for details). Dashed lines denote corotation at the boundaries.

If the perturbation pressure is to vanish at both boundaries, we find

$$\left\{\bar\sigma_{in}^2 - \frac{\pi}{Mm/2}(2n_{in} - 1/2)\right\} \left\{\bar\sigma_{out}^2 - \frac{\pi}{Mm/2}(2n_{out} - 1/2)\right\} = \varepsilon(\omega) \qquad (4.2)$$

where $\varepsilon \sim O((mM)^{-2})$ and $_{in}$ and $_{out}$ refer to the inner and outer boundary respectively. To first order the modes corresponding to the dispersion relation (4.2) can be regarded as a superposition of the neutral modes belonging to the inner and outer boundary where mode crossings (resonances) unfold into bands of instabilities, if $\bar\sigma_{in}$ and $\bar\sigma_{out}$ of the corresponding neutral modes have opposite sign, i.e. if they corotate somewhere within the flow, and into avoided crossings otherwise. This behaviour is found indeed, if we compare Fig. 1 with Fig. 2, where the eigenfrequencies according to equation (4.2) are plotted for the same boundary positions and parameters as in Fig. 1.

It has been shown by Glatzel (1987b) that for neutral modes a pseudoenergy - not to be confused with the perturbation energy as defined in Section 3 - can be defined which is proportional to $\bar\sigma$ evaluated at the boundary where $\tilde p$ is required to vanish. This suggests a physical interpretation of the resonance instabilities in terms of the energy of a mode (see also Cairns (1979)): If two modes interact whose energy has different sign the amplitude of both of them can grow by an exchange of energy thus producing an instability even if the energy of the entire system is to be conserved.

Also the radiation instability can be understood in terms of the energy of a mode. In this case energy loss due to acoustic radiation causes the amplitude of a mode having negative energy to grow. However, it is not the same pseudoenergy as above which is appropriate to describe this process (see Glatzel (1987b) for details).

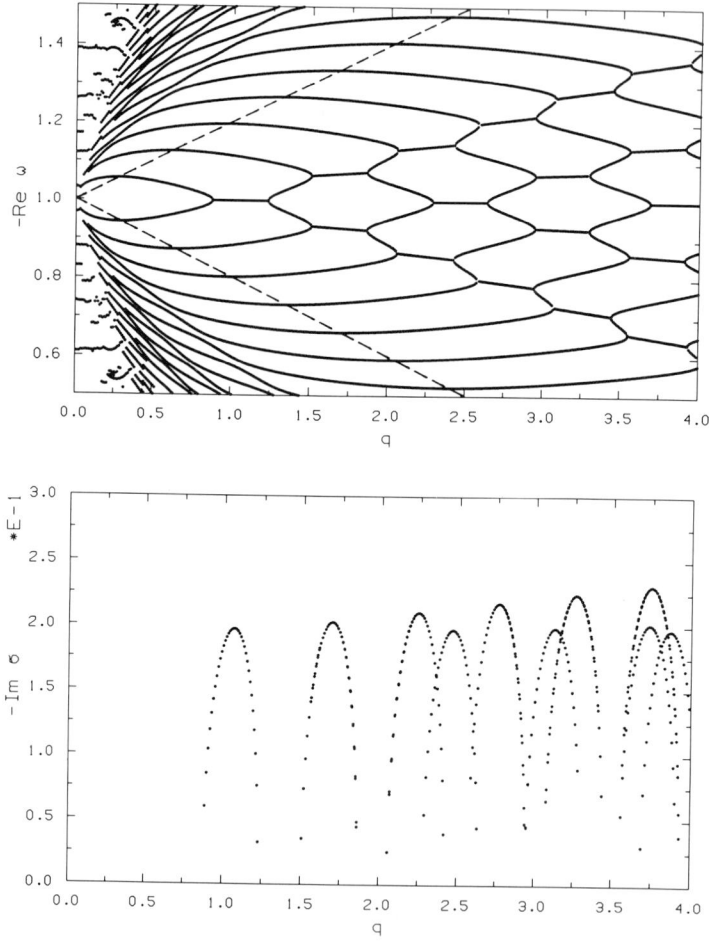

Figure 2. Eigenfrequencies of a rotating shear layer for zero pressure perturbation at the boundaries (see text for details). Dashed lines denote corotation at the boundaries.

REFERENCES

Cairns, R.A., 1979. *J. Fluid Mech.*, **92**, 1.
Glatzel, W., 1987a. *Mon. Not. R. astr. Soc.*, **225**, 227.
Glatzel, W., 1987b. *Mon. Not. R. astr. Soc.*, in press.
Papaloizou, J.C.B. & Pringle, J.E., 1984. *Mon. Not. R. astr. Soc.*, **208**, 721.
Papaloizou, J.C.B. & Pringle, J.E., 1987. *Mon. Not. R. astr. Soc.*, **225**, 267.
Rees, M.J., 1984. *Ann. Rev. Astron. Astrophys.*, **22**, 471.

ON THE STABILITY OF SUPERSONIC SHEAR LAYERS

W. Glatzel
Max - Planck - Institut für Astrophysik
Karl - Schwarzschild - Straße 1
D - 8046 Garching bei München
FRG

ABSTRACT. The stability of a linear supersonic shear layer is reexamined. It is shown that the solution of the perturbation equation can be given analytically in terms of confluent hypergeometric functions. The dispersion relation is derived for various boundary conditions. In particular, the effect of a different sound speed outside and within the shear layer is studied. An additional kind of instability is found if considerable reflection of sound waves occurs at the edges of the shear layer.

1. INTRODUCTION

The stability of linear supersonic shear layers, i.e. the influence of compressibility on the stability of a simple shear layer has been investigated by several authors (e.g. Ray (1982), Choudhury & Lovelace (1984) and references given there). Apart from possible applications to laboratory flows there is a variety of astrophysical situations in which the stability of supersonic shear layers may play an important role, e.g. at the boundaries of galactic and extragalactic jets. In particular, the wiggles observed in radio jets may be due to instabilities occurring at the boundaries.

Except the investigation by Ferrari et al. (1982) the instability of the boundary between two flows having a supersonic velocity difference has been treated either in the vortex sheet approximation or for smooth velocity profiles with equal sound velocities inside and outside the shear layer. Here we shall reconsider the case of a linear shear layer using various kinds of boundary conditions for the perturbation equation. In particular, these boundary conditions will be chosen to allow for a different density of the shear layer and the surrounding medium.

2. AN ANALYTICAL SOLUTION OF THE PERTURBATION EQUATION

We consider the linear stability of a shear layer, where the velocity $\bar{V}$ is taken in the x - direction and varies linearly with z $(\bar{V} = z)$ from $z = -1$ to $z = +1$. Lengths and velocities are measured in units of half of the thickness of the shear layer and the flow speed at the edge of the shear layer respectively. The Mach number M is defined as the ratio of the flow speed at the edge of the shear layer and the (constant) sound velocity. By Squire's theorem it is sufficient to consider twodimensional perturbations $\sim \exp(ik(x+\omega t))$ only, where k is the wavenumber and $\sigma = \omega k$ is the complex eigenfrequency.

The perturbation equation as derived from the Euler equation for an adiabatic flow is then easily solved numerically (see Ray (1982) and Choudhury & Lovelace (1984)). However, by choosing $Q = \bar{\sigma}^{-1/2} \tilde{p}$ (with $\bar{\sigma} = \omega + \bar{V}$) as the dependent variable, where $\tilde{p}$ is the Eulerian pressure perturbation, and $\varsigma = ikM\bar{\sigma}^2$ as the independent variable the perturbation equation is reduced to the standard form of Whittaker's equation, whose solution can be given analytically in terms of confluent hypergeometric functions $M_{\kappa,\mu}$ (cf. Abramowitz & Stegun (1970)). With $\kappa = \frac{i}{4} \frac{k}{M}$ and $\mu = 3/4$ the general solution of the perturbation is then given by

$$Q = c_1 M_{\kappa,\mu}(\varsigma) + c_2 M_{\kappa,-\mu}(\varsigma) \tag{2.1}$$

where c_1 and c_2 are integration constants. The ratio c_2/c_1 and the pattern speed ω is determined by the boundary conditions applied at $z = \pm 1$.

3. BOUNDARY CONDITIONS

For the derivation of the dispersion relations we consider three types of boundary conditions. If the sound speed within the shear layer is equal to the sound speed outside the appropriate boundary condition is a radiation condition (cf. Ray (1982)) given by $\frac{1}{Q}\frac{dQ}{d\varsigma} + \frac{1}{4\varsigma} = \pm\frac{1}{2}(1 - \frac{4\kappa}{\varsigma})^{1/2}$, where the upper sign has to be taken at $z = -1$ and the lower sign at $z = +1$. If the density in the surrounding medium is much larger than in the shear layer, the normal velocity component at the boundaries has to be zero: $\frac{1}{Q}\frac{dQ}{d\varsigma} + \frac{1}{4\varsigma} = 0$. In the opposite case the pressure perturbation must vanish: $Q = 0$.

4. DISPERSION RELATIONS

Since the results turn out to be - even quantitatively - very similar to the exact treatment and more physical insight is gained we discuss here the dispersion relations derived on the basis of an asymptotic expansion of the functions $M_{\kappa,\mu}$ for $|\kappa| \to 0$ and $|\varsigma| \to \infty$. In the following quantities with subscript $_-$ have to be evaluated at $z = -1$, with subscript $_+$ at $z = +1$.

If the pressure or the velocity perturbation is required to vanish at both of the boundaries we find the dispersion relation:

$$\exp(\varsigma_-) = \frac{\mp 1 + i\sqrt{2}\exp(-\varsigma_+)}{i\sqrt{2} \pm \exp(-\varsigma_+)} \tag{4.1}$$

where the upper sign has to be taken for a vanishing pressure perturbation and the lower for a vanishing velocity perturbation. The dispersion relation (4.1) is equivalent to

$$\left\{\bar{\sigma}_-^2 - \frac{\pi}{kM}(2n_- \pm 1/2)\right\}\left\{\bar{\sigma}_+^2 - \frac{\pi}{kM}(2n_+ \pm 1/2)\right\} = \varepsilon(\omega) \tag{4.2}$$

where $n_\pm$ are integers satisfying $2n_\pm \pm 1/2 \geq 0$ and ε is small $(\varepsilon(\omega) \sim O((kM)^{-2}))$.

Since ε is small, an approximate solution of equation (4.2) is given by the zero of each of the expressions in curly brackets and we obtain for each of the boundaries a set of neutral modes. This approximation fails, if the expressions in curly brackets vanish at the same time, i.e. around a mode crossing. As ε is positive at a mode crossing $(\varepsilon = (\pi/(kM))^2)$ we obtain a band of instability with a maximum growth rate of about

$$\Im\sigma = \frac{1}{2}(\frac{\pi k}{M})^{1/2}[(2n_- \pm 1/2)(2n_+ \pm 1/2)]^{-1/4} \qquad (4.3)$$

if $\bar{\sigma}_-$ and $\bar{\sigma}_+$ of the corresponding neutral modes have different sign and an avoided crossing of neutral modes otherwise (for the treatment of mode crossings see also Cairns (1979)).

If the radiation condition is applied at both of the boundaries, we obtain

$$\varsigma_- \exp(-\varsigma_-) = \frac{i\sqrt{2} - 8\varsigma_+ \exp(\varsigma_+)}{8 + i16\sqrt{2}\varsigma_+ \exp(\varsigma_+)} \qquad (4.4)$$

or

$$\left\{\bar{\sigma}_-^2 - \frac{2\pi n_-}{kM} + \frac{i}{kM}\log(kM\bar{\sigma}_-^2)\right\}\left\{\bar{\sigma}_+^2 - \frac{2\pi n_+}{kM} - \frac{i}{kM}\log(kM\bar{\sigma}_+^2)\right\} = \varepsilon(\omega) \qquad (4.5)$$

where $n_\pm \geq 0$ are integers and $\varepsilon \sim O((kM)^{-2})$ is small. As above we obtain two sets of modes which are damped due to the logarithmic term for $n_\pm \geq 1$ but unstable for $n_\pm = 0$. The effect of resonant mode interactions which is responsible for the instability bands above is still present but does not change the stability properties of the modes. The two modes having $n_\pm = 0$ can be identified with the modes described by Ray (1982) and Choudhury & Lovelace (1984) and are the compressible counterparts of the ordinary Kelvin - Helmholtz - modes.

5. DISCUSSION

Comparing the modal structure of the shear layer for reflecting and transmitting boundary conditions we find apart from the Kelvin - Helmholtz - modes, which do not exist for the zero velocity boundary condition, a similar behaviour of the real parts of the eigenfrequencies. However, the stability properties are very different indicating different physical mechanisms for the instabilities. In the case of perfectly reflecting boundaries we have pure resonance instabilities occurring at mode crossings while for the (not perfectly) transmitting boundary condition acoustic radiation is essential and can cause - depending on the particular mode - damping or destabilization. If acoustic radiation is dominant, the instabilities occur in a continuous way rather than having a band structure as in the case of resonance instabilities (for a detailed discussion see Glatzel (1987)).

6. CONCLUSION

The stability of a supersonic shear layer, in particular the physical mechanisms leading to an instability, sensitively depend on the boundary conditions adopted for the perturbations. If the sound velocities within the shear layer and in the surrounding medium are equal, the two Kelvin - Helmholtz - modes are at high Mach numbers unstable due to acoustic radiation, while all other modes are damped. If the sound velocities differ significantly and sound waves are reflected at the edges of the shear layer, mode resonances lead to bands of instability, even if the existence of the Kelvin - Helmholtz - modes is excluded by the boundary conditions.

REFERENCES

Abramowitz, M. & Stegun, I., 1970. *Handbook of Mathematical Functions*, Dover Publications, New York.
Cairns, R.A., 1979. *J. Fluid Mech.*, **92**, 1.
Choudhury, S.R. & Lovelace, R.V.E., 1984. *Astrophys. J.*, **283**, 331.
Ferrari, A., Massaglia, S. & Trussoni, E., 1982. *Mon. Not. R. astr. Soc.*, **198**, 1065.
Glatzel, W., 1987. in preparation.
Ray, T.P., 1982. *Mon. Not. R. astr. Soc.*, **198**, 617.

SELF-SUSTAINING GALACTIC WINDS IN SPIRAL GALAXIES WITH STARBURST NUCLEI

John E. Beckman,
Instituto de Astrofisica de Canarias,
38200 - La Laguna,
Tenerife,
Spain.

ABSTRACT. In the circum-nuclear zones of the discs of some spirals there is an absence of neutral interstellar hydrogen. Here, a model is devised in which gaseous outflow is sustained by the combined action of massive stars in these zones. The energy requirement, which implies driving out the gas against gravity, can be satisfied if 10% or more of the gas condenses into massive stars. A problem is efficient coupling of energy into outflow without disrupting the molecular gas. It is shown that combined stellar winds can just satisfy the energy and momentum requirements for outflow, as can supernovas also. Dust in the molecular clouds would permit radiation to couple strongly to the gas. All these effects are present. Observable consequences should be: (a) Expansion with velocities of tens of km. per second near the nuclei of some galaxies. (b) Absence of neutral gas near the centres of post-starburst galaxies. (c) Metallicity gradients, with residual gas near the nucleus being metal-rich and a peak of stellar metallicity in a ring whose distance from the nucleus depends on the age of the burst.

1. INTRODUCTION

Observational evidence for outflow round the nuclei of spirals has been presented by Muñoz-Tuñon and Beckman in a paper in the present proceedings. The aim of this paper is to show that once star formation has been triggered at the centre of a giant molecular clouds around the nucleus in the disc of a spiral, physical conditions permit the outward propagation of a zone of enhanced star formation. The presence of massive stars in large concentrations gives rise to a set of processes: high radiative fluxes, stellar winds and supernova explosions, whose dynamical effects combine to produce a galactic wind. The wind sweeps out the neutral gas from the centre: that which remains is supernova-product enriched. Star formation proceeds in a steadily expanding zone, given the triggering effect of the outflowing wind.

The energy and momentum provided by the stars must be sufficient to sustain the expansion against gravity, and the expansion time scale short enough to ensure successive gravitational instability collapse before the gaseous ring can dissipate. Here we briefly examine quantitatively the physical constraints of such a model. We assume cylindrical symmetry for simplicity. Star formation might be triggered by an active nucleus, or by a central gas density peak: the model is viable in either case. An initial association of O stars or of even more massive objects is assumed to form. We examine the condition that the winds from this association compress the surrounding gas so that a new generation of stars is formed before the compression wave can dissipate. Elmegreen (1983) gave an analogous picture for spiral arms, and here some of his formalism has been adopted. In the second section we look at the instability condition without reference to the causes of the wind, which are examined in sections 3 (energetics) and 4 (momentum). In the final section, 5, we outline observable predictions for the model.

2. GRAVITATIONAL INSTABILITY

For the case of a giant molecular cloud under self-gravitation round the nucleus we adopt a density distribution

$$\rho = \rho_0 \left(\frac{R_0}{R}\right)^2 \quad (1)$$

and assume a momentum-conserving impulse of expansion such that the product of the mass M in a shell with its expansion velocity V is always equal to an initial value $M_0 V_0$. In general V will fall with time, and the condition that the shell does not dissipate before a new generation of stars can form out of its gas is expressed as

$$(t + \frac{R_0}{V_0}) > t_{grav} \quad \text{before } V<c' \quad (2)$$

where t is the time, t_{grav} is the gravitational collapse time of the cloud ("Jeans time") and c' the sound speed in the gas. Condition (2) translates to a requirement on the initial mach number $\eta_0 = \frac{V_0}{c'}$,

$$\eta_0 > \eta_{crit} = [1 + \frac{3}{2} \frac{V_0}{R_0} (t_{grav} - \frac{R_0}{V_0})]^{1/2} \quad (3)$$

for a ring geometry. The collapse time is given by: $t_{grav} = 1/(G\rho)^{1/2}$ (4) and in a reasonably dense cloud, with density $n_0 = 10^3$ mol. cm^{-3}, t_{grav} takes a value of 2.5×10^5 years. For $R_0 = 50 pc$, $V_0 = 50$ km s^{-1}, values measured in NGC 6946, we find $\eta_{crit} = 1.75$. For sound speeds characteristic of clouds with these densities and temperatures no higher than 10^5 K, values of η_0 are at least 3 and more probably of order 10, and hence condition (3) is generally satisfied. Self-sustaining propagation of star formation is thereby permitted under physical conditions in molecular clouds surrounding galactic nuclei.

3. ENERGETICS

With the initial density distribution in (1) we compute the balance between the energy generated in the stars and the kinetic plus potential energy required to drive out the gas. To push gas out from the shell between galactic radii R_1 and R_2 requires potential energy W:

$$W = \frac{16 \pi^2 G}{9} \rho_{to} \rho_o (1-\alpha)(R_2^3 - R_1^3) R_o^2 \tag{5}$$

where α is the fraction of gas which goes into massive stars, and ρ_{to} is the density at radius R_o of gas plus stars (the stellar density is for simplicity also given by a distribution law of form (1). The kinetic energy K of gas outflow is

$$K = \frac{1}{2}(1-\alpha) M_R V^2 = 2\pi (1-\alpha) R_o^2 R_2 \rho_o V^2 \tag{6}$$

The energy E available from stellar sources can be expressed as

$$E = \alpha \beta M c^2 = 4\pi \alpha \beta R_o^2 R_2 \rho_o c^2 \tag{7}$$

where β is a conversion factor: the efficiency of conversion of mass into energy coupled to the outflow. Equation (7) does not show the time dependence of this conversion. For outflow to be maintained

$$E > K + W \tag{8}$$

and this translates into a limiting condition on β, i.e.

$$\beta > \frac{(1-\alpha)}{\alpha} \left[\frac{V^2}{2c^2} + \frac{4\pi}{9} \frac{\rho_{to}}{c^2} \cdot R_2^2 \right] \tag{9}$$

where we have taken $R_2 >> R_1$ for simplicity. Using V=100 Km s^{-1} for $R_2 = 1$ kpc and assuming $\rho_{to} = \rho_o$ i.e. that the underlying stellar density is of the same magnitude as the gas density we find

$$\beta > 7.5 \times 10^{-7} \tag{10}$$

which is separable into $\beta_K > 5 \times 10^{-7}$ and $\beta_W > 2.5 \times 10^{-7}$ for kinetic and potential energy respectively.

This condition on β may not appear severe. The total radiative plus mechanical energy emitted by an O star during its lifetime is (roughly) 1% of the stellar mass; if only 1% of this were coupled to the galactic wind we would have $\beta = 10^{-6}$ which satisfies (10). A more realistic calculation based on known properties of winds from O stars (see eg. Abbott et al., 1975) yields the relation

$$\beta_{wind} = \frac{1}{2M_*} \left(\frac{\Delta M}{\Delta t}\right)_w \left(\frac{V_w}{c}\right)^2 t_w \tag{11}$$

where M_* is the stellar mass, $(\Delta M/\Delta t)_w$ is the mass loss rate, V_w the wind speed, and t_w the lifetime. For $M_* = 50\ M_\odot$, $t_w = 10^6$ yr, and

$(\Delta M/dt)_w = 5 \times 10^{-6}$ $M_\odot$ yr^{-1} at $V_w \simeq 1500$ km s^{-1} we find $\beta_{wind} = 1.5 \times 10^{-6}$. As half of any stellar wind is directed towards the galactic nucleus, we can place an upper limit to the effective β from winds as

$$\beta_{wind} < 7.5 \times 10^{-7} \qquad (12)$$

Given turbulent losses, winds are only marginally able to sustain the flow. There are additional energy sources available and we have been too conservative in several ways. (a) If 20% of the gas went into massive stars, β need be < 1/2 of the estimate in (10); (b) a velocity of 100 km s^{-1} at R_2 = 1kpc is high; for 50 km s^{-1} β_k could be 4 times less. (c) The radiative output from an O star in its lifetime is 10^{52} ergs which represents an equivalent for β of $\beta_r = 1.5 \times 10^{-4}$. Hence winds represent a coupling of radiative energy to outflow energy of only 1%. We examine below a mechanism for coupling some of the other 99% of the radiation to impel outflow in the gas. (d) O stars end their lives as type II supernovas. The total mass to energy conversion in a supernova is of order 1% of the energy output during the star's lifetime, but the fraction directed outwards as kinetic energy in a supernova remnant is 90% of the energy in the explosion (Chevalier, 1977). The effective β for supernovas β_s, is thus similar in scale to β_{wind}. A total expression for β is then

$$\beta = \gamma \ (\beta_{wind} + \beta_s + \delta \beta_r) \qquad (13)$$

where δ is a coupling factor for radiative to kinetic energy conversion (see § 4), and γ is a factor which allows some streamline flow from individual stars to be degraded into turbulence. Since $\beta_{wind} < 7.5 \times 10^{-7}$ $\beta_s < 7.5 \times 10^{-7}$, and $\beta_r \simeq 7.5 \times 10^{-5}$, the dynamics of the galactic outflow will be governed by γ and δ.

4. MOMENTUM COUPLING AND THE ROLE OF DUST

Since β_r is so large, any mechanism which can couple the radiation to the gas with δ values of only a few percent, can play a role in producing galactic winds. Dust is a ubiquitous feature of molecular clouds, due to the prompt explosion of the most massive stars, and offers a suitable mechanism. Considering only dust interaction we can set a requirement for a coupling factor ϵ_r which conserves linear momentum as:

$$\epsilon_r \beta_r = \frac{(1-\alpha)}{\alpha} \frac{V_R}{c} \qquad (14)$$

Here, ϵ_r is the ratio by which momentum transfer to the molecular gas must exceed the initial stellar photon momentum $L_r/c = \alpha \beta_r \ (M_R c^2/c)$. To yield expansion velocities V_R of order 150 km s^{-1} we need $\epsilon_r \simeq 10$. Faulkner (1970) showed how multiple scattering can yield ϵ_r values greater than unity in an axisymmetric expanding system. For absorbers of integrated optical depth τ_r we find

$$\varepsilon_r \simeq \tau_r \qquad (15)$$

In NGC 6946, which has a very dense cloud, τ_r for thermalized infrared photons was estimated (Lebovsky and Ricke, 1979) at 50 and VR of 50 km s^{-1} was observed by Muñoz-Tuñon et al. (1987). In general for a cloud of mass 10^9 M$_\odot$ and density distribution similar to that in (1), the hydrogen column density will be 5x10^{23} cm^{-2} and with a dust to gas ratio from Bohlin et al. (1978) within our galaxy we find $\tau_r \simeq 250$ at visible wavelengths, τ_r of order 10 in the infrared. As molecular clouds are generally more opaque than the clouds used by Bohlin et al. to estimate dust extinction in the UV near the Sun, the value of 10 is a lower limit, and we would predict dust to be a significant coupling agent in producing galactic winds.

To complete the analysis we look at the momentum equations for winds and supernovas. For stellar winds

$$\alpha\, M_R\, \beta_W\, \eta_W\, \frac{c^2}{V_W} = (1-\alpha)\, \frac{1}{2}\, M_R\, V_R \qquad (16)$$

where η_W is the fractional mass of the star expelled during its lifetime as a wind. Given $V_R \simeq 50$ km s^{-1}, $V_W \simeq 1000$ km s^{-1} and $\beta_W \simeq 1.5 \times 10^{-6}$ we find $\eta_W \simeq 0.1$. For a star of $M_* = 50 M_\odot$ this means $5 M_\odot$ is needed to drive the galactic wind. This is in accord with measured mass loss rates of 5×10^{-6} $M_\odot$ over 10^6 years. These figures, though only illustrative, show that winds can satisfy the momentum as well as the energy requirements. For supernovas the equivalent equation is

$$\alpha\, M_R\, \beta_S\, \eta_S\, \frac{c}{V_S} = (1-\alpha)\, \frac{1}{2}\, M_R\, V_R \qquad (17)$$

For typical supernova outflow velocities of order 10^4 km s^{-1} this yields for $\beta_S \simeq 10^{-6}$, a value of $\eta_S \simeq 0.01$. A $50 M_\odot$ star needs to emit $0.5 M_\odot$ in rapid outflow to balance the galactic wind momentum. This is characteristic of mass outflows in type II supernovas.

5. DISCUSSION

I have shown that a giant molecular cloud surrounding the nucleus of a spiral galaxy can be the site of a self-sustaining axisymmetric burst of star formation in which the gas is driven outwards by the combined effects of stellar winds, supernova explosions, and radiation pressure on dust. If 10% of the gas goes into massive stars, the process can be self-sustaining, in terms of energy and momentum. Of the 90% of the radiation left uncoupled, a significant fraction will be left to heat the gas along the lines originally proposed by Mathews and Baker (1971) as a stripping mechanism for ellipticals. Spirals are different as their disc geometry leads to higher local gas densities, and toroidal structures implied here. Whatever the trigger mechanism for the burst we would not expect much gas to be blown out of the spiral by the mechanisms described.

Among predictions of this model are : (a) A ring of enhanced gas density and residual outflow velocity within a couple of kpc of the nucleus. (b) Metallicity gradients: with turbulent conditions in the supernova-enriched gas near the nucleus ("LINERS"), and higher stellar metallicities in a toroid of young stars whose radius depends on the age of the burst. (c) Radial expansion in the gas near the nucleus. All these phenomena are observed, but much work remains to assess the frequency of the phenomenon in normal spirals.

References

Abbott, J.I., Castor, J.I., and Klein, R.I.: 1975, Ap. J. 195, 157.
Bohlin, R.C., Savage, B.D., and Drake, J.F.: 1978, Ap.J. 224, 132.
Chevalier, R.: 1977, Ann. Rev. Astr. Ap. 15, 175.
Elmegreen, B.: 1983, in 'Birth and Infancy of Stars' (Luca, Dmont, Stora Eds.), North Holland, p. 215.
Faulkner, D.J.: 1970, Ap. J. 162, 523.
Lebovsky, M.J., and Rieke, G.H.: 1979, Ap. J. 229, 111.
Mathews, W.G., and Baker, J.C.: 1971, Ap. J. 170, 241.
Muñoz-Tuñon, C., and Vilchez, J.M.: 1987, Astron. Astrophys. (In Press).

OBSERVATIONS OF OUTFLOW AND ITS CONSEQUENCES IN CIRCUM-NUCLEAR ZONES OF SPIRALS

C. Muñoz-Tuñon and John E. Beckman,
Instituto de Astrofisica de Canarias,
38200 - La Laguna,
Tenerife,
Spain.

ABSTRACT. The radial hydrogen H_2 + HI) distribution in spirals takes two forms: with a density peak at the nucleus or with a circular "hole" around it. Here we look at spectral and photometric properties of selected spirals and find cases of expansion in the circum-nuclear gas, as well as ring structures in gas and stars. The correlation between absence of central gas and size of nuclear bulge (Sa's have a lower bulge gas:stars ratio than Sc's), plus the presence of LINERS in the interstellar medium close to the nucleus indicate probable outflow during starbursts. In NGC 6946 we observe the outflow in H_α and NII, at velocities ~1000 km s^{-1}. In NGC 1068 and NGC 4736 expanding rings are found between 1 or 2 kpc, and in NGC 7331 a central LINER plus a static ring at 4 kpc. Outflow from the nucleus is probably a common feature of the formation of central regions of spirals.

1. INTRODUCTION

Galactic winds were proposed by Mathews and Baker (1971) to account for the lack of observable gas in elliptical galaxies, in spite of the undeniable continual feeding of the interstellar medium by stellar mass loss. Faber and Gallagher (1976) suggested that winds may also occur in spirals, because nuclear bulges have similar populations and velocity dispersions to ellipticals. They could not tell at that stage whether the "holes" in HI distribution might be filled with H_2. Subsequent CO measurements suggest that this is often not the case. Here we have collected evidence for the presence of winds near the nuclei of spirals. In two cases we observe the wind "blowing", i.e. expansion in the gas near the nucleus. In a third case a clear expanding ring of CO has been detected (Myers and Scoville, 1987), and in a fourth we look at a galaxy where the aftermath of a wind is detectable. Cumulative evidence points to a significant role for galactic winds in the development of bulges of spirals.

2. NGC 6946

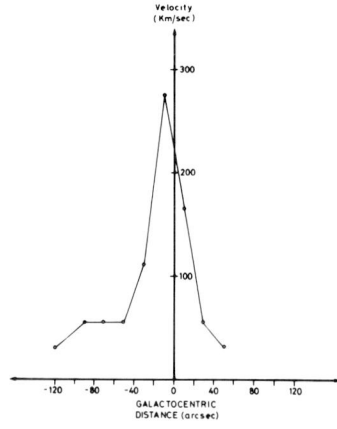

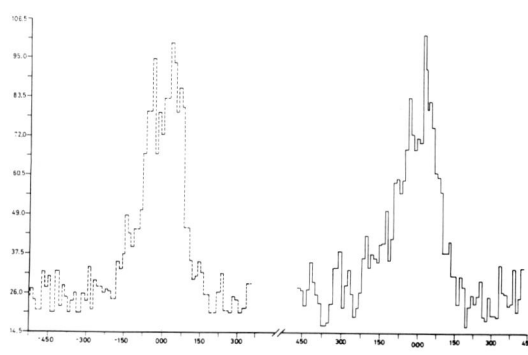

Fig. 2: Profiles of Hα and NII towards the nucleus of NGC 6946

Fig. 1: Velocity dispersion with radius for gas in NGC 6946.

In papers (Beckman et al., 1986; Muñoz-Tuñon et al., 1987; Muñoz-Tuñon, 1987) we show that for this Sc galaxy: (a) Velocity dispersion in the gas peaks on the nucleus (see Fig. 1), a sign of radial motion. (b) Line profiles in Hα and NII towards the nucleus show congruent splittings (see Fig. 2) with a stronger blue peak than red peak. This indicates expansion with measured velocities of order 50 km s^{-1}. (c) A flat-topped CO spectrum on the nucleus (Young and Scoville, 1982) with velocity dispersion 200 km s^{-1}, and enchanced blue wing. This, too indicates expansion, shared by the molecular gas as well as the atomic. We have shown in the works cited that the observed galactic wind in NGC 6946 could be fed by the 10^4 O stars known to be embedded in the central cloud through supernovas, or combined stellar winds or a combination of both. We showed that 10% of the central placental 10^9 $M_\odot$ of H_2 have been converted to stars in a starburst lasting 10^7 years, and we speculated on the future of the burst.

3. NGC 4736

In this galaxy there is a molecular ring some 0.7 kpc from the centre (Garman and Young, 1986) an expanding ring of HII regions some 1.5 kpc from the centre (Van der Kruit, 1976) and a dip in the HI density near the nucleus. This suggests NGC 4736 to be at a later stage of a similar phenomenon to NGC 6946. Sanders and Bania (1976) estimate a time interval of 10^7 years since an initial central explosion, but a refined calculation with steady flow fueled by star formation (see Beckman, this volume) yields 10^8 years. The fact that only 1.5% of the mass in the nuclear bulge NGC 4736 is gas, (cf. 13% for NGC 6946, Young and Scoville 1982, Garman and Young, 1986) and that it is Sab, suggests a more evolved object. Is there an evolutionary sequence implied here?

4. NGC 7331

This galaxy will be dealt with in detail elsewhere (Muñoz-Tuñon et al., 1987 b) and here we summarize our evidence. As shown in Fig. 3, (U-V) photometric mapping gives a steep colour gradient outwards from the nucleus, peaking at 3.5 kpc radius. The HII regions away from the nucleus show "normal" $H\alpha$/NII 6584 A intensity ratios and sharp lines, whereas on the nucleus there is a strong broad NII emission but no $H\alpha$ (see Fig. 5). In Fig. 6 we show that the ratios NII/$H\alpha$ and OIII/$H\alpha$ are much higher than in previous LINER spectra surveyed by Keel (1983), but this is due to our increased spatial and spectral resolution. The spatial distribution of NII and OIII is shown in Fig. 6 and in Fig. 7 we show a stellar metallicity index (Mgb). High interstellar metallicity coincides with low stellar metallicity and vice versa. All this is consistent with a starburst which left the circumnuclear gas rich in primary supernova products, and increased the metallicity of the newly formed stars in the blue ring ~4 kpc from the centre. We pose the question of whether all LINERS are old starbursts.

5. NGC 1068

Myers and Scoville (1987) recently presented evidence that in this SB Seyfert the H_2 as detected via CO, is in an expanding ring with radial velocity 70 km s^{-1} and radius 1.6 kpc. They speculate whether the bar can feed gas into the centre to fuel a starburst. Their observations fit well our framework of an evolutionary sequence starburst (outflow) LINER. Maybe here, as in NGC 6946 the active nucleus triggered the original starburst.

6. CONCLUSIONS

A model incorporating the observations summarized in this paper is presented in these proceedings (Beckman, 1987. The role of outflow in the development of galactic bulges appears to be important. The temporal sequence starburst-liner may be a common feature of galaxies and may help to understand the dependence of evolution rates of galaxies on mass (Sa's more massive, larger bulges, less central gas, more evolved).

References

Beckman, J.E.: 1987, Proceedings of the 2nd Torino Workshop "Mass outflows from stars and galactic nuclei" (This volume)

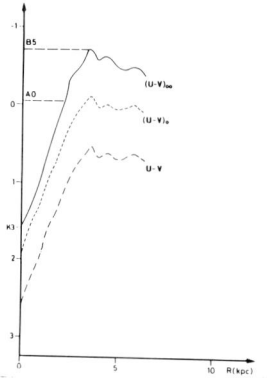

Fig. 3: (U-V) v. radius for NGC 7331 corrected for internal extinction

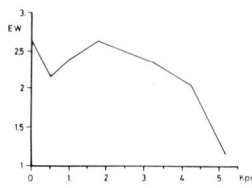

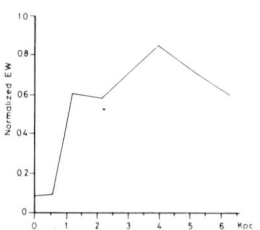

Fig. 7: Radial distributions of EW of Mgb in NGC 7331 uncorrected for stellar temperatures (upper) and corrected (lower).

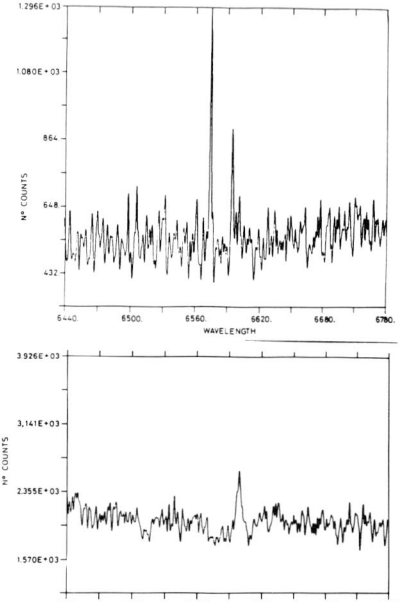

Fig. 4: spectrum of nuclear zone (no Hα, broad NII) and off-nucleus for NGC 7331.

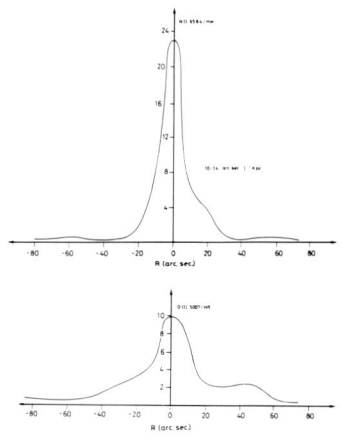

Fig. 6: Angular distributions of OIII/Hβ and NII/Hα in NGC 7331.

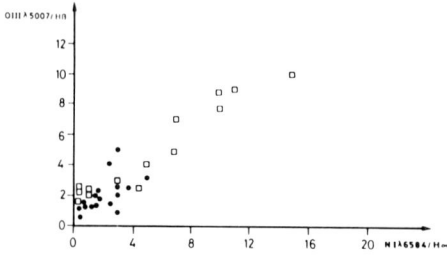

Fig. 5: OII/Hβ versus NII/Hα for NGC 7331 (squares) compared with other LINERS (circles)

Beckman, J.E., Muñoz-Tuñon, C., Battaner, E., Prieto, M., Sanchez-Saave-
    dra, M.L.: 1986, Astron. Astrophys. 161, 55.
Faber, S.M., Gallagher, J.S.: 1976, Ap. J. 204, 365.
Garman, L.E., and Young, J.S.: 1986, Astron. Astrophys. 154, 5.
Keel, W.C.: 1983, Ap. J. 269, 486.
Mathews, W.G., Baker, J.C.: 1971, Ap. J. 170, 241.
Muñoz-Tuñon, C., Beckman, J.E., Battaner, E., Prieto, M.: 1987 b,
    Astron. Astrophys. (In Press).
Muñoz-Tuñon, C., Beckman, J.E., Prieto, M.: 1987 a, Rev. Soc. Astr. Mex.
    Vol. 14 (In Press).
Muñoz-Tuñon, C., Vilchez, J.M.: 1987, Astron. Astrophys. (In Press).
Myers, S.T., Scoville, N.Z.: 1987, Ap. J. 312, L39.
Sanders, R.H., Bania, T.M.: 1976, Ap. J. 204, 341.
Van der Kruit, P.C.: 1976, Astron. Astrophys. 52, 85.
Young, J.S., Scoville, N.J.: 1982, Ap. J. 260, L41.

EVOLUTION OF PERTURBATIONS AND SHOCK FORMATION IN STELLAR WINDS AND JETS

E. Trussoni[1], A. Ferrari[2], R. Rosner[3], K. Tsinganos[4]

(1) Istituto di Cosmogeofisica del CNR, Torino, Italy
(2) Universita' di Torino and Osservatorio Astronomico, Torino, Italy
(3) Harvard-Smithsonian Center for Astrophysics, Cambridge, USA
(4) Department of Physics, University of Crete, Heraklion, Greece

ABSTRACT. We study the temporal evolution of disturbances in spherically symmetric and polytropic winds from a central source. Such disturbances may be due to localized momentum addition in the outflow, or localized deviation from spherically symmetric expansion. We follow the evolution of an initial steady state which is perturbed to a continuous or discontinuous final equilibrium state, as predicted by previous calculations of stationary flows. We show that some of the predicted discontinuous equilibrium solutions are not physically accessible, while the attainment of the other equilibrium solutions depends on both temporal and spatial parameters characterizing the perturbation.

1. INTRODUCTION

Shocks occur in several astrophysical environments. They form in the interplanetary medium, supernova remnants, star formation regions inside molecular clouds, active galactic nuclei and jets. In this short note we discuss a possible way to model the formation of shocks in astrophysical outflows, occurring in stellar or extragalactic scales. For our purposes we start from the classical theory of a thermally driven wind (Parker 1958), which via the interplay of the gravitational and pressure gradient forces in a polytropic atmosphere, leads to supersonic acceleration by means of a "nozzle" effect.

Recent detailed studies, in the framework of stationary analysis, have extended this theory, pointing out that, in a flow undergoing non-spherical area expansion and/or momentum addition, multiple critical points and corresponding solutions may be found. These solutions may be connected by standing shocks at the points where the Rankine-Hugoniot conditions are satisfied; and these shocks, where heating of gas and particle acceleration is likely to occur, have been proposed to be associated with morphological features observed in jets (Ferrari et al. 1986) and H-H objects (Silvestro et al. 1987).

When the stationary analysis predicts multiple degenerate critical solutions, the only way to differentiate among these solutions is through a time dependent analysis of the problem. Here we present some results showing the evolution of perturbations for a polytropic wind subject to localized non radial areal expansion and momentum addition.

2. HYDRODYNAMIC EQUATIONS

In writing the equations which govern the motion of the plasma, we follow the "quasi-two-dimensional" approximation, wherein the transverse stresses are included via a suitable form of the cross sectional area expansion (Trussoni et al. 1987). Then the continuity and momentum conservation equations for a polytropic gas can be written in dimensionless units:

$$\frac{\partial \rho'}{\partial t'} + \frac{1}{R^2}\left[\frac{\partial(\rho' R^2 V)}{\partial R}\right] = -\rho' B(R, t')$$

$$\frac{\partial(\rho' V)}{\partial t'} + \frac{1}{R^2}\left[\frac{\partial(\rho' R^2 V^2)}{\partial R}\right] = -\rho' V B(R, t') - \frac{\partial P}{\partial R} - \rho' \frac{Gm_o}{r_o v_{so}^2 R^2} + \rho' D'(r, t)$$

$$B(R, t') = \frac{V}{f}\frac{\partial f}{\partial R} + \frac{1}{f}\frac{\partial f}{\partial t'}$$

with

$$R = r/r_o, \quad \rho' = \rho/\rho_o, \quad t' = t/t_{cr}, \quad V = v/v_{so}, \quad P = p/(\rho_o v_{so}^2)$$

$$D'(R, t') = r_o D(r, t)/v_{so}^2, \quad t_{cr} = r_o/v_{so}, \quad v_s = 128(\alpha T_6)^{1/2} \text{ km s}^{-1}.$$

The numerical integration has been performed using the Flux Corrected Transport (FCT) algorithm, by using a spatial grid of about 250 mesh points to cover a distance range from $R = 1$ to $R = 5$.

3. RESULTS

We discuss solutions for a choice of the physical parameters typical of the solar wind: $R_o = R_\odot$, $m_o = m_\odot$, $T = 2 \times 10^6 K$, $\alpha = 1.1$, and analyze the case of an outflow subject to a nonradial expansion of the following form:

$$f(R, t) = \left\{f_m(t) \exp\left[\frac{(R - R_1)}{\sigma_1}\right] + c(t)\right\} \left\{\exp\left[\frac{(R - R_1)}{\sigma_1}\right] + 1\right\}^{-1},$$

with

$$c(t) = 1 - [f_m(t) - 1]\exp\left[\frac{(1 - R_1)}{\sigma_1}\right], \quad f_m(t) = 1 + (F_m - 1)[1 - \exp(-t/\tau_1)]$$

where F_m is the asymptotic value of the areal expansion, and τ_1 is the time scale of expansion. In addition, we also include the effect of momentum addition given by the following expression:

$$D(R, t) = D_o\left[1 - \exp(-t/\tau_2)\right]\exp\left[-\frac{(R - R_2)^2}{\sigma_2^2}\right].$$

We briefly discuss two interesting cases.

a) *Non-spherical area expansion.* In the previous formulae it has been assumed that $F_m = 12$, $R_1 = 1.5$ and $\sigma_1 = 0.1$. The stationary analysis predicts in this case three

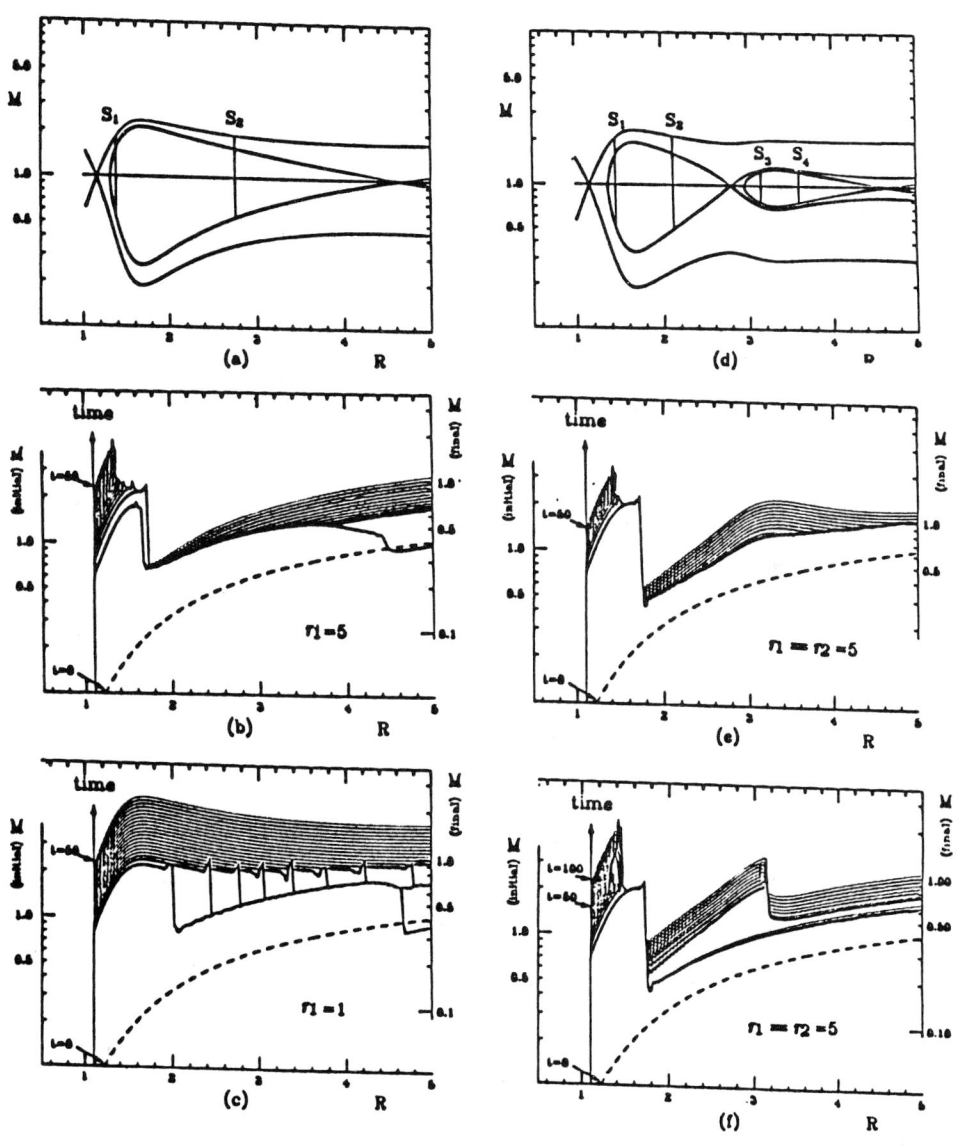

Figure 1. Topologies (Mach number vs distance) of wind-type solutions of the governing hydrodynamic equations with non-spherical area expansion (1a-c), and non-spherical area expansion plus momentum addition (1d-f). In 1a and 1d the stationary solutions are shown, while the time dependent evolution of perturbations is plotted every $\Delta t = 2.5 t_{cr}$ (1b,c), and $\Delta t = 5 t_{cr}$ (1e,f).

critical points and three solutions: one continuous solution through the inner critical point, and two discontinuous solutions involving two standing shocks at positions S_1 and S_2, respectively (see Fig. 1a). The time dependent results remove this degeneracy of the solutions, by showing the physical accessibility of the discontinuous solution through S_1 (Fig. 1b), and the continuous solution (Fig. 1c), while they show the inaccessiblilty of the discontinuous solution through S_2. The alternative among the two accessible solutions is ruled by the time scale τ_1 of the area expansion: for slow expansion time scales, the solution with the inner standing shock S_1 is obtained, while for faster expansion time scales, a shock is formed that is convected downstream, and the continuous solution is reached. We were unable to find a value of τ_1 such that we obtain asymptotically the discontinuous solution with the shock at S_2; this solution has been in fact shown to be unstable (Trussoni et al. 1987).

b) *Non-spherical area expansion and momentum addition.* The momentum addition has been assumed to peak downstream of the point of the localized non-spherical area expansion, with the following parameters: $D_o = 0.36$, $R_2 = 3$, and $\sigma_2 = 0.3$. In this case the stationary analysis predicts five critical points with four possible shocks: those already found in the previous case (S_1 and S_2) plus an additional shock pair downstream at positions S_3 and S_4 (Fig. 1d). The temporal evolution of solutions is always governed by the values of the time scales τ_1 and τ_2; the behaviour of solutions involving shocks S_1 and S_2 is similar to the previous case. Concerning the solutions connected with the second pair of shocks, we stress that there is again no possibility to obtain a discontinuous solution at this position, whichever are the time scales τ_1 and τ_2 (see Fig.1e). The only possibility to obtain a discontinuous solution with an additional shock at S_3 is to "switch on" the perturbation of the momentum addition at a later time (Fig. 1f): this position looks as if it is stable at least for relatively long enough time scales, such as $t = 100 t_{cr}$.

4. SUMMARY

Our two main conclusions from this temporal analysis are the following:
i) some discontinuous wind solutions predicted by stationary results have been found to be physically inaccessible;
ii) the possibility to have multiple shocks along the wind flow depends critically on the detailed time history of the imposed perturbation on the flow.

These results, although obtained for relatively simple configurations, are important to interpret the morphologies of outflows. The presence of stationary and convected shocks is observed in more complicated numerical simulations also: here we are in a position to discuss which physical parameters rule different types of solutions. An application of these conclusions to specific astrophysical outflows is rather suggestive, and will be presented elsewhere.

REFERENCES

Ferrari, A., Trussoni, E., Rosner, R., and Tsinganos, K., 1986, *Ap. J.* **300**, 577
Habbal, S.R., and Tsinganos, K., 1983, *J.G.R.* **88(A3)**, 1965
Holzer, T.H., 1977, *J.G.R.* **82**, 23
Parker, E.N., 1958, *Ap. J.* **128**, 664
Silvestro, G., Ferrari, A., Trussoni, E., Rosner, R., and Tsinganos, K., 1987, *Nature* **325**, 228
Trussoni, E., Ferrari, A., Rosner, R., and Tsinganos, K., 1987, *Ap. J.*, in press

THE FLOWING INTERSTELLAR MEDIUM IN THE HOST GALAXIES OF QUASARS

J. E. Dyson   and   J. J. Perry
Department of Astronomy,   Institute of Astronomy,
University of Manchester,   Madingley Road,
Manchester M13 9PL,   Cambridge CB3 9HD,
U. K.   U. K.

ABSTRACT. QSOs are associated with galaxies whose stellar populations provide an interstellar medium in the circumquasar environment. Arguments relating to continuum variability effectively rule out the standard model of broad emission line clouds confined by the thermal pressure of a hot plasma. We argue that these clouds are confined by the dynamic pressure of a mass loaded flowing interstellar medium which has a flat density profile resulting from the mass loading process itself. The mass loading is also responsible for the conversion of a small fraction of the flow into the emission line clouds.

1. INTRODUCTION

There seems little doubt that QSOs are associated with galaxies. As a result, a dynamically active interstellar medium is inevitable around any central black hole in a galaxy, partly as a result of stellar disruption during the fuelling process and of stellar ablation by the radiation field (e.g., Edwards 1980) and partly as a result of normal (e.g., stellar wind, supernovae) mass loss phenomena. A flowing medium is produced by these processes and the flow must be modified by these processes. Perry and Dyson (1985 - henceforth PD1) proposed that this interaction produces strong shocks in the flow. The superheated post-shock gas is out of equilibrium with the radiation field and cools quasi-isobarically by inverse Compton scattering, ultimately leading to the formation of cool (T $\approx$ 10^4K) line emitting clouds.

Much attention has been focussed on the so-called 'standard' model for these clouds (e.g., Krolik, McKee and Tarter 1981 - henceforth KMT) where an interstellar medium pressurizes these clouds thermally. Collin-Souffrin (1987) has queried the existence of the two phase thermal equilibrium discussed by KMT from considerations relating to the spectral shape of the radiation field. We now briefly demonstrate that, regardless of the details of the model, thermal pressurization is not consistent with observational - specifically continuum variability - data. Further, we demonstrate that constraints on the density profile of the ISM link in a natural way to the cloud formation model proposed in PD1.

2. OBSERVATIONAL CONSTRAINTS ON THE ISM

The ISM is kept fully ionized by the central radiation field at the Compton temperature $T_c \approx \langle h\nu \rangle / 4\pi$ which lies in the range $T_c \approx 10^6 - 10^8 K$. Current work (Fabian et al 1986; Collin-Souffrin 1987) favours the lower end of this range. The opacity is dominated by electron scattering and the optical depth down to the radius, R_c, at which the observed continuum is produced (the QSO 'photo-sphere') must be less than unity. Characteristically $R_c \approx 0.01 R_H$, where R_H is the radius at which the broad line clouds exist (Perry et al 1987). Ionization equilibrium calculations imply that $R_H \approx L_{47}^{\frac{1}{2}}$ pc (where L_{47} is the QSO bolometric luminosity in units 10^{47} erg s^{-1}).

The optical depth to electron scattering along a radius is (Perry and Dyson 1987 - henceforth PD2) $\tau \approx 4 \cdot 10^{-6} n_o (R_H/pc) \underline{S}'$, where n_o is the ISM density at $r = R_H$ and S' is a geometrical factor which is about equal to unity for spatially flat ($n \sim r^0$) flows and is considerably greater than unity for steep ($n \sim r^{-2}$) flows (PD2).

Thermal pressurization demands that the ionization parameter Ξ ($\equiv$ 2.3 x (ionizing radiation pressure/gas pressure)-KMT) be the same in both ISM flow and cloud. The required optical depth is (PD2) $\tau \approx 5(R_H/pc)(T/10^8 K)^{-1} S'$ which is always much greater than unity under any conceivable circumstances. Thus, regardless of model details, thermal pressurization can be ruled out.

On the other hand, dynamic pressurization (ISM ram pressure $\simeq$ cloud thermal pressure) can tolerate a higher ISM ionization parameter because the ionization parameter decreases by a large factor across a strong shock as a result of the post-shock increase of pressure (PD1). This translates into a much less stringent requirement on the flow, namely that $\tau \approx S'(R_H/pc)$ (for characteristic ISM flow velocity $\approx$ broad emission line velocity dispersion). However S' cannot be appreciably greater than unity, thus ruling out steep flows (PD2).

Flat flows arise naturally as a result of mass loading which decelerates supersonic flow, thus flattening the velocity profile (e.g., Perry 1986). Since, as argued in PD1, considerable mass loading of the ISM throughput must occur to produce the cool clouds in the post-shock gas, the requirement of a flat density profile fits naturally into the shock production cloud mechanism.

3. CONCLUSIONS

We have argued that observational restrictions on the electron scattering optical depth lead to the rejection of the thermal pressurization model for broad emission line clouds, but that dynamic pressurization is consistent with these restrictions. The ISM must have a much flatter density profile than normally assumed and this arises as a result of mass injection which in turn produces the shock waves that give rise to the cool clouds. This injection results from the presence of the stellar component of the galaxy, and the injected material is heavy element enriched as required to produce the observed line emission. Thus the stellar component of the galaxy fuels the central radiation source, produces the ISM gas, and is responsible for the flow interactions which ultimately produce cool clouds. A satisfyingly self-consistent picture seems to emerge.

REFERENCES

Collin-Souffrin, S., 1987, Proc. RAL Conference on Broad Emission Line Regions in QSO's and AGN (ed. P. M. Ghandalekhar), in press.
Edwards, A. C., 1980, Mon. Not. R. astr. Soc., 190, 757.
Fabian, A. C., Guilbert, P. W., Arnaud, K. A. Shafer, R. A., Tennant, A. F. and Ward, M. J., 1986, Mon. Not. R. astr. Soc., 218, 457.
Krolik, J. H., McKee, C. F. and Tarter, C. B., 1981, Astrophys. J., 249, 422.
Perry, J. J., 1986, in IAU Symposium 119, Quasars (eds. G. Swarup and V. K. Kapahi), D. Reidel Publ. Co., Dordrecht, p.307.
Perry, J. J. and Dyson, J. E., 1985, Mon. Not. R. astr. Soc., 213, 665.
Perry, J. J. and Dyson, J. E., 1987, Mon. Not. R. astr. Soc. (submitted).
Perry, J. J., Collin-Souffrin, S., Dyson, J. E. and McDowell, J. C., 1987, Mon. Not. R. astr. Soc. (submitted).

NEW ARGUMENTS SUPPORTING THE INTRINSIC ORIGIN OF BAL QSOs:
FORMATION OF THE COMPLEX Lyα+NV LINE PROFILE

D. Hutsemékers[1], J. Surdej[2]

Institut d'Astrophysique, Liège, Belgium.

Also, [1] aspirant and [2] chercheur qualifié au Fonds
National de la Recherche Scientifique (FNRS)

ABSTRACT. In order to further elucidate the question whether
all quasars are affected by the BAL phenomenon, we
investigate the possibility that observed BAL profiles are
formed in spherically symmetric expanding atmospheres.

1. EXAMPLES OF BAL QSOs

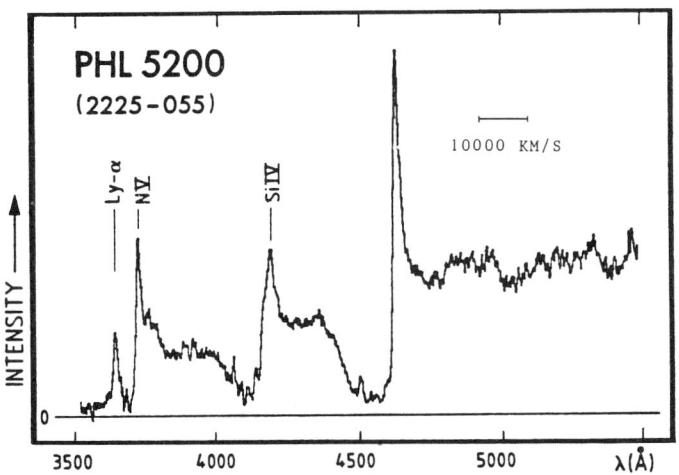

Figure 1.a : The spectrum of PHL 5200 (from Wampler, 1983)

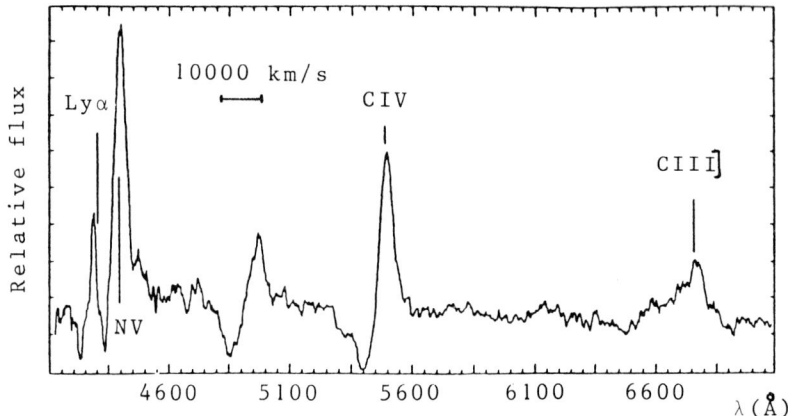

Figure 1.b : The spectrum of H1413+113 (from Turnshek, 1981)

Note that:

- i) on the average, the equivalent width EW_{abs} of the blueshifted absorption component of a BAL profile is substantially larger than the equivalent width EW_{em} of the scattered emission-line

- ii) the residual intensity observed in some highly saturated absorption troughs can be unusually small

- iii) the Lyα+NV line profile presents a complex structure.

Because of i) and ii), models involving jet-, fan- and/or disk-like geometries have been proposed in order to account for the BAL phenomenon. According to such models, BAL gas could be present around most, if not all QSOs.

2. REVIVAL OF THE SPHERICALLY SYMMETRIC BAL ATMOSPHERES

Surdej and Hutsemékers (1987) have recently shown that both observational constraints (i) and (ii) might be reproduced by the resonance scattering of line photons across spherically symmetric BAL regions, taking into account the possible effects due to turbulence (cf. the following figure taken from Lamers et al. 1987), occultation and/or electron collisions.

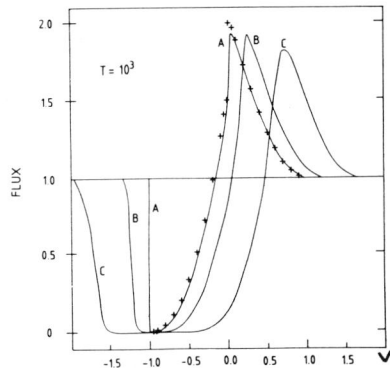

Figure 2 : Turbulence effects on the formation of P Cygni line profiles (from Lamers et al., 1987). Turbulence increases from A to C.

3. EMISSION-LINE PROPERTIES OF BAL QSOs: THE CASE OF NV

If the distribution of emission-line properties would appear to be the same for BAL and non-BAL QSOs, it would give further support to the view that all quasars undergo the BAL phenomenon. Let us compare the NV emission-line properties in BAL and non-BAL QSO spectra

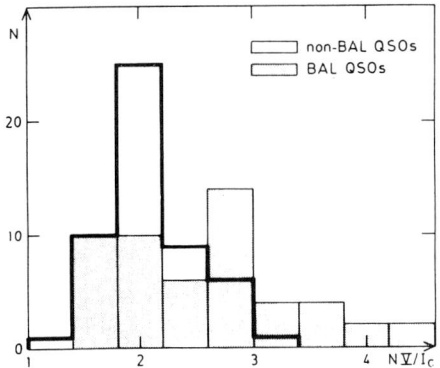
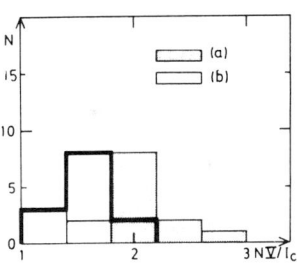

Figure 3.a: Normalized distributions of the NV/Ic emission-strength determined for 52 non-BAL and 26 BAL QSOs.

Figure 3.b: Referring to the sample of non-BAL quasars published by Young et al. (1982), this figure represents the NV/I_c distributions derived from the NV emission peak being measured from just above the continuum (a) as well as from the extrapolated red wing of the Lyα emission-line (b).

We confirm that the NV emission strength appears to be generally higher in the spectrum of BAL quasars.

4. FORMATION OF THE COMPLEX Lyα +NV LINE PROFILE

Using Sobolev-type approximations for the transfer of line radiation in spherically expanding envelopes, we find that an overall good agreement can be achieved between the theory and

- the observed attenuation of the Lyα emission-line
- the observed enhancement of the NV emission strength
- and the possible presence of a shoulder-like feature in the red wing of the NV emission profile.

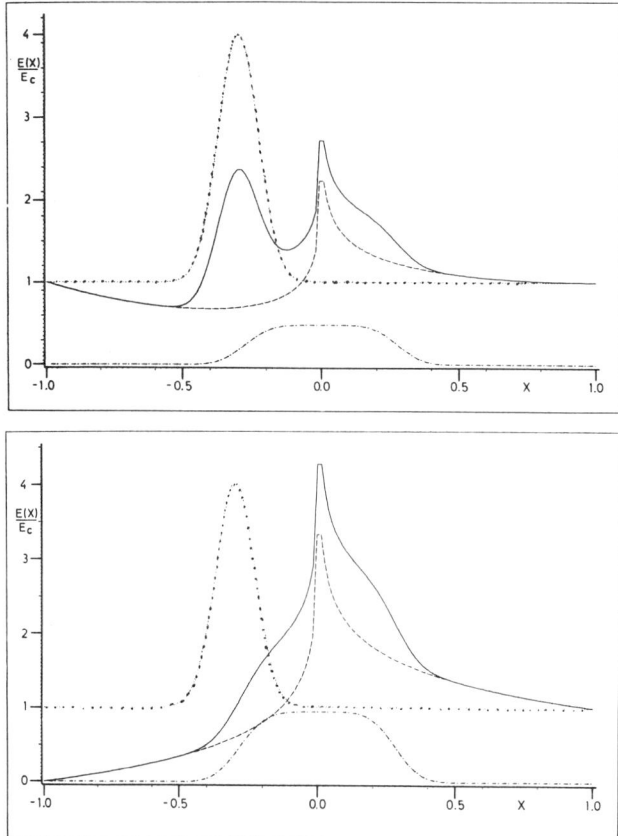

Figure 4 : Computed Lyα+NV line profiles for different choices of parameters. The full line represents the complex profile $E(X)/E_c$ versus the dimensionless frequency X. The dotted line corresponds to the underlying emission due to Lyα. The dashed line depicts the NV line profile that would be observed if there were no Lyα-N^{4+} interactions. Finally, the dotted-dashed line illustrates the contribution of Lyα photons to the NV emission profile.

5. CONCLUSIONS

All the above results strengthen the view that BAL and non-BAL QSOs form two distinct classes of quasars.

REFERENCES

- Lamers, Cerruti-Sola, Perinotto, 1987, Ap. J., <u>314</u>, 726
- Surdej, Hutsemékers, 1987, Astron. Astrophys., <u>177</u>, 42
- Turnshek, 1981, Ph. D. Thesis, University of Arizona
- Wampler, 1983, Astron. Astrophys., <u>122</u>, 54.

PART III: FUTURE SPACE MISSIONS

THE LYMAN MISSION

M.Grewing
Astronomisches Institut der Universitaet, D-7400 Tuebingen

1. INTRODUCTION

Around the World, scientists interested in Ultraviolet Astronomy have come to the conclusion that there is need for a spectroscopic satellite to observe at wavelengths below the hydrogen Lyman-alpha line. Indeed, several mission proposals have been received bythe European Space Agency (ESA), by NASA, and by agencies in other countries (e.g. Australia and Canada). In joint discussions started by study teams set up by ESA and NASA a single definition of the scientific objectives and their relative priorities, a single instrument approach and a basic concept for a mission called LYMAN was agreed upon already in 1983. More recently scientists and technicians from Australia and Canada have joined the study activities.

2. GOALS OF THE LYMAN MISSION

The prime goal of the LYMAN mission is high resolution spectroscopy in the 91-125 nm range, i.e. from the hydrogen absorption edge to just above the Lyman-alpha line. This range is complementary to both the IUE and the HST spectral ranges, and it will be observable with higher spectral resolution and roughly four orders of magnitude greater sensitivity than that achieved by the COPERNICUS (OAO-3) satellite in the 1970s.

The particular interest in this primary range results from the fact that it contains many of the most important atomic, ionic and molecular transitions, including the resonance lines of atomic hydrogen and deuterium, the molecules H2 and HD, which are all powerful diagnostic tools (cf. Fig. 1). This applies to dilute gases in the interstellar medium as well as to the atmospheres of stars and to extragalactic systems, especially the Active Galactive Nuclei and QSOs.

In addition, two further spectral ranges will be covered by the LYMAN mission, the 120-200 nm FUV range, and the 10-35 nm EUV range, the later one with an extension to 91 nm if feasible. While a resolution of

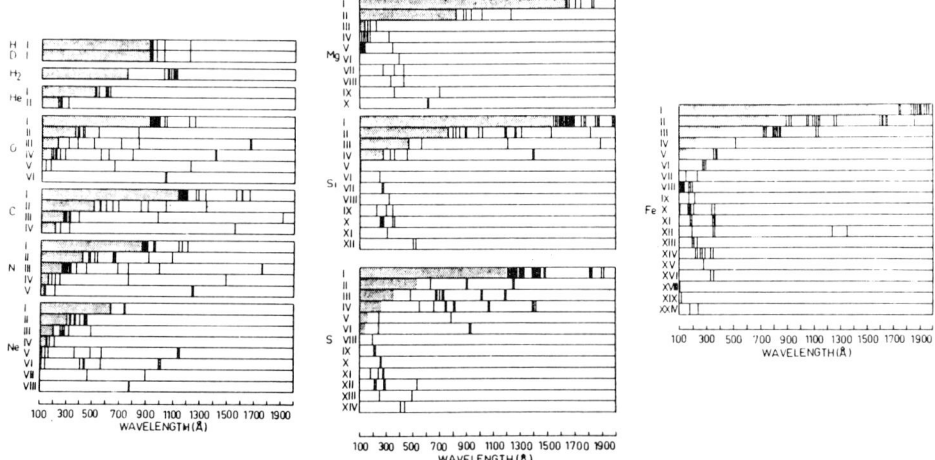

Figure 1: The distribution of atomic, ionic and molecular transitions throughout the spectral range accessible to the high and medium resolution spectrographs onboard the LYMAN observatory.

30.000 is required for the primary wavelength range, for the secondary ranges resolutions of 10.000 and >300, respectively, have been specified.

The scientific objectives behind these specifications are summarized in Table 1 taken from the ESA Report. A detailed description of the scientific aims has been given in the ESA Assessment Phase report (ESA Sci(85)4) and in the final report of the NASA Science Working Group for the Far Ultraviolet Spectroscopic Explorer (FUSE), and the reader is referred to these documents.

As shown in Table 1, the original requirement was to be able to observe objects as faint as 16th mag at the highest resolution, and even fainter objects at somewhat degraded resolution. This translated into an overall instrument efficiency equivalent to an effective photon collecting area of 100 cm . It turned out that due to the specific problems with reflective coatings at wavelengths below 115 nm (which are the reason for the short-wavelength cut-offs of the IUE and the HST), this is a very ambituous and correspondingly expensive requirement. For the ongoing ESA Phase A study activity for LYMAN it was therefore decided to take as a baseline a descoped version of the LYMAN mission as described in the document ESA Sci (85)4 which allows to do almost all of the science originally foreseen by trading orbit efficiency and integration time against effective collecting area. There is, of course, a limit to this set by the background noise, and about two magnitudes will be lost at the faint end of the target list.

Table 1: Lyman Science Goals

Science Objective	Target	Magnitude
Interstellar Medium:		
Galactic Halo	OB stars	4-15
Warm phase of Interstellar Medium	OB stars	4-15
Local Interstellar Deuterium	WD and sdO	12-16
H_2 and HD in Galactic Disk	OB stars	4-15
Extragalactic:		
Intergalactic Deuterium	AGNs ($Z \simeq 0$)	12-14
Intergalactic Helium	QSOs ($Z \simeq 2$)	16-17
Hot Plasma ($10^5 - 10^6$ K) in AGNs	AGNs	12-14
Halos around Galaxies	QSOs	16-17
Interstellar Medium in LMC and SMC	OB stars	12-14
Interstellar Medium in M31 and M33	OB stars	17-18
Hot Stars:		
Hot Plasma in Stellar Winds	OB stars	4-15
Abundances and T of highly evolved stars	sdO and WR	6-15
Mass-loss in Stars of different Galaxies	OB in LMC,M31	10-18
Cool Stars:		
Stellar Chromosheres	Cool stars	4-10
Winds and Mass-loss	Cool stars	4-8
Activity in Pre Main Sequence Stars	A, F, G stars	4-10
Interacting Binaries:		
Flows of Warm Gas in Binaries	OB X-binaries	6-15
Accretion of Low Mass Binaries	Sco X-1 stars	8-12
Mass transfer in Cataclismic Binaries	AM Her stars	14-16

Note: Missing from this compilation are solar system objects which as in the case of the IUE will clearly be observed with LYMAN. They are, however, not primary drivers of the design except that moving targets must be tracked (and a long slit capability of the spectrograph is desirable).

3. INSTRUMENT CONCEPT

The payload concept is based on a grazing-incidence telescope which is the only means to provide a reasonable collecting area throughout the 10-200 nm range. Such telescopes have already successfully been built, in particular in Europe, for imaging at x-ray wavelengths. The high spectral resolution to be achieved with LYMAN would, however, require further development of the imaging quality of such telescopes in order to achieve about 1 arcsec on-axis.

The light from the grazing-incidence telescope will be fed into one of three spectrographs : the prime, the FUV, or the EUV spectrograph. We shall focus our attention here on the first one. Given the need to resolve about 0.03 Angstrom, and taking into account the pixel sizes of currently (or soon) available array detectors with good quantum efficiency at 91-125 nm, one quickly comes to the conclusion that a very delicate decision has to be made due to the already mentioned fact that the normal incidence reflectivities of known metal coatings at these short wavelengths are quite poor (e.g. 20-30 %). To minimize the reflection losses, the obvious spectrograph concept is the Rowland circle design, with only one 'normal incidence' reflection. The price to pay ,however, is its large geometrical size which is driven by the resolution/pixel size requirements, and which implies a heavy, voluminous piece of equipment. Also, a very long curved detector is needed which would probably have to be built as a mosaic.

In the ongoing ESA study, an alternative spectrograph design based on the Echelle concept (much like in the IUE spectrographs) is therefore also considered. While this has many obvious attractions, including the easy to make compact-size detectors, it still remains to be demonstrated that the accumulation of reflection losses at the several 'normal incidence' surfaces leaves enough overall efficiency to meet the scientific requirements.

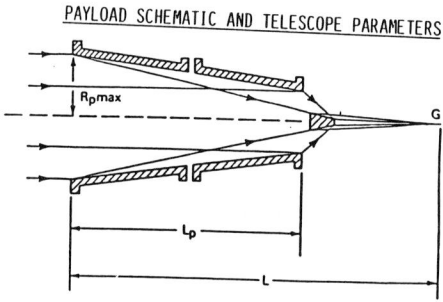

Fig.2: Schematic layout of the LYMAN grazing-incidence telescope.

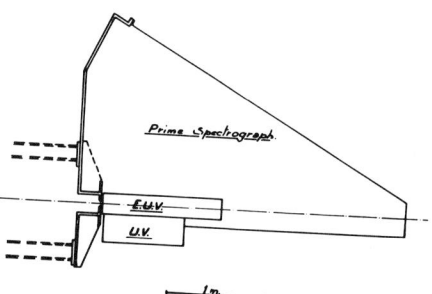

Fig.3: Schematic layout of the three spectrographs foreseen in the case of the Rowland option.

4. SPACECRAFT AND MISSION CONCEPT

A mission aiming at high spectral resolution necessarily requires a spacecraft capable to accurately point to, and stabilize on a given object. Given the faintness of many of the foreseen targets, the attitude information needed for this must come from the main instrument itself, i.e. from the actual field of view of the grazing-incidence telescope because no available star tracker could reach this faint.

This remark should be enough to indicate that the LYMAN mission requires quite a capable spacecraft bus. On the other hand, every attempt will be made to make use of existing (or currently developed) spacecraft busses in order to minimize the cost.

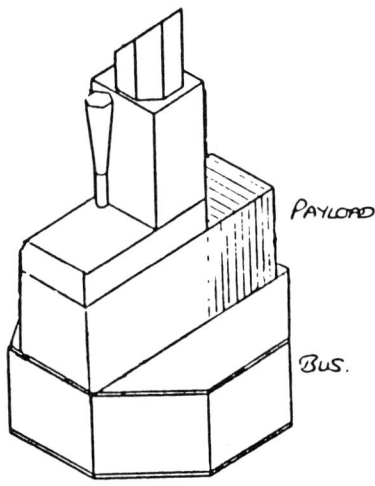

Fig.4:Possible satellite configuration showing the bus and the payload section.

One further important element in these considerations comes from the desire, expressed in particular by European scientists, to put the scientific instruments into a high Earth orbit which for LYMAN will be much more efficient than a low Earth orbit, and which can be reached by the ARIANE launcher. Such a choice would basically allow IUE-type, i.e. true observatory-type operations of LYMAN, which would make it a scientifically extremely promising mission indeed.

THE SOHO PROJECT

E. Antonucci

Istituto di Fisica Generale, University of Torino, Italy

ABSTRACT. The model payload of the Solar and Heliospheric Observatory (SOHO), part of the first cornerstone of the long-term ESA scientific programme, includes a package of instruments those primary scientific goal is the investigation of the processes leading to the formation of a hot solar atmosphere and the generation of the solar wind. The SOHO instrumentation will allow to establish in great detail the density and temperature structure of the solar atmosphere and the velocity fields, including the flow velocity of the expanding coronal plasma, from the base of the corona up to 6 solar radii. The combined remote and 'in situ' observations are expected to have as a result the identification of the processes at the base of the solar wind generation.

1. INTRODUCTION

More than two decades of 'in situ' observations of the solar wind, dating back to the beginning of space experiments, and more than one decade of detailed observations of the solar corona in the XUV domain have brought a wealth of information on the solar wind. Nevertheless, the basic mechanisms leading to the formation of a hot corona and to its supersonic expansion cannot yet be considered fully identified and understood.

The 'in situ' observations have been obtained up to now in the ecliptic plane beyond 0.3 AU; while, the inner heliosphere remains not yet explored. For what concerns coronal observations, it was possible to obtain, for the first time, systematic full disk images of the solar corona in XUV out to 1.5 R_θ and extend the limb observations of the white light corona out to 6 R_θ, in the years 1972-1973, during the Skylab mission. During solar cycle 21, new space experiments have extended the imaging of the solar corona to the shorter wavelength region of the spectrum by obtaining hard X-ray images up to 40 GeV, and achieved capabilities to obtain high resolution spectra also in the soft X-ray domain down to 2 A, with the US Solar Maximum Mission and Airforce P78 satellites and the japanese Hinotori mission.

The observations have clearly shown the highly inhomogeneous character of the corona which is in turn found in the solar wind structure in the heliosphere (e.g. Sime, 1985). The configuration of the corona is in fact completely dominated by magnetic fields, which are markedly inhomogeneous, and the coronal large-scale structure is evolving through the solar cycle according to the large-scale magnetic field evolution.

The topology of the coronal fields has been recognized as a determinant factor in the expansion of the coronal plasma. The brighter, denser regions of the corona are not considered to play an important role in the wind formation, although they can be responsible for mass ejections when they become active. They in fact consist of magnetic coronal loops, those closed field lines confine the plasma at temperatures of 2-3 10^6 K. On the other hand, the coronal holes (e.g. Zirker, 1977), corresponding to large photospheric unipolar regions, are characterized by an open diverging magnetic structure, and along their open field lines the solar wind is preferentially accelerated, giving origin to the fast solar wind streams (Hundhausen, 1977). Coronal holes are extended regions at lower density and lower temperatures (1.5 10^6 K), with pressure of the order of 10^{-2}-10^{-1} dyne cm^{-2}; although in the lower atmosphere such regions do not differ appreciably from the quiet solar regions. The effect of the diverging magnetic field configuration becomes, however, important in the transition region, which is believed to have a reduced pressure and a temperature gradient five times smaller than in the quiet sun.

The flow velocity of the fast streams originating in coronal holes is about 700 km s^{-1} at 1 AU, almost twice the value found for the slow wind streams. The source of the slow solar wind streams, characterized also by lower temperatures and higher densities, is more uncertain. Fast and slow streams are also differing for the helium abundance as measured at 1 AU (Borrini et al. 1983). The evolution and latitudinal distributions of coronal holes through the solar cycle has a direct influence in the three-dimensional velocity structure of the solar wind (Gosling et al., 1976, Coles et al., 1980).

2. SCIENTIFIC OBJECTIVES OF THE CORONAL EXPERIMENTS OF THE SOHO PROJECT.

Clearly a full understanding of the solar wind sources and expansion is directly linked to the identification of the basic non-radiative processes responsible for coronal heating, which may be different , or be effective in different degree, in closed and open magnetic field regions. Solar wind and corona, in fact, are to be considered as one physical system. The layers of primary heating in the solar atmosphere are identified as the lower chromosphere and the corona, while in the upper chromosphere-transition region layer the total radiation flux, of the order of 10^6 erg cm^{-2} s^{-1}, is balanced by the energy flowing back from the hot corona (e.g. Athay, 1985). Coronal heating processes, however cannot be confined just in the lower corona, since energy inputs are required far out in the extended corona to maintain the fast streams observed in the solar wind at 1 AU. In fact, although a solar wind in continuos expansion is a natural consequence of the hot solar corona, with temperatures above 10^6 K, thermally driven solar wind models (for density and temperature values at the base of the corona compatible with the observations) are not sufficient to explain the observed high flow speed in the fast solar wind streams. The observed and predicted velocities become consistent only when energy is added to the supersonic region of the wind flow. Energy input in the subsonic region has the effect of increasing the mass flux, but not appreciably the flow speed (Leer and Holzer, 1980; Leer et al., 1982).

Heating in the solar atmosphere is fluctuating both in space and time and the identification of the heating processes requires high resolution observations of the solar atmosphere. The plasma-magnetic field interactions appear to occur at a scale not resolved with the spatial resolution of the present instrumentation. This is the case of magnetic field dissipation induced by the photospheric motions of the footpoints of the coronal loops, proposed by Parker (e.g. 1972, 1983), and studied by many other authors, as a possible candidate of heating of the closed field line regions. In addition the velocity fields associated for instance to waves which are probably playing an important role in coronal heating, primarily in the open field line regions, are still unresolved. Hence, the observations required to study heating processes have to achieve, as well, a high spectral resolution to allow detailed measurements of profiles and Doppler shifts of XUV lines. Rocket and Spacelab experiments that have achieved 1 arcsec resolution have shown a wealth of small-scale features and phenomena in the transition region (Bonnet et al. 1980; Brueckner and Bartoe, 1983), which have not yet been systematically observed. The supersonic jets observed by Brueckner and Bartoe (1983), for instance, may have a role in the generation of the solar wind due to the high momentum associated.

The study of coronal heating and solar wind expansion does not only require the understanding of the fine structure and small-scale phenomena of the solar atmosphere. Other crucial measurements are needed. First of all, we lack of a complete mapping of the flow speed throughout the corona and inner heliosphere. The few measurements available are limited to the base of the corona where the flow velocity is still quite low (Rottman et al., 1982). Furthermore, measurements beyond 1.5 R_θ are limited up to now to the electron density and the magnetic field topology, which are obtained with the white light coronographs.

The coronal experiments included in the model payload of the Solar Heliospheric Observatory (SOHO), proposed in the ESA scientific programme, have been conceived to perform 'ad hoc' observations in these specific areas. They will provide the capability of mapping the physical conditions and dynamics, and in particular the flow velocity, throughout the solar atmosphere from its base out to 6-10 solar radii in the extended corona. Features of the chromosphere-transition region-corona will be observed at spectral, spatial and temporal resolutions suitable to study their fine-structure, dynamics and evolution, in order to establish the transport and deposition of energy through the solar atmosphere. In addition, the coronal observations will be linked to solar wind observations at 1 AU, obtained with in situ experiments.

3. SOHO CORONAL EXPERIMENTS

The SOHO and Cluster missions toghether form the first cornerstone of the long-term ESA scientific programme and are included in the ESA/NASA Solar Terrestrial Science Programme (Horizon 2000, ESA SP-1070, 1984). In addition to the study of the fundamental processes of the solar corona and solar wind, the SOHO mission is devoted to the investigation of the solar interior by using helioseismology. Cluster consists of an ensemble of four spacecraft, carrying homogeneous payloads, for the study of three-dimensional turbulence and the small scale structure in the magnetosphere. As a cornerstone, the two missions allow a coordinate investigation of the solar, heliospheric and magnetospheric plasma physics.

SOHO will be operating in privileged conditions, as a solar observatory, since it will orbit at the Lagrangian point L1, at $1.5 \; 10^6$ km from Earth in the Sun direction. The choice of the Lagrangian point has been suggested by the need of continuos pointing at the Sun to obtain uninterrupted time series of intensity and velocity data for helioseismology studies. The satellite is three-axis stabilized with a pointing stability of 1 arcsec/ 15 minutes. The launch is forseen in 1984, and the satellite will operate for at least 2 years. The selection of the actual payload will be completed during 1987.

The model payload of SOHO is formed by three packages of experiments devoted to coronal and solar wind remote observations, solar wind 'in situ' observations, helioseismology (ESA SCI(85)7).

The helioseismology package consists of a High Resolution Spectrometer to measure solar velocity oscillations in integrated sunlight, a Solar Oscillations Imager to record full-disk Dopplergrams with 2 arcsec resolution, and a Solar Irradiance Monitor to measure continuously the solar 'constant'. These instruments will probe the solar conditions in the core by measuring with high precision the compressional p-modes and the gravity g-modes of the Sun.

The solar wind 'in situ' package consists of a number of experiments to measure particles and magnetic fields in the solar wind at 1 AU, obtaining the three-dimensional distribution of electrons, protons and alfa particles, the solar wind composition, the particle energy spectrum.

The coronal instruments can be divided in two groups: one, formed by the Ultraviolet Coronal Spectrometer and the White Light Coronagraph, for observations of the extended corona out to at least 6 R_θ; the second, formed by the Grazing Incidence Spectrometer, the Normal Incidence Spectrometer and the EUV Imaging Telescopes, devoted to the observations of the atmosphere within 1.4 R_θ.

As mentioned above, the extended corona beyond 1.5 R_θ has been only partially explored up to now, since measurements have been limited to white light observations, which give mainly information on the electron density and the three-dimensional geometry of the corona. The SOHO instrumentation will allow for the first time the determination of the electron and ion temperatures and of the flow velocity throughout the solar atmosphere. This is achieved by measuring the profiles of UV spectral lines with the Ultraviolet Coronal Spectrometer. This instrument uses new diagnostic techniques (Kohl and Withbroe, 1982, Withbroe et al. 1982) based on the fact that coronal ions resonantly scatter the strong UV emission of the chromosphere and transition region. The combined measurements of the electron and resonantly scattered components of the radiation emitted from hydrogen (Lyman α and Lyman β emission) and heavier ions such as OVI, can provide information on both the electron and ion temperatures and on the magnitude of non-thermal velocities. Furthermore, it is possible to perform the crucial measurement of the flow velocity of the coronal plasma in its expansion by using a Doppler dimming technique. In fact, the radiation resonantly scattered by the coronal ions is reduced due to the Doppler shift of the coronal absorption profile relative to the chromospheric, transition-region line profile, introduced by the solar wind velocity. The instrument can be sensitive to velocities in the range from 25 to 300 km s^{-1}.

The instruments devoted to the observation of the chromosphere, transition-region and inner corona will allow the study of the fine structure at 1 arcsec angular resolution and of its temporal fluctuations with a resolution of a few tens of seconds by measuring the emission of UV lines. The Grazing Incidence Spectrometer and the Normal Incidence Spectrometer are complementary. The Normal Incidence Spectrometer achieves a better spectral resolution, reaching the capability to detect wavelength shifts corresponding to line of sight velocity of 1 km s^{-1}. The spectral domain of this instrument give access to lines formed in the 10^4-10^6 K temperature range. The Grazing Incidence Spectrometer, with a reduced spectral resolution, aims primarly to measure UV line pairs which allow to determine the density and temperature over the range 10^7-10^{13} cm^{-3} and 10^4-10^7 K respectively.

The combined observations of the coronal instruments on the SOHO satellite are then expected to provide a sufficient information to achieve a new understanding on the physics of the solar corona and the solar wind sources. This will be achieved:

by determining the role in the coronal heating processes of small-scale phenomena observed at the base of or in the inner corona, such as spicules or transition-region jets or microflares;

by identifying the signatures of the energy transport and deposition processes;

and by determining for the first time the density and temperature structure and the velocity fields from the base of the corona out to at least 6 solar radii and the evolution of such structure and fields through a considerable part of the solar cycle.

REFERENCES

Athay, R.G., 1985, Solar Phys., 100, 257.

Bonnet, R.M., Bruner, E.C., Acton,L.W., Brown, W.A., and Decaudin,M., 1980, Astrophys. J., 237, L47.
Borrini,G., Gosling,G.T., Bame,S.G.,and Feldman,W.C., 1983, Solar Phys. 83, 367.
Brueckner, E.G., and Bartoe,J.D., 1983, Astrophys. J., 272, 329.
Coles, W.A., et al., 1980, Nature 286, 239.
Hundhausen, A.J., 1977, in J.B. Zirker, 'Coronal holes and high speed wind streams.' Colorado Ass. Univ. Press, Boulder.
Gosling,J.T., Asbridge,J.R., Bame,S.J., Feldman,W.C., 1976, J. Geophys. Res., 81, 5061.
Leer,E., and Holzer,T.E., 1980, J. Geophys. Res., 85, 4681.
Leer,E., Holzer,T.E., and Fla,T., 1982, Sp. Sci. Rev. 33, 161.
Kohl,J.L., and Withbroe, G.L., 1982, Astrophys. J., 256, 263.
Parker,E.N., 1972, Astrophys. J., 174, 499.
Parker,E.N., 1983, Geophys. Ap. Fluid Dyn., 24,79.
Rottman,G.J., Orral,F.Q., and Klimchule,J.A., 1982, Astrophys. J., 260,326.
Sime,D.G., 1985, 'Future missions in solar, heliospheric, space plasma physics.' Proceedings ESA SP-235.
Withbroe,G.L., Kohl, J.L., Weiser,H., Noci,G., Munro,R.H., 1982, Astrophys. J., 254,361.
Zirker,J.B., 1977, "Coronal holes and high speed wind streams.' Colorado Ass. Univ. Press, Boulder.

QUASAT - A SPACE VLBI OBSERVATORY

R.T. Schilizzi
Netherlands Foundation for Radio Astronomy
Dwingeloo
The Netherlands

ABSTRACT: QUASAT was proposed to ESA as a joint ESA-NASA medium scale mission for space VLBI in December 1982. It has just entered an ESA Phase A study in European industry, supported by JPL and Australian and Canadian space interests. If QUASAT is selected to fly at the end of the Phase A study, it will be launched in 1995-1996. The scientific goals, mission concept, and current status are reviewed in this paper.

1. INTRODUCTION

   The quest for higher angular resolution has taken radio astronomy a long way since the first two element interferometric observations were made in 1946 with baselines of a few hundred metres. Current VLBI networks and the planned VLBA in the USA utilise the whole Earth as a radio telescope to achieve not only sub-milli-arcsecond angular resolution, but also good quality images. The next obvious step for the interferometry technique is to go beyond the confines of the Earth and position an element or elements in Earth orbit.
   Apart from the considerable technical challenge of going into space, there are sound scientific reasons for doing so, foremost amongst these is high quality imaging at increased angular resolution to investigate those sources and source components still unresolved with ground-based arrays. In addition, prospects for imaging in the equatorial region of the sky and in the southern hemisphere are significantly enhanced by the presence of a space element which can provide adequate U-V coverage for the first time. "Rapid" imaging on timescales of a few hours for variable galactic sources is also made possible, again because of enhanced U-V coverage compared to ground based arrays.
   The following sections outline the history of the QUASAT project, briefly summarise the scientific goals, and review the mission concept. The current status of QUASAT is discussed in the concluding section.

2. HISTORY OF THE QUASAT PROJECT

QUASAT was proposed to ESA as a space VLBI mission in November 1982 by an international group of astronomers. It was approved by ESA for an Assessment Study as a joint ESA-NASA medium-scale mission in early 1983 and this was carried out from October 1983 to June 1985 in parallel with a pre-Phase A study in NASA and on the basis of science requirements established jointly by science teams on both sides of the Atlantic. The early concepts for the mission and the scientific return to be expected were discussed at a Workshop on QUASAT held in Gross Enzersdorf near Vienna in June 1984, and published by ESA (Document SP-213) and reviewed again at a second Workshop held in Charlottesville, Virginia in May 1985. The results of the ESA Assessment Study were finally published in November 1985 as Document SCI(85)5, and presented to the wide scientific community in Darmstadt (FRG) in January 1986 as part of a competitive selection for Phase A studies. QUASAT was successful in passing this hurdle and was formally approved for Phase A in Europe in February 1986.

The afternath of the Challenger crash in February 1986 forced NASA to delay consideration of new projects in the Explorer programme, including QUASAT, and caused ESA to re-evaluate the cooperative framework under which QUASAT had been studied up to that time. This re-evaluation resulted in a mission concept with ESA as the leading agency but with substantial support from JPL, and Australian and Canadian space interests (ESA Document SCI(86)8). This new concept was approved for an ESA Phase A in February 1987.

3. SPACE VLBI DEMONSTRATED USING THE TDRSS

In July 1986, and again in January 1987, succesful interferometric observations of radio sources were made at 2.3 GHz using an orbiting spacecraft, the Tracking and Data Relay Satellite System (TDRSS), and radio telescopes in Japan and Australia (Levy et al. 1986, 1987). Twenty three out of 24 sources showed fringes on baselines up to 2.2 Earth diameters. This clearly demonstrates that space VLBI is a viable, and potentially very valuable, discipline in space astronomy.

4. SCIENTIFIC GOALS OF QUASAT

With QUASAT we expect to obtain high quality images at resolutions of ~60 micro arcsec at 22 GHz in the "final" orbit and lower quality images at angular resolutions of 40 micro arcsec in the "initial" orbit, (see discussion on orbits in the next section). The primary goal will be to probe the nuclei of radio galaxies and quasars more deeply and in greater detail than is possible with ground-based networks alone (see Table 1). The increased angular resolution will, for example, allow us to resolve the radio emission on scales of the Broad Line Region in many galaxies; the combination of good U-V coverage with increased resolution in the final orbit will bring this radio emission clearly into focus. In addition, the measurement of polarised emission at several frequencies will allow us to examine the physical characteristics of the emission and deduce the nature of the medium through which it passes. One other prime area worth mentioning is the use of astrometric techniques on H_2O masers in nearby galaxies to directly measure distances to those galaxies (Reid, 1987).

The main scientific goals are listed in Table 2 and described in detail in ESA SCI(85)5 and ESA SP-213. The most recent VLBI results can be found in the Proceedings of IAU Symposium 129 on The Impact of VLBI on Astrophysics and Geophysics (Eds. M.J. Reid and J.M. Moran) which is in the press. Some examples of these results can be found in my accompanying paper in these Proceedings.

Table 1: Spatial resolution at 1.35 cm wavelength.
(H_o=100 km^{-1} Mpc^{-1}, q_o=0.5)

Object	Distance	Linear size for 60 micro-arcsec beam	
Galactic Centre	7.1 kpc	6.7×10^{12} cm	0.4 A.U.
Centaurus A	5 Mpc	4.6×10^{15}	1.8 light days
NGC 1275 (3C84)	53 Mpc	4.6×10^{16}	0.6 light months
3C273 (z=0.158)	491 Mpc	3.4×10^{17}	4 light months
3C345 (z=0.595	1992 Mpc	7.4×10^{17}	0.8 light years

Table 2: QUASAT Scientific Goals

Measurement of high brightness temperature phenomena in:

- **Active Galactic Nuclei**
  Quasars, RG's, BLLacs/Blazars, Seyfert I+II's, Liners
  Images on scales of accretion disk, broad line region, narrow line region, collimated flow
- **Newly-formed stars**
  masers, flow patterns, magnetic fields, densities
- **Evolved stars**
  masers, mass-loss rates, densities
- **Binary flare stars**
  images of giant flares, energy generation
- **X-ray binary stars**
  images of mini-jets, collimated flow

Extragalactic masers
  properties of neutral gas in active nuclei

Scattering properties of the interstellar medium

Distances within our galaxy

Distances to nearby galaxies - Hubble Constant

5. MISSION CONCEPT

5.1. General

QUASAT is a free-flying satellite carrying a 15 m class radio telescope in elliptical orbit around the Earth. The radio telescope will be deployed in space, and used to make interferometer observations of radio sources in conjunction with the major ground-based VLBI arrays in Europe, USA, Australia and the USSR. The mission design lifetime is 2 years, but an operational lifetime of 5 years is expected.

The space-borne antenna will be capable of observing in both hands of circular polarisation simultaneously at any two of the frequency complement of 22, 5, 1.6 and 0.3 GHz, and will relay the received signals via a digital or analogue link directly to telemetry stations (NASA DSN and/or ESA) on the ground (see Figure 1). The bandwidth of the dual polarisation link will be at least 32 MHz in each hand of polarisation. A phase/frequency reference for the antenna in space, stable to about 1×10^{-14}, will be based on hydrogen maser oscillators on the ground and relayed directly to the satellite via a two-way link from the telemetry stations in turn. All communication with the space element will be through one or more telemetry stations in the network.

After transmission to the ground, the signal will be recorded on magnetic tape in digital form, and transported to the central processing facility of the European or US VLBI array for correlation with similar tapes from the ground VLBI arrays. After correlation and calibration the data will be sent to the principal investigators for further analysis.

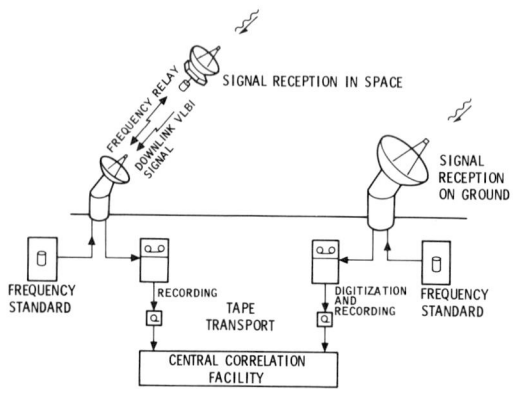

Figure 1. Schematic diagram of a space/ground VLBI system.

5.2. Orbits

Two orbital situations have been considered:
1) an "imaging" orbit, optimised for very high quality imaging with the European and US grounds arrays. The perigee altitude is 5700 km, the apogee altitude 12500 km, and the inclination 63°. This orbit would require a dedicated launch since no passengers to this orbit have been identified. Reduced levels of funding projected for the next medium scale mission in ESA, make the cost of getting to this orbit unattractive.
2) a "two-tier" orbit to provide high angular resolution with lower quality imaging (the "initial" orbit) before bringing the satellite down to lower altitudes for high quality imaging at somewhat lower angular resolution (the "final" orbit) after perhaps 1 or 2 years. The <u>initial orbit</u> has perigee and apogee altitudes of 5000 and 36000 km respectively, an inclination of 30° and a period of $12^h.2$. The <u>final orbit</u> has its apogee altitude reduced to 22000 km and the period to $7^h.75$. The advantage of this concept is that QUASAT would share a launch to geostationary transfer orbit, and thus share the costs. Figure 2 depicts the two-tier orbit.

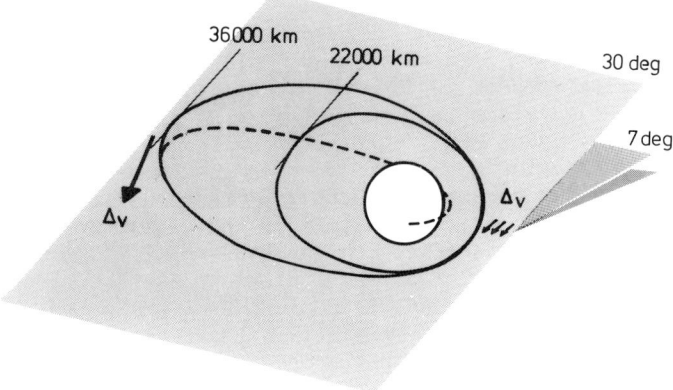

Figure 2.   The orbit injection sequence, and two-tier orbit.

Computer simulations of the imaging potential of QUASAT in the "imaging" orbit have demonstrated that superb results are obtained (ESA SCI(85)5). Simulation of the imaging potential of the "initial" and "final" orbits is in progress at Jodrell Bank as part of the Phase A study. Table 3 lists the angular resolution and sensitivity expected for QUASAT in the "final" orbit at its observing wavelengths of 1.35, 6, 18 and at 92 cm.

Table 3: Angular resolution in the "final" orbit and sensitivity of QUASAT

λ (cm)	Angular resolution* (micro-arcsec)	RMS noise per resolution element (continuum)**	RMS noise per spectral channel+	Frequency (GHz)
1.35	60	1 mJy	35 mJy	22.0-22.50
6	270	0.5	-	4.72-5.02
18	800	0.3	20	1.60-1.72
92	4000	3	-	0.324-0.327

* this is the dimension of the synthesized beam. Information on size and position angle is in fact available on scales 2 to 3 times smaller than this.
** for a 48 hour observation assuming there is a sufficiently strong feature (~150 mJy at 1.35 cm and ~50 mJy at 6 and 18 cm) present in the map to allow self-calibration.
+ for a 48 hour observation assuming there is a spectral feature present of sufficient strength (~20 Jy at 1.35 cm and ~5 Jy at 18 cm) to allow self-calibration.

5.3. Frequencies

The maser line emissions at 22 GHz for H_2O and 1.6-1.7 GHz for OH dictate two of the frequency complement. 22 GHz is at the same time the highest operating frequency of the ground networks for which there are telescopes of large collecting area and for which good sensitivity for both continuum and line can be achieved with a 15m antenna in space.

The third frequency of 5 GHz, approximately the geometric mean of the other two, has been chosen to fill the gap between 22 and 1.6 GHz, and provide a different combination of resolution and surface brightness sensitivity. On the ground, 5 GHz is one of the prime frequencies for high quality imaging since it is least affected by problems due to variations in atmospheric propagation delay. The sensitivity will be 2 to 3 times higher than at 22 GHz, thereby widening the range of objects studied to include the less luminous galaxies.

QUASAT at 5 GHz will have similar resolution to the global ground network at 22 GHz, and at 1.6 GHz a similar resolution to the ground network at 5 GHz. It will thus be possible for the first time to compare spectral and polarisation effects at different wavelengths at these extreme resolutions. Spectral index measurements will also be possible between QUASAT at 22 GHz and millimetre VLBI networks at 89 GHz, since their angular resolutions are similar. This will however only be possible for the stronger compact sources since the sensitivity of millimetre VLBI will be an order of magnitude lower than for QUASAT at 22 GHz.

5.4. Science Payload

The Assessment Studies have shown that the overall mission, which is based on current VLBI and spacecraft engineering practice, is technically feasible. As far as the spacecraft is concerned the major new aspect is the deployable antenna. Two antenna configurations have been studied; in ESA a 15 metre class inflatable space rigidised structure developed by Contraves, Switzerland (see Figure 3), and in NASA a 15 metre wrap-rib deployable mesh structure developed by the Lockheed Corporation, California. Both appear to be satisfactory for the mission. Other areas of development are the radio astronomy feeds and an active cooling system for the radio astronomy receivers. An industrial study of the feeds has shown good performance for a compact three frequency co-axial feed at prime focus which can also accomodate a dipole array for 92 cm. The radio astronomy receivers will be based on HEMT amplifiers, cooled actively by Stirling cycle coolers.

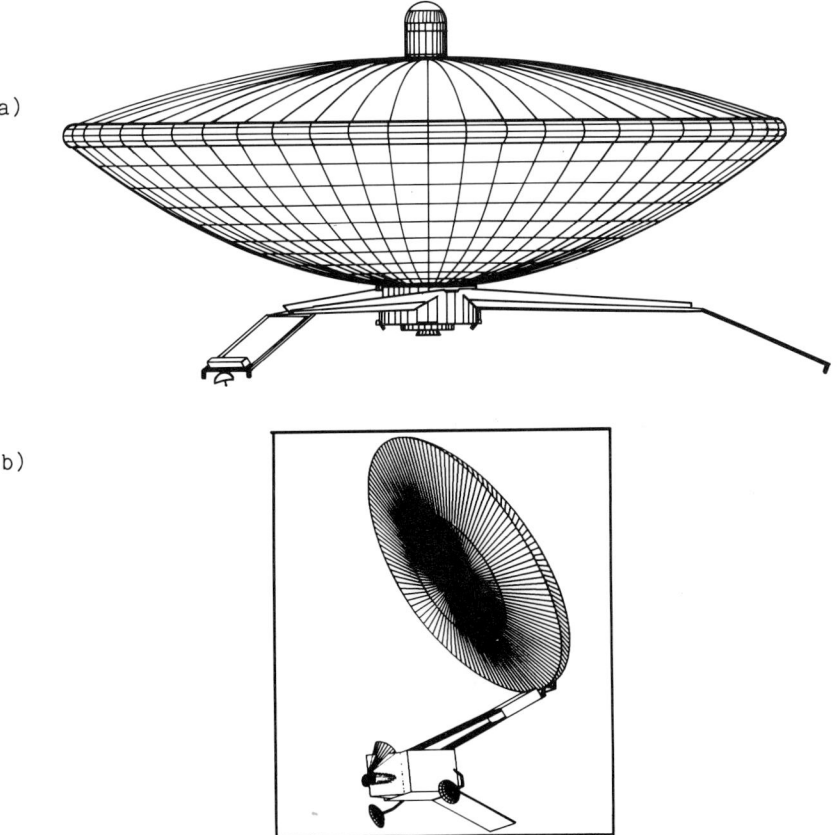

Figure 3 ESA concepts for the orbiting antenna: a) centre fed, b) offset fed

6. CURRENT STATUS AND FUTURE DEVELOPMENT

The industrial Phase A study finally began on 12 May 1987 and is being conducted by a European consortium led by Aeritalia in Turin, Italy. NASA has at the same time allocated funds to JPL for support of the ESA Phase A, particularly in the area of antenna technology, links, tracking, and operation. Details of the Australian and Canadian participation are currently under discussion.

The "science payload" - feeds, receivers, coolers, synthesisers etc but excluding the antenna - is to be studied by the radio astronomy community in parallel with the Phase A, with the aim of establishing the costs and effort required in constructing space qualified payload elements. Institutes in Europe, USA, Australia and Canada have been identified for the various elements.

QUASAT will enter a competitive selection round in ESA at the end of 1988. Assuming it is selected, the project will then move into a detailed design phase in 1989-90, followed by the construction and testing phase from 1990-1994. Final integration of the spacecraft will occur in 1995 with launch in that same year or in 1996.

REFERENCES

Levy, G.S. et al (1986) Science 234, 187

Levy, G.S. et al (1987) Proc. IAU Symposiuim 129 on The Impact of VLBI on Astrophysics and Geophysics, (Eds. M.J. Reid, J.M. Moran), May 1987, in the press

Reid, M.J. (1987) Proc. IAU Symposium 129 on The Impact of VLBI on Astrophysics and Geophysics, (Eds. M.J. Reid, J.M. Moran), May 1987, in the press

THE INTERNATIONAL EUV/FUV HITCHHIKER (IEH)

IEH Team

Requests for further information may be made to: Prof. M.L. Malagnini, DAT.

FOREWORD. The Lunar and Planetary Laboratory of the University of Arizona, the Astronomical Observatory of Trieste and the Department of Astronomy of the University of Trieste are collaborating in the preparation of a space flight experiment which has been approved by NASA and the Italian Piano Spaziale Nazionale (PSN) for Hitchhiker and Spartan flight opportunities on the STS.
Principal investigators for the experiment, International EUV/FUV Hitchhiker (IEH), are A.L. Broadfoot of the Lunar and Planetary Laboratory and R. Stalio of the Department of Astronomy of the Trieste University. A.L. Broadfoot and R. Stalio represent groups of scientists working in the fields of planetary science and ultraviolet astronomy respectively.
The preparation of the experiment is progressing as a collaborative effort with each of the P.I. groups sharing in the design, construction, flight sequencing, and data analysis effort.

The members of the IEH team are:

A.L. Broadfoot, PI, planetary programs	LPL
R. Stalio, PI, astronomy programs	DAT
G.L. Chincarini, Chm. extragalactic programs	OAB
C. Cosmovici, Chm. solar system programs	IFS
M. Hack, Chm. stellar atmosphere programs	DAT
J. Holberg, Chm. instr. calibration	LPL
D.L. Judge, Chm. absolute solar (EUV) flux programs	SCU
C. Morossi, Chm. star fields	OAT
R.S. Polidan, Chm. Software	LPL
B. R. Sandel, Chm. planetary atmosphere programs	LPL
L.M. Berens, software	LPL
M. Comari, software	OAT
S. Ferluga, instr. calibrations	DAT
T. Forrester, software	LPL

M. Franchini, software and instr. calibrations   ISAS
S. Furlani, instr. configuration   OAT
L. Lampi, star fields   OAT
R. Longo, star fields and instr. calibration   DAT
A. Magazzu', software and instr. calibrations   OAC
M.L. Malagnini, star fields and instr. calibration   DAT
L. Rossi, star fields and instr. calibration   IAS
P.L. Selvelli, instr. calibrations   GNA

DAT   Dipartimento di Astronomia, Universita' degli Studi di Trieste, via G.B. Tiepolo 11, I-34131 Trieste, Italy.
GNA   Gruppo Nazionale di Astronomia del CNR, Unita' di Ricerca di Trieste, via G. B. Tiepolo 11, I-34131 Trieste, Italy.
IFS   Istituto Fisica dello Spazio Interplanetario, CNR, via G. Galilei, I-00044 Frascati, Italy.
ISAS International School Advanced Studies, Strada Costiera 11, I-34014 Trieste, Italy.
IAS   Istituto Astrofisica Spaziale,CNR, I-00044 Frascati, Italy.
LPL   Lunar Planetary Laboratory, University of Arizona, 9th floor, Gould-Simpson Building, Tucson, AZ 85721.
OAB   Osservatorio Astronomico Brera, via Brera 28, I-20121 Milano, Italy.
OAC   Osservatorio Astronomico di Catania,
OAT   Osservatorio Astronomico Trieste, via G.B. Tiepolo 11, I-34131 Trieste, Italy.
SCU   Center for Space Sciences, University of Southern California, Los Angeles, California 90089-1341, U.S.A.

1. THE IEH EXPERIMENT

The experiment is an imaging spectrographic facility which will be an efficient tool for obtaining astronomical and planetary observations in the 400 to 1300 A region. One of the characterizing features is that it will permit spectral imaging of extended planetary and astronomical sources. The experiment complement includes a companion instrument that has been proposed by D. Judge (University of Southern California) to measure the absolute solar flux at selected UV bands.

The IEH is an enlarged version of the Voyager ultraviolet spectrometers (UVS), instruments of demonstrated capability. Like the UVS, the instrument is an objective grating spectrograph employing a mechanical collimator to restrict the field of view. Its most important advance over the Voyager instruments are, (1) the use of an area detector giving good spatial resolution capabilities in the cross dispersion direction, (2) significantly higher spectral resolution (approx. 0.8 A), (3) ten times larger aperture, and (4) greater sensitivity.

Figure 1 illustrates the configuration of the instrument and Table 1 gives the instrument characteristics. The IEH operates in the 400 to 1300 A region, where refractive materials have no transmission and even reflections are very inefficient. With a single reflection the concave grating images the target on the detector, dispersing in wavelength and forming monochromatic images of an extended emission line source within the field of view (3'x15') of the mechanical collimator. That is, an image of the source will be formed in the light of each of its emission lines.

Two grating will be used to optimize the throughput over the whole spectral range 400-1300 A. The best performance will be obtained by optimizing the blaze, reflective and quantum efficiencies in bands 400 A wide about 600 and 1000 A.

Table 1. IEH characteristics

	Spectrometer 1	Spectrometer 2
Mirror Size	15 x 15 cm²	15 x 15 cm²
Central Obscuration	25 cm²	25 cm²
Effective Area	200 cm²	200 cm²
Meccanical Collimator		
Field of View	.05°x.25°	.05°x.25°
Point Source Transmission (Tc)	.59	.59
Solid Angle of Acceptance	3.8×10^{-6} sr	3.8×10^{-6} sr
Spatial Resolution	4.5''x 4.5''	4.5''x 4.5''
Spectral Range	400-860 A	840-1300 A
Spectral Dispersion	0.4 A/pixel	0.4 A/pixel
Concave Gratings (parabolic, holografic)		
Radius of Curvature	2 m	2 m
Ruling Density	600 l/mm	600 l/mm
Blaze Angle	1.03°	1.85°
Blaze Wavelength	600 A	1075 A
Angle of Incidence	2.06°	3.70°
Angle of Diffraction (Center)	0°	0°
Plate Dispersion	16.7 A/mm	16.7 A
Reflective Coating	Os(?)	Os(?)
Estimated Efficency (η) on Blaze	7%	7%
Intensifier		
Cathode	LiF	CsI
Quantum Efficency (Center Wav.)	.40	.30
Active Area	25 mm D.	25 mm D.
Wavelength Transducer (Ph)	P20	P20
Area of Fiber Coupler (Approx)	4.25 x 25 mm²	4.5 x 25 mm²
CCD (EEV, P8602)		
Pixel Size	.022 mm²	.022 mm²
Array of Pixels	385 x 576 px.	385 x 576 px.
Area	8.47 x 12.67 mm²	8.47 x 12.67 mm²

The adoption of an area detector in place of the photographic plate provides single photon counting capability. The detector system is an intensified charge coupled device, ICCD. The intensifier is a windowless microchannel plate with phosphorus/fiber optic output coupled coherently to a CCD readout.

Point source fluxes of approximately 3×10^{-13} erg s^{-1} cm^{-2} Å^{-1} at 600 Å and 1×10^{-13} erg s^{-1} cm^{-2} Å^{-1} at 1075 Å will be observable with S/N=10 in 30 minutes. For extended emission sources we must multiply the previous figures by Ad/Ω F^2 where Ad is the pixel area, F the focal length of the instrument and Ω the solid angle subtended by the source.

Numerous applications of the spectrometers to planetary and astronomy research are envisioned. Among the most important planetary targets are the Io plasma torus and comets. The Voyager UVS discovered the presence of the hot plasma (electronic temperature of approximately 300,000 K) consisting of singly and doubly ionized sulphur and oxygen ions forming a complete toroidal cloud about Jupiter near Io's orbit. The EUV measurements of torus emission are a crucial link in understanding plasma torus physics.

The research from IEH will cover both galactic and extragalactic astronomy. In the FUV spectral region IEH will give information on variability, activity in binary stars, FUV interstellar extinction, mass loss etc. In addition the spectral imaging capabilities of IEH will allow the study of extended sources such as planetary nebulae, HII regions, supernova remnants, star formation regions, bright galaxies.

The EUV astronomy program will be fundamentally different from that at longer wavelengths because of the opacity of neutral hydrogen in the interstellar medium. In general we will be able to observe only within a few tens of parsecs from the Sun; however in those directions where the interstellar medium has low density we will be able to go farther (to about 50 parsecs) and possibly obtain a map of the patchiness of the local interstellar medium more detailed than those presently available.

Figure 1

SUBJECT INDEX

Algol systems	95
Extinction (interstellar)	231
Galaxies:	
AGN: accretion discs	377,381
---: jets	163
QSO: their galactic environment	401
---: BAL QSOs	151,405
Spirals : winds from nuclei	385,391
HH objects	147,263,293
Jets: see also Young stars and Galaxies	
----: models	163,317,321,373
----: instabilities	397
----: superluminal sources	151,163
Luminous Blue Variables	39
Mass loss: see Stars, Galaxies	
Outflows: Theoretical requirements for models	177
Nebulae:	
Planetary nebulae: dynamics	123
-----------------: evolution towards	123,137
-----------------: formation	123,291
-----------------: imaging	259,289,343
-----------------: masses, sizes	109
-----------------: nuclei (mass loss)	63,109,123,267
-----------------: nuclei, lifetimes	109
-----------------: shells(diameters,multiple)	257,259
-----------------: spectroscopy	257,273,285
-----------------: velocities	63,257,273,285
Reflection nebulae:	263
Wind bubbles	223
Stars:	
Binaries: interacting	95
--------: symbiotic,RS CVn	325,333
--------: X-ray	151

Cool stars: chromospheres   25
----------: dust shells   325,337
----------: magnetic fields   353
----------: mass loss   3,25,353
----------: wind velocities   25,79,353
Evolution: nucloesynthesis   279
----------: post-AGB   137
----------: thermal pulses   137
Hot stars: abundances   215,231
----------: atmospheres   231
----------: Be, disks   205, 211
----------: mass loss, effect of rotation   39
----------: mass loss, rates   39,95,215,227,253
----------: mass loss, time scales   177
----------: mass loss, variability   39,205
----------: winds, acceleration   3,63,79,177,195,239
----------: winds, emission(IR,radio,X-ray)   219,227
----------: winds, heating   79,195
----------: winds, hydrodynamics   239
----------: winds, interaction with i.s.matter   123,223
----------: winds, magnetic fields   235
----------: winds, multiphase   177
----------: winds, velocities   63,79,231
----------: Wolf-Rayet   231,199
Young stars: bipolar outflows   147
-----------: stellar winds   295,299,309
-----------: related structures(mol.clouds etc.)   147,263,293,299,303,
                                                        309,313,317,329,333,
                                                        337,347
Sun: solar wind   3,11
---: corona, mass loss   11
Supernovae: progenitors   357,365
----------: remnants, morphology   357
----------: spectroscopy   365
----------: neutrino outflows   367
White Dwarfs: accretion   245
------------: LTNR   245

Zeta Aurigae systems   25

INDEX OF OBJECTS:

A 78   257
R Aqr   333

BD+60 2522   223

RU Cancri	325
Gamma Cas	205
CED110	303
CH Cyg	333
P Cyg	39
V645 Cygni	263
59 Cyg	205
Cyg X-3	151
CRL2591	347
AG Draconis	333
G35.2-0.74	347
H1413+113	405
HB5	343
HD 316285	215
HD 170634,170739,170784	309
He2-274,2-131	273
HH 34	313
HH 24	317
IC 351,2165,289	259
IC 418	267,273
IC 2448,4642,4776,4846	273
IRAS sources	329,337
K3-50	347
L1551	303
VV Monocerotis	325
M3-28	343
M1-75	259
M2-9	273
M31:OB78-277	231
M31:OB69-WR2	231
NGC 2169	303
NGC 7563	223
NGC 7009,6891,6629,2440,5882,3918,3242,1535	273
NGC 6884,6886	259
NGC 6537,6445,6741	343
NGC 1068,6946,4736,7331	391
NGC 2264,2024,1333	347
PHL 5200	405
UV Piscium	325
TY Pixidis	325

R 50, 126   211

Mu Sgr   95
S 88, 140   347
S 106   285
S 22   211
SSV 3   317
SS433   151,163
SN 1987A   365,367
SNR Kepler   359
SwSt1   273

W3   347
WR70,80,96,95,69,48,140,137   197

DEC 19 1990